AP* Q&A BIOLOGY

600 QUESTIONS AND ANSWERS

David Maxwell, M.S.
Science Department Chair and AP Biology Teacher
The Pingry School
Basking Ridge, New Jersey

About the Author

David Maxwell has taught AP Biology at The Pingry School, an independent college preparatory school in New Jersey, for 16 years, and he has served as the Science Department Chair for four years. He has helped grade the AP Biology exam every year since 2004. David holds degrees from the University of Wisconsin (M.S. in Plant Pathology) and Juniata College (B.S. in Ecology). He lives with his family in a small town in rural New Jersey.

Acknowledgments

I would like to thank Nat Conard and The Pingry School for encouraging me to take on this project. I would like to thank my editor, Samantha Karasik, and my copyeditor, Michele Sandifer, for editing my work and making it clearer. Thanks to my wife, Krista, and my daughters, Katie and Libby, for the time they allowed me to take away from the family to complete this book. I would like to thank the faculty of The Pingry School Science Department for providing inspiration and feedback on my material. I would also like to thank my fellow AP Biology Readers for their help. Finally, I would like to thank all of my students, who have consistently inspired me to learn more by asking new questions that I simply had to learn the answers to, and the biological world for being so interesting.

All inquiries should be addressed to:
Barron's Educational Series, Inc.
250 Wireless Boulevard
Hauppauge, New York 11788
www.barronseduc.com

Library of Congress Control Number: 2018938612

ISBN: 978-1-4380-1120-2

PRINTED IN CANADA

9 8 7 6 5 4 3 2 1

Contents

ANSWERS

Introduction

Learn to think scientifically. Do not just memorize facts. These statements epitomize the framework of the AP Biology exam and curriculum. Since 2013, the AP Biology exam and the curriculum framework, as provided by the College Board, have focused on "big ideas" and "science practices" that students need to understand fully rather than quick scientific facts that can be memorized in preparation for the test. The field of biology is changing so rapidly that memorizing facts, which are now easily accessible on the Internet, is no longer enough. The test makers want you to learn the practices of science so that you will develop the tools and skills necessary to understand, analyze, and interpret new information as it is discovered and developed and so that you can apply these analytical skills well beyond your biology classroom.

How can *you* learn to think scientifically? How can *you* develop and practice the skills necessary to achieve success on your AP Biology exam? You've already set yourself on the path to success by purchasing this study guide, which contains 600 practice questions that you will need to solve by thinking scientifically. Many of the situations described in these questions are based on real-life scenarios. For example, is there a connection between military sonar and whale beachings? Does being infected with a parasite influence people to own cats? Can providing bears with a place to cross the road prevent genetic inbreeding? All of these questions and more will be answered as you learn to think scientifically and as you practice answering all of the various question types on the AP Biology exam.

THE STRUCTURE OF THE AP BIOLOGY EXAM

The AP Biology exam consists of 77 questions that you will have 3 hours to answer. The exam is broken down into two sections as follows.

Section I		
Question Type	**Number of Questions**	**Timing**
Part A: Multiple-Choice	63	90 minutes
Part B: Grid-In	6	
Section II		
Question Type	**Number of Questions**	**Timing**
Long Free-Response	2	80 minutes + 10-minute reading period*
Short Free-Response	6	

* You must monitor the length of the reading period yourself. The exam proctor will not do it.

BIG IDEAS

There are four big ideas (BI) in the AP Biology curriculum and covered on the AP Biology exam, each of which has several learning objectives (LO). Those four big ideas are as follows.

BIG IDEA 1: The process of evolution drives the diversity and unity of life.

BIG IDEA 2: Biological systems utilize free energy and molecular building blocks to grow, to reproduce, and to maintain dynamic homeostasis.

BIG IDEA 3: Living systems store, retrieve, transmit, and respond to information essential to life processes.

BIG IDEA 4: Biological systems interact, and these systems and their interactions possess complex properties.

QUESTION TYPES

Multiple-Choice

For these questions, you will be presented with a question and four possible answer choices. You must select the best answer to the question. The multiple-choice questions typically present you with some background information in the form of a text, a graph or graphs, or a diagram or diagrams that will be necessary to answer the question. These will not be simple "fact recall" questions. You will need to think and reason. By exam day, you will want to be able to answer these questions fairly rapidly. Remember, you will only have 90 minutes to answer the 63 multiple-choice questions (plus the 6 grid-in questions) on this exam. That means you have approximately 1 minute, 20 seconds to answer each question. Make sure that you do not leave any of these questions blank. You have a 1 in 4 (25%) chance of guessing the right answer if you're not sure which one is correct, and there is no penalty for incorrect answers or for guessing.

Grid-In

For these questions, you will read the question, calculate the correct answer, and bubble your answer on the grid-in section of your answer sheet. The grid-in questions require you to apply mathematics to biological questions. In preparation for this part of the exam, you should familiarize yourself with the AP Biology Equations and Formulas sheet and know how to use each equation and formula. Note that these equations and formulas are displayed in the Reference Tables at the end of this book. Pay attention to the number of digits in your answer. If the question asks for your answer to the nearest *tenth*, be sure that you do not give your answer to the nearest *hundredth*. Even though the grid-in questions come at the end of the machine-scored first section of the exam, you should consider answering the grid-in questions first. It is much more difficult to guess correctly on the grid-in questions (where no answer choices are provided) as opposed to guessing on the multiple-choice questions (where you have a 1 in 4 chance of selecting the correct answer). Remember, you should plan to answer each of these questions in no more than 1 minute, 20 seconds.

Long Free-Response

The long free-response questions require you to provide a long, detailed response to the question. You can earn up to 10 points for each question. Pay careful attention to the wording of each question. The exam writers put the action words in bold type.* The writers do this to help you figure out exactly what they want in your response. Always directly answer the questions that the test makers instruct you to address. You may find it helpful to check off each bold word as you answer each question to guarantee that you have answered everything and have not missed any important points. You will not be awarded any points for style, for rewriting the question, or for providing a thesis statement, introductory paragraph, or concluding paragraph. You will only have approximately 20 minutes to brainstorm and write your response to each long free-response question. As a result, it is important to keep your response concise and be sure to demonstrate your understanding of the topic. The long free-response questions also include some specific instructions, such as "give ONE example." If the question specifically states "ONE example," then make your one example a good one. If you give more than one example in your response, only your first response will be graded, regardless of whether or not your second example is correct. If you do come up with a better example after you've already written your first one, simply cross out your first example and begin writing the one you'd rather submit. Diagrams are often very helpful for illustrating specific information. You should use them when possible and appropriate. If you do draw a diagram, be sure that you reference the diagram in your response, providing it with a title and a label, and incorporate the diagram into your response. For example, you could write, "See Figure 1—The Cell" within your response. Diagrams that are not referred to in the response are not graded. When answering long free-response questions, be aware of how much time you have left. Since these questions come first in Section II, students tend to spend a lot of time writing their responses. Students often do not leave themselves enough time to answer the short free-response questions. Answer each long free-response question fully, and then move on to the next question.

---

*For the purposes of distinguishing the bold words in the long free-response and short free-response questions in this book, the editors have taken the extra measure of also underlining these action words to help readers recognize them more easily.

Short Free-Response

The short free-response questions require you to provide a short, less involved response to the question. Like the long free-response questions, the short free-response questions have key words bolded so that you'll know exactly what to include in your response. On recent AP Biology exams, the first 3 short free-response questions have been worth 4 points each, while the last 3 short free-response questions have been worth 3 points each. Plan to spend a little over 6 minutes on each of these questions. Remember, don't spend all your time on the long free-response questions. Some students leave several of the short free-response questions blank because they used all of their time responding to the long free-response questions. Keep this in mind so that you do not fall into that trap.

SCORING

The AP Biology exam is graded on a scale from 1 to 5, with 5 being the highest score. The following table is based on the students who sat for the AP Biology exam in May 2017.

AP Grade	Percentage of Students Who Scored That Grade	Description	Equivalent Grade	Composite Score	Percentage Correct
5	6.4%	Extremely well qualified	A+ or A	86–120	71.7%
4	21%	Well qualified	A–, B+, or B	68–85	56.7%
3	36.7%	Qualified	B–, C+, or C	49–67	40.8%
2	27.5%	Possibly qualified	C–, D+, or D	30–48	25.0%
1	8.4%	No recommendation	D– or F	0–29	0%

As you can see, 6.4% of all students who took the test in May 2017 earned a 5, so those who earned a 5 were in a fairly select group. Students who earn a 5 are described by the College Board as "extremely well qualified." It is expected that students who earn a 5 on the AP Biology exam would likely earn an A+ or an A in a typical college Biology course. The raw scoring for this exam is completed on a 120-point scale. The final column shows the percentage of questions that needed to be answered correctly in May 2017 to earn the lowest composite score for that AP grade. For example, to earn a 5, students needed to earn at least 71.7% of the points available on that exam.

Section I of the exam, which includes both multiple-choice questions and grid-in questions, is scored by a machine. Section II of the exam, which consists of long free-response questions and short free-response questions, is scored by exam readers. These readers will be looking for you to provide clear, concise responses in which you answer every part of each question in order to earn all of the points available. The two long free-response questions will be worth 10 points each. The first three short free-response questions are typically worth 4 points each. The last three short free-response questions are typically worth 3 points each.

Note about Scoring for the Grid-In, Long Free-Response, and Short Free-Response Questions

When working through the grid-in, long free-response, and short free-response questions in this book that involve math, your answer may vary slightly from the given answer in the Answers section at the end of this book. This may be due to rounding. In most cases, it is best to keep all of the digits you can during your calculations (just keep them in your calculator) and then round off at the very end. This will likely minimize rounding errors. It is also possible that you may reach a different numerical answer if you read a graph differently. On the actual exam, there will be an acceptable range of answers that takes into account these differences. Usually, if you are within 5% of the actual answer, then your answer will be marked as correct.

Also, when working through these three question types in this book, be sure to keep scratch paper handy to work out any calculations.

TIPS FOR SUCCESS ON SECTION I

When working your way through the multiple-choice questions and the grid-in questions, keep the following strategies in mind:

- Pace yourself so that you have enough time to answer all of the questions in this section.
- Remember to always bubble in your responses neatly. Responses that are not marked clearly or are not properly erased could possibly be graded as incorrect by the scoring machine.
- Guess whenever you're not sure because there is no penalty for incorrect answers. Answer all questions even if you have to make a wild guess.
- Answer the grid-in questions first (even though they are at the end of this section) because it is difficult to guess a numeric response correctly whereas you have a 1 in 4 chance of guessing correctly on each multiple-choice question.
- Don't second guess yourself. Trust your instincts and the knowledge you've learned from your AP Biology course.

TIPS FOR SUCCESS ON SECTION II

When working your way through the long free-response questions and the short free-response questions, keep the following strategies in mind:

- Write *legibly*. The exam readers go to great lengths to try to understand what you have written, but if they cannot read your response, then it will not matter how well you knew the answer.
- Budget your time carefully so that you can answer all 8 questions in this section. Try to spend only 20 minutes on each long free-response question and a little over 6 minutes on each short free-response question.
- Organize your thoughts before you begin writing to ensure that your response will be clear and concise.
- Remember to answer all parts of the question. Keep an eye out for bold words that tell you exactly what the test makers are looking for in your response to each question. Check off each part as you answer it.
- Pay attention to words in all capitals. If the question asks for ONE example, only give one example. If you give more than one example, only the first one will be graded.
- Don't waste time writing "fluff." Don't restate the question. Don't provide a thesis statement or a concluding paragraph. You won't have time to add this information, and including this information won't help you earn points.
- Remember that you will earn points as you state correct information that answers the question. You will not lose points for stating incorrect information. However, be sure not to contradict yourself, as that could cause you to lose points.
- Keep an eye on your spelling, and always try to spell things correctly. If a word is misspelled but is close to the correct spelling, you will generally be awarded the point. However, if it is unclear to the reader what word you were trying to say based on your misspelling, you likely won't earn the point. For example, the nucleus and the nucleolus are two different structures in a cell. If you meant to say nucleolus, but you wrote nucleus, you would not be awarded that point.
- Pictures are worth a thousand words, but they are worth zero points unless you refer to them in your response. Be sure to reference the figure in your answer (i.e., "As you can see in Figure 1 below, the F1 cross results in a 3:1 ratio, demonstrating that the red eye trait is dominant.")

HOW TO USE THIS BOOK

The four big ideas that make up the conceptual framework for the AP Biology curriculum and exam also make up the framework for this book. Each chapter contains practice questions that revolve around one big idea. The proportion of questions in each chapter is approximately the same as the proportion of learning objectives outlined for each big idea by the College Board. Chapter 3, which focuses on Big Idea 3, contains the most practice questions because Big Idea 3 has the most learning objectives.

There are enough questions in this book to make up 7 full AP Biology exams and approximately 80% of an 8th exam. Therefore, you should have more than enough practice material to master each content area and question type. If you do need additional practice beyond what is in this book, please visit the AP Biology page on the College Board's website:

https://apstudent.collegeboard.org/apcourse/ap-biology

Where should you begin? First, find out what your strengths and weaknesses are. One strategy is to choose randomly, from each chapter, 6 multiple-choice questions, 1 grid-in question, and 1 short free-response question. See what questions you answer correctly. Then determine what content areas and question types you need to focus your efforts on. As you'll see when you review the Answers section of this book, the majority of the answer explanations, especially those for multiple-choice questions, explain why one answer choice is correct and why the remaining answer choices are incorrect. Each answer explanation also identifies the learning objective that is covered in each question. Use this information to search your class textbook or online resources for more practice with that learning objective topic if you are still confused. Another great resource is the *Barron's AP Biology* test prep guide, which reviews every subject topic by topic.

Even if there are topics and question types that you feel confident about, you should still take the time to work through every practice question in this book. Each question is framed after modern-style AP Biology practice questions. Remember, all the questions focus on higher-order thinking skills rather than on information recall. In fact, if you have enough time to review a question carefully, you will almost always be able to figure out the correct answer from the information in the question. If your general strategy is to memorize information, you will quickly find that this is not the most efficient use of your efforts. Focus on the mental processes behind scientific reasoning, and become proficient at thinking like a scientist. Doing so will not only make the AP Biology exam that much easier, but it will also help you retain more biology information and develop thinking habits that will serve you in all areas of your life.

As a student, your life is likely always on the go, so keep this book tucked in your backpack at all times. Life has lots of idle moments when you can answer even just 1 or 2 questions. For example, you might be waiting for your sibling to finish athletic practice, or you might be waiting for a parent to pick you up after your music lesson. Rather than twiddling your thumbs, take those few moments to work through as many questions as you can. If you can complete just 5 questions per day, you will finish this book in about 4 months. Spreading out your learning over time, rather than cramming at the last minute, is the most effective way of learning new things.

Above all, remember that by working through this book, you've taken the appropriate steps to score well on this exam. Don't let your test-taking anxieties cause you to second-guess yourself. Students tend to have a significant amount of anxiety about answering questions incorrectly. One idea that might help ease your anxiety is to remember that part of thinking scientifically often means being uncertain about answers. Scientists work to learn things that are not already known. In other words, they deal with trial and *error*. Science only moves forward when individuals look at the current understanding, find something that doesn't quite fit, and find new ways of looking at an idea or concept. To think scientifically, you need to become comfortable with not being certain about the correct answer 100% of the time. You don't have to know everything, and you don't have to be 100% right 100% of the time. You don't need to be perfect—you just need to study hard, practice frequently, and do your best.

STUDY GUIDE

Your individualized study plan will largely depend on how much time you have to devote to preparing for this exam. The following study guide assumes that you are beginning your review with 5 weeks left until the exam. If that is the case, you will need to answer fewer than 20 questions per day in order to answer all of the questions in this book. If you spend only 45 minutes practicing each day, this is an attainable goal. Answer the questions that look simple or even interesting at first since your interest will motivate you. As you get better at answering questions, the ones that initially appeared to be hard will no longer seem as difficult. Prioritize the multiple-choice questions since they make up the bulk of the first section of the exam. Then work on the short free-response questions since there are more short free-response questions than long free-response questions. In addition, the skills you will develop when answering the short free-response questions will translate to when you answer the long free-response questions.

5 Weeks Before the Exam

As soon as possible, set aside four study periods, one for each question type. You will need approximately 25 minutes for the multiple-choice questions, 15 minutes for the grid-in questions, 15 minutes for the short free-response questions, and 30 minutes for the long free-response questions. If you don't see yourself as a numbers person, complete the short free-response questions before the grid-in questions.

Use the first 25 minutes to answer 2 multiple-choice questions from each chapter, spending 5 minutes per chapter. Flip through the chapter until you find a problem that looks interesting to you. If it looks too hard, skip it for now. Once you've answered all 8 multiple-choice questions, check the answer explanations. You should have several goals at this point. The first is to familiarize yourself with all of the question types on the exam and with the layout of this book. The second is to develop confidence when answering AP Biology–style questions. Right now, the questions should seem hard. Do not worry about working quickly. With the last 5 minutes left, think about your life schedule and when you can find time to study. Set a goal for how many questions you will answer each day. Make your goal as reasonable as possible but small enough that you can actually accomplish it. Motivating yourself is easier if you first set realistic goals. Then set new, bigger goals once you accomplish the smaller ones. You will get faster as you practice. Therefore, your initial goal does not need to be more than 20 questions per day.

Use a similar process for the remaining question types. For the 15 minutes that you spend answering the grid-in questions, answer 1 question from each chapter (4 questions total). For the 15 minutes that you spend responding to the short free-response questions, answer a question (from any chapter) that looks interesting. Focus on the bolded action words, and answer only what is asked. For the 30 minutes that you spend responding to the long free-response questions, attempt 1 question from any chapter. Make sure that you answer the question fully, and then check the answer explanation.

Collect data on the number of questions that you answer each day. Adjust your goals depending on your average per day. If you find that you can attain a higher goal, go for it.

4 Weeks Before the Exam

Create a graph showing the data you collected regarding the number of questions that you answered each day during your first week of study. Calculate the percentage of questions that you answered correctly per minute for each type of question. Use this information to create a bar graph, and place the question types on the *x*-axis and the response variable (the percentage of questions that you answered correctly per minute) on the *y*-axis. Label the graph correctly, and add a title. Creating a graph will allow you to see your progress. Displaying information graphically is also a vital scientific skill. Then test your answering rate by seeing how many multiple-choice questions you can answer in 16 minutes.

2 to 3 Weeks Before the Exam

Create a graph showing the percentage of questions that you answered correctly per minute each day during your first two weeks of study. Test yourself to see how quickly you answer multiple-choice questions. Compare last week's speed to this week's speed. If you practiced every day so far, you should be faster this week. If you haven't, keep calm and add a few more minutes of study time to each day next week. Complete a test of your speed answering grid-in, short free-response, and long free-response questions if you have time and add these calculations to your graph.

1 Week Before the Exam

Make sure that you have a four-function calculator with a square root button. Check the calculator policy on the College Board's website to ensure that your device is on the approved list. Ensure that you have fresh batteries. Make sure that you know how to get to your exam location. You don't want to waste time thinking about these things on the day of the exam.

Graph the data again, and complete another round of pace testing. If you are on target, congratulations! If you are significantly behind where you would like to be, remain calm. You have undoubtedly learned a lot about how to think scientifically as well as a lot of really interesting biology information in your AP Biology class. That is valuable in and of itself. Ultimately, the AP Biology exam is a snapshot of your progress toward mastering scientific skills. Even if you don't earn a high score on this one exam, you can still become a doctor, a veterinarian, a cancer researcher, or whatever you choose as long as you persevere. If you have been working hard and following the plan described in this book, you should go into the exam confident that you can solve all of the puzzles that the test makers will present to you.

1 Day Before the Exam

Exercise, eat a good meal, and get to sleep at a reasonable time. By this point, you have put in a lot of effort preparing for this exam. You are ready. Lay out your clothes for the next day as well as your approved calculator, pencils, and sharpener. Put your materials for the exam into your car. Plan when you will leave the next morning. Then look at the graphs that you made charting your progress. Practice a few more questions if you feel the need to.

The Morning of the Exam

Get up in time to have breakfast. Your brain runs on glucose, and you will need your brain to be fueled. Leave early enough for the exam so that you will get there even if the traffic is bad and the parking lot is full. If you are often late, give yourself an extra 15 minutes of travel time on top of the estimated time for bad traffic and parking.

Above all, remember to breathe and take comfort that you know this material. Remember that you have mastered the tools needed to succeed on the AP Biology exam.

QUESTIONS

Big Idea 1—Evolution

CHAPTER 1

Answers and explanations can be found on page 275.

BIG IDEA 1

The process of **evolution** drives the diversity and unity of life.

For the complete list of big ideas, learning objectives, enduring understandings, essential knowledge, and science practices, refer to the "AP Biology Course and Exam Description" from the College Board:

https://secure-media.collegeboard.org/digitalServices/pdf/ap/ap-biology-course-and-exam-description.pdf

MULTIPLE-CHOICE QUESTIONS

Questions 1–6

A biology student collected seed samples from pigweed (*Amaranthus palmeri*) from two neighboring farms in Arkansas. One farm planted glyphosate-resistant corn and soybeans for the past 10 years and regularly sprayed the pesticide glyphosate to control weeds (conventional farming). The other farm was an organic farm that did not use pesticides. The student grew seedlings to the six-leaf stage. The student also exposed each group to varying levels of pesticides to determine the dose that killed 50% of the seedlings (LD_{50}). The data are summarized in the table below.

Table I Pesticide Resistance in Conventional and Organic Fields

	LD_{50}
Glyphosate-treated (conventional) fields	2,820 g/ha
Nontreated (organic) fields	30 g/ha

In a follow-up experiment, the student developed inbred populations of pigweed that were pure-breeding resistant and pure-breeding susceptible strains. The student hybridized the pure strains. The student then tested the offspring for their level of resistance using a tissue culture method that allowed the student to determine the minimum concentration of glyphosate that caused cell death (threshold toxicity level). The F1 offspring showed resistance levels that were halfway between those of the resistant and susceptible strains. The F1s were then self-crossed to produce an F2 generation. The data that follow show the results of the F2 generation.

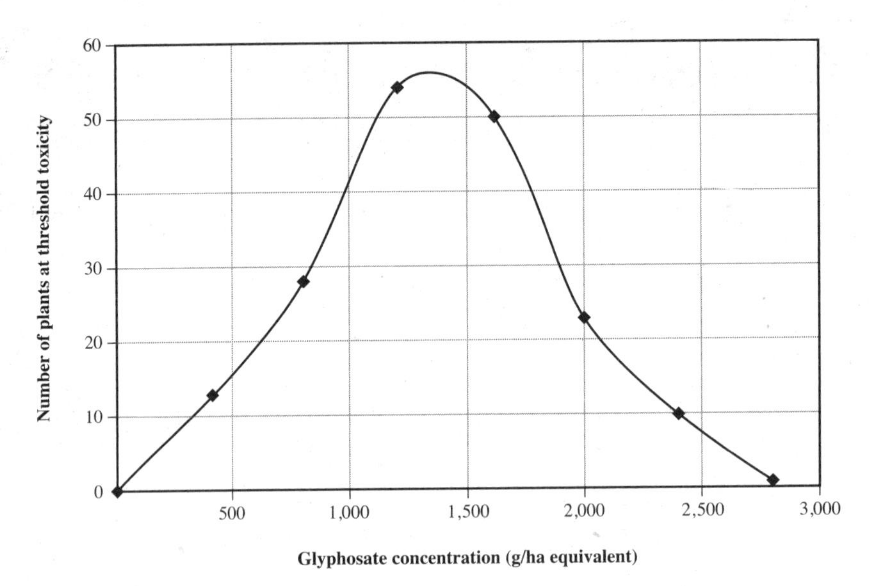

Figure I Pesticide resistance in the F2 generation

1. **Which of the following is most consistent with a modern understanding of evolution?**

   (A) The pigweed population in the glyphosate-treated fields gradually acquired resistance because plants exposed to the herbicide changed in order to survive.
   (B) The pigweed population in the nontreated fields contains individuals that are genotypically resistant to glyphosate at levels above 30 g/ha.
   (C) Glyphosate-resistant plants from the treated fields must have interbred with the organic plants.
   (D) When exposed to pesticides, plants modify their DNA in order to survive the pesticide.

2. **The student took plants that had been exposed to 2,820 g/ha glyphosate in the experiment described, grew them to maturity, allowed them to cross-pollinate, and collected seeds. The student then grew those seeds to the six-leaf stage and exposed them to 2,820 g/ha glyphosate. Which of the following would be the most likely result?**

   (A) None of the plants would be resistant to that concentration of pesticide.
   (B) About 25% of the plants would be resistant to that concentration of pesticide.
   (C) More than half of the plants would be resistant to that concentration of pesticide.
   (D) All of the plants would be resistant to that concentration of pesticide.

3. The student took plants that had been exposed to 30 g/ha glyphosate in the experiment described, grew them to maturity, and collected seeds. The student then grew those seeds to the six-leaf stage and exposed them to 2,820 g/ha glyphosate. Which of the following would be the most likely result?

(A) Very few, if any, of the plants would be resistant to that concentration of pesticide.
(B) Half of the plants would be resistant to that concentration of pesticide.
(C) Most of the plants would be resistant to that concentration of pesticide.
(D) All of the plants would be resistant to that concentration of pesticide.

4. The student attempted to repeat the experiment the following year. However, the student discovered that some of the pigweed samples from the organic farm showed levels of glyphosate resistance similar to the pigweed samples from the conventional farm. The organic farmer was insistent that she had not used pesticides on the farm. Which of the following is the most likely biological explanation for the presence of glyphosate-resistant pigweed samples on the organic farm?

(A) Random mutation occurred in the pigweed samples on the organic farm.
(B) There is no biological explanation. The organic farmer must not be telling the truth.
(C) Transduction by viruses that were transported by aphids spread genes from nearby conventional farms.
(D) There was a spread of resistant seeds from neighboring farms.

5. In the follow-up experiment described in the prompt, all of the F1 offspring showed resistance levels that were halfway between those of the resistant parental strain and the susceptible parental strain. Which of the following mechanisms of inheritance would most likely explain this result?

(A) Glyphosate resistance is sex-linked.
(B) Glyphosate resistance is recessive.
(C) Glyphosate resistance is controlled by a dominant allele.
(D) Inheritance of glyphosate resistance is incompletely dominant.

6. In the follow-up experiment described in the prompt, the F2 offspring showed the distribution of levels of resistance shown in Figure I. Which of the following is most consistent with the pattern of inheritance shown by the F2 offspring?

(A) Glyphosate resistance is controlled by many gene loci.
(B) Glyphosate resistance is controlled by a dominant allele and a recessive allele at a single gene locus.
(C) Glyphosate resistance is controlled by two codominant alleles at a single gene locus.
(D) Glyphosate resistance is controlled by sex-linked genes.

Questions 7–9

The Trans-Canada Highway runs through Banff National Park. Prior to 1996, many animals died crossing the highway. Beginning in 1996, the government began installing wildlife overpasses and underpasses and fencing along the highway. Researchers obtained stored tissue samples from black bears in the park and performed genetic tests to determine the average heterozygosity in the bear populations (a measure of genetic diversity). The data are presented in the graph below.

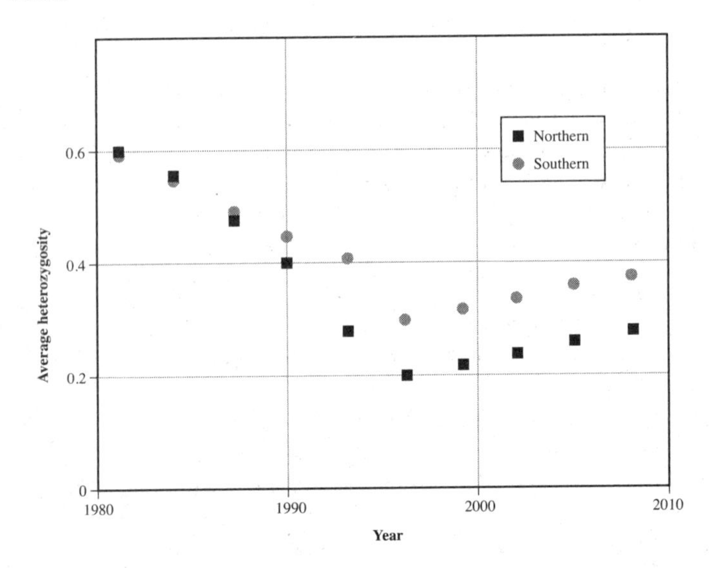

7. **Which of the following is the most likely explanation for the trend in the data before 1996?**

    (A) The immigration of southern bears to the north was a result of climate change.
    (B) Natural selection against bears that crossed the road led to increased gene flow between the populations.
    (C) Increased mutation rates, as a result of the pollution from the highway, decreased genetic diversity.
    (D) Genetic drift led to declining diversity in both populations because the road effectively created two small populations of bears.

8. **Which of the following is the most likely explanation for the trend in the data after 1996?**

   (A) Wildlife crossings were ineffective because bears can't read signs.
   (B) Wildlife crossings increased the likelihood that bears from different populations mated. As a result, the mutation rate decreased.
   (C) The introduction of wildlife crossings increased the amount of immigration and emigration between the two populations, leading to an increase in diversity in both populations.
   (D) The introduction of wildlife crossings increased the bears' ability to obtain a greater variety of food. This led to more sexual selection and greater diversity.

9. **Which of the following would be the most effective way to restore the levels of genetic diversity to 1980 values?**

   (A) Capture and raise black bear cubs in captivity, and release them when they are mature.
   (B) Install bear-proof garbage cans to prevent the animals from becoming reliant on human food sources.
   (C) Provide sources of food for the bears to increase their population size.
   (D) Capture black bears from areas with high levels of diversity, and release them in Banff National Park.

Questions 10–11

Students in New Jersey at a school that is at latitude 40.6° N used a DNA sequencing method to detect the diapause minus allele in wild populations of *Drosophila melanogaster* every month from May through October for five years (2011–2015). Homozygotes for this allele are not able to enter a cold-resistant state. The average allele frequency for each month is shown in the figure below. Error bars show the standard error of the mean.

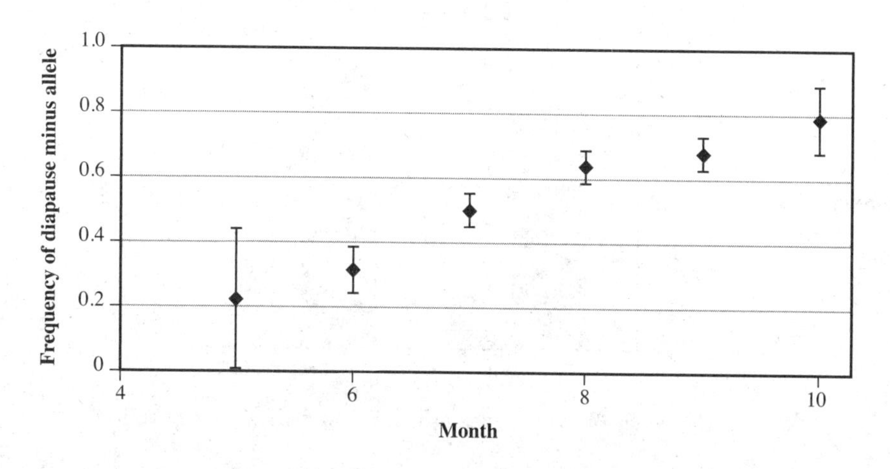

10. **Which of the following is the most likely evolutionary explanation for the larger standard error of the mean for the samples taken during May?**

    (A) Inbreeding among fruit flies during the winter led to greater variability.
    (B) Harsher winters more strongly selected against diapause minus alleles. The variability in winter severity led to more variable allele frequency.
    (C) Nonrandom mating of the fruit flies led to higher variability in May.
    (D) Genetic drift was higher in May.

11. **Which of the following does NOT account for the trend shown in the graph?**

    (A) Cold winter weather selects against the diapause minus allele from October through May (during the winter).
    (B) Immigration of fruit flies from southern regions introduces diapause minus alleles into the population from May through October (during the summer).
    (C) Natural selection favors diapause minus mutants from May through October (during the summer).
    (D) Genetic drift dictates that one allele will disappear from the population.

Questions 12–17

The snail *Poamopyrus antipodarum* can reproduce either sexually or asexually. It is infected by a parasite that is carried by a species of duck that feeds only in shallow water. As a result, the parasite is not present in deep water. Researchers sampled snails from deep and shallow water in two lakes and determined the frequency of males (Figure I). Next they collected snails from all four locations and isolated parasites from the two lakes. They performed a laboratory experiment in which snails from each location were challenged with parasites from each of the lakes (Figure II).

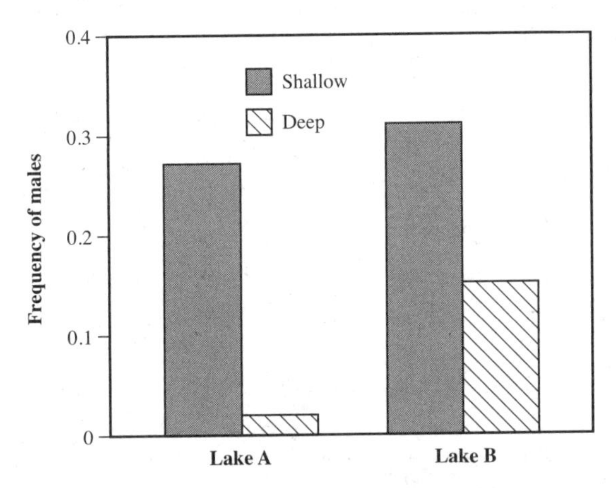

Figure I Frequency of male snails in shallow and deep water in two lakes

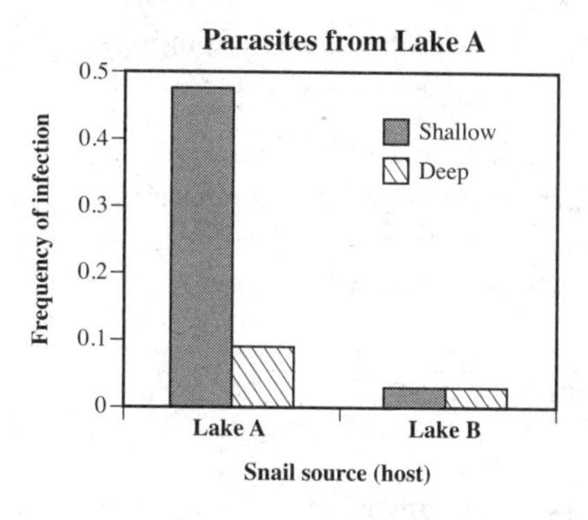

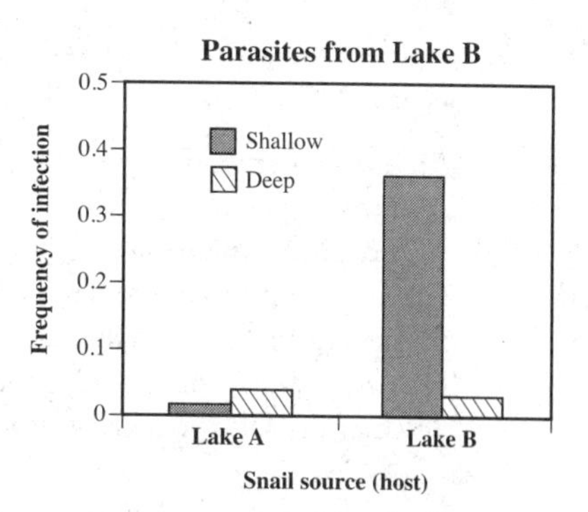

Figure II Frequency of infection in a laboratory assay—snails from shallow and deep water in two lakes when challenged by parasites from each lake

12. Which of the following is the most likely explanation for the distribution of males in the two lakes?

(A) Random chance accounts for the distribution of males.
(B) Males are selected for in deep water.
(C) Males are selected against in deep water.
(D) Sexual selection makes males more abundant in deep water.

13. Which of the following best explains the pattern of infection shown in Figure II?

(A) Snails move to deep water to avoid parasites.
(B) Snails from Lake B have built up an immune response to parasites from Lake A.
(C) Parasites from Lake B are more likely to infect snails from Lake A because Lake A snails have no resistance genes.
(D) Parasites from each lake coevolved to infect snails that inhabit the lake in which the parasites live.

14. Two diploid snails mate. What fraction of each snail's genes are passed on to the offspring?

(A) 100%
(B) 75%
(C) 50%
(D) 25%

15. **Which of the following statements about evolutionary fitness is most accurate?**

    (A) The most athletic animal will survive because it is the most fit.
    (B) Fitness has to do with the healthiness of the organism. Longer-lived organisms that resist disease have higher fitness.
    (C) An organism that passes the highest proportion of its genes to future generations has the highest fitness.
    (D) An organism that obtains the most food and controls the largest territory has the highest fitness.

16. **An asexual snail has 100 offspring. How many offspring would a sexually reproducing snail need to have in order to have equal fitness with the asexual snail, assuming all other factors are equal?**

    (A) 200
    (B) 100
    (C) 50
    (D) 25

17. **Which of the following statements is consistent with the data from Figure I and Figure II?**

    (A) Sexual reproduction produces organisms with higher fitness in all environments.
    (B) Sexual reproduction is more likely in habitats where parasites are present because a diverse population of snails presents a variety of challenges to the parasite.
    (C) Asexual reproduction is more likely in habitats where parasites are present because a strong snail genotype is resistant to all parasites.
    (D) Asexual reproduction produces organisms with higher fitness in all environments.

Questions 18–22

In order to study animal domestication, researchers divided a wild population of rats (*Rattus norvegicus*) into two groups (tame and aggressive) based on their response to handling by humans. In the tame group, researchers selected the tamest 30% of the rats for breeding. In the aggressive group, the most aggressive 30% of the rats were permitted to breed. This process was repeated for 60 generations. Corticosteroid levels were measured in the aggressive and tame rats before and after being restrained by a gloved human hand. The average corticosteroid levels and the standard errors of the mean (SEM) for both treatment groups are shown in the following table.

	Tame (ng/mL)	Aggressive (ng/mL)
Baseline	0.5	1.5
After Handling	3	5
Baseline SEM	0.2	0.34
After Handling SEM	0.3	0.51

18. Which group of rats is more fit in this experiment?

(A) The aggressive rats are more fit because they are more likely to obtain food from the environment and survive.
(B) The tame rats are more fit because they are more likely to obtain food from humans.
(C) The aggressive rats are more fit in the aggressive group, and the tame rats are more fit in the tame group because these individuals are most likely to have offspring.
(D) The aggressive rats are more fit because they are more likely to compete for mates.

19. The tame rats exhibited a large variety of coloration patterns that were not seen in the aggressive rats, in addition to exhibiting droopy ears and a decreased brain size. Which of the following is the most likely explanation for these observations?

(A) The researchers deliberately selected for animals with these other traits.
(B) Genes that were directly responsible for tameness were also involved in creating these other phenotypes or were linked to these other genes.
(C) Since artificial selection leads to mutations, these traits arose by random chance.
(D) Female aggressive rats are not attracted to traits like droopy ears.

20. Which of the following would be a reasonable conclusion from the data presented in the table?

(A) High levels of corticosteroids cause aggressive behavior.
(B) Domestication selects for high corticosteroid levels.
(C) Although tame rats had higher baseline corticosteroid levels, aggressive rats had higher levels after handling.
(D) Aggressive rats had higher baseline corticosteroid levels and higher levels after handling.

21. **If the aggressive rats were released into a natural environment that prevented interbreeding with other rats and then were sampled for aggressiveness after approximately 50 generations, which of the following would most likely result?**

    (A) The rats in the population would become more aggressive on average because their genes would be concentrated.
    (B) There would be no change in aggressiveness without human intervention.
    (C) Aggressiveness would decrease to the levels typical of rats in natural environments because the rats would be subjected to the same selective pressures as that of a wild population.
    (D) There would be no way to predict what would happen because natural selection is an entirely random process.

22. **Which of the following would be the most appropriate display for the data in the table?**

    (A)

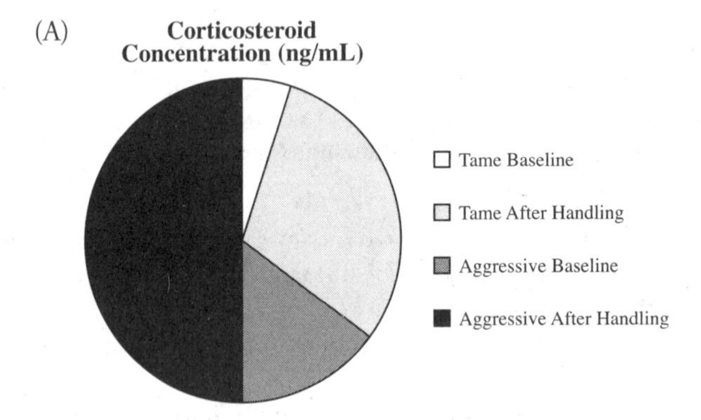

    (B)

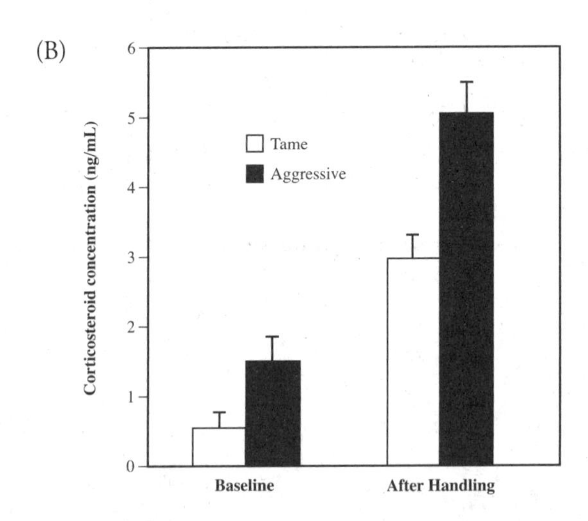

(C)

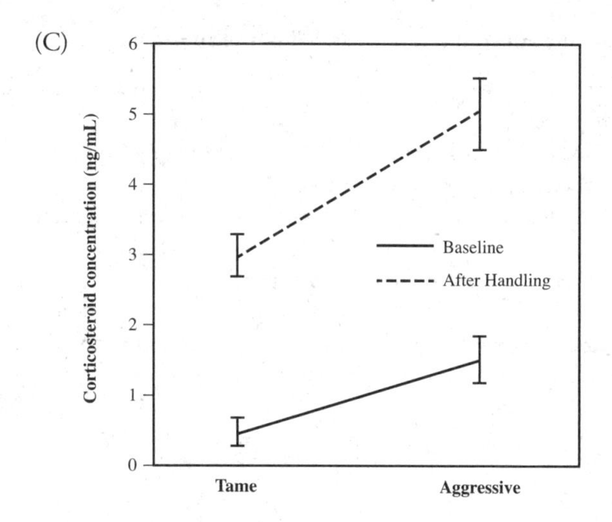

(D)

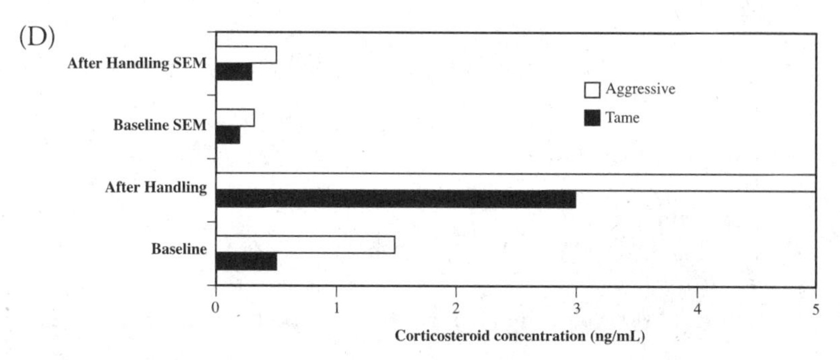

Questions 23–26

Scientists compared DNA sequences for the cytochrome b gene from two existing species and from two fossil species of elephantids in order to determine their phylogenetic history. The table below shows the number of differences in amino acids among the different species. The dugong was used as an outgroup. An outgroup is a species that is known to be more distantly related than the group being studied.

	African Elephant	Asian Elephant	Mammoth	Mastodon	Dugong
African Elephant (*Loxodonta*)	0	7	6	11	24
Asian Elephant (*Elephas*)	7	0	2	12	25
Mammoth (*Mammuthus*)	6	2	0	12	23
Mastodon (*Mammut*)	11	12	12	0	26
Dugong (*Dugong*)	24	25	23	26	0

23. **Why is the dugong (the outgroup) included in the analysis?**

(A) It provides a reference point for comparing the different sequences in the other species.
(B) It is included so that the scientists can see if the dugong is more closely related to the mammoth or the mastodon.
(C) Dugong DNA was easily obtained. Any animal species would have sufficed.
(D) There is no rationale for including the dugong.

24. **According to the data, which species is most closely related to the Asian elephant?**

(A) African elephant
(B) mastodon
(C) mammoth
(D) dugong

25. **According to the data, which species is most closely related to the African elephant?**

(A) Asian elephant
(B) mastodon
(C) mammoth
(D) dugong

26. **Which of the following phylogenetic trees best matches the data presented?**

(A)

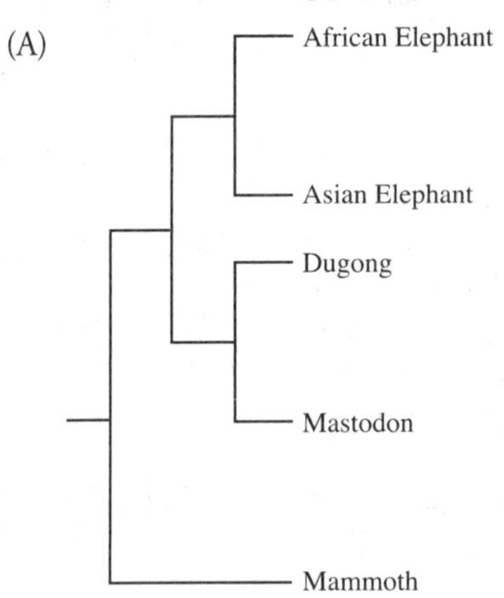

(B)

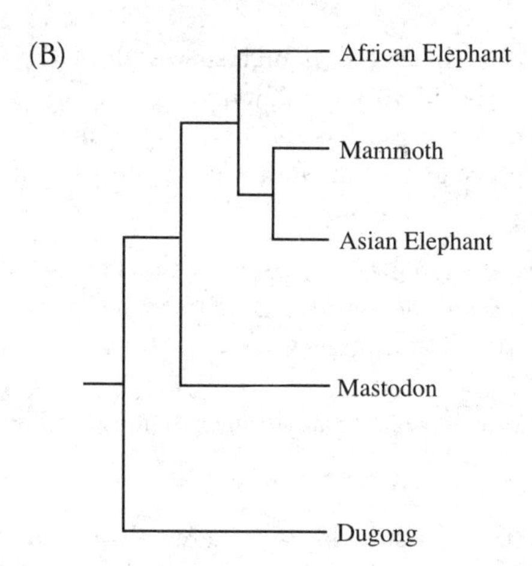
African Elephant
Mammoth
Asian Elephant
Mastodon
Dugong

(C)

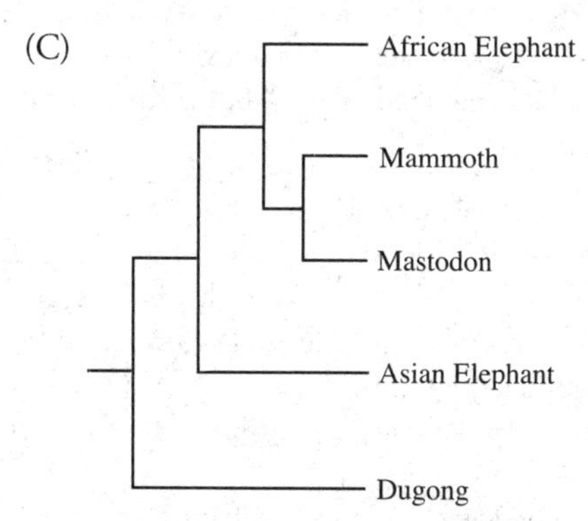
African Elephant
Mammoth
Mastodon
Asian Elephant
Dugong

(D)

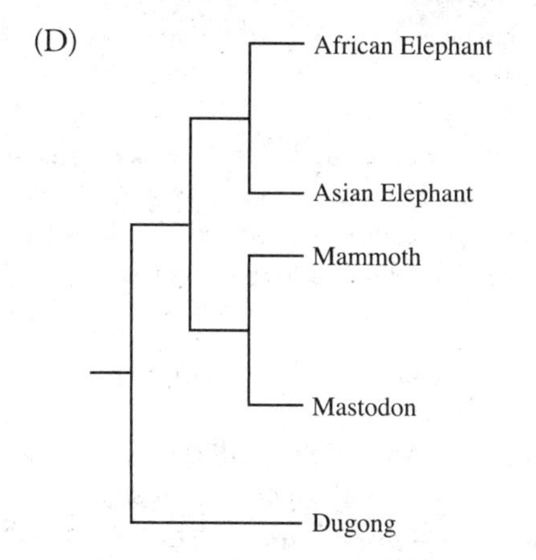
African Elephant
Asian Elephant
Mammoth
Mastodon
Dugong

27. ***Archaefructus liaoningensis*** **is a fossil species of angiosperm (flowering plant) found in rock strata that is 125 to 130 million years old. Molecular data suggest that the divergence of angiosperms from earlier plants occurred 215 million years ago. Which of the following is a reasonable scientific approach to resolve this discrepancy?**

    (A) Manipulate the molecular data to make it match the fossil record.
    (B) Search for fossil angiosperms in rocks older than 215 million years old.
    (C) Search for fossil angiosperms in rock strata between 130 and 215 million years old.
    (D) Search for fossil angiosperms in rock strata younger than 125 million years old.

28. **Northern jacanas (*Jacana spinosa*) are shorebirds that are polyandrous. Female jacanas compete for territory and access to males. When one female defeats another and claims her territory, the first thing she does is to kill all the existing chicks and break the unhatched eggs. What is the most direct selective advantage of this behavior?**

    (A) Since the chicks and the unhatched eggs are not hers, the conquering female increases her fitness by destroying them. She is also using all the resources of the territory and the males to raise chicks that are hers.
    (B) The conquering female gains a nutritional advantage for herself by eating the chicks and the unhatched eggs.
    (C) The conquering female avoids potential competition from the next generation of jacanas.
    (D) By severing the connection between the previous female and her offspring, the conquering female removes an additional reason for the conquered female to attempt a counterattack.

29. **Some viruses, such as the dengue virus and the tobacco mosaic virus, have RNA genomes that are replicated by RNA-dependent RNA polymerases (RdRPs). These RdRPs have no proofreading ability. As a result, RdRPs have very high mutation rates. Which of the following explains the adaptive significance of this?**

    (A) The high mutation rate is inconsequential because viruses produce enough offspring to maintain a stable genome.
    (B) Since their genomes are large, RNA viruses typically encode several additional proteins to proofread the RNA before packaging.
    (C) Since RNA viruses use the host's DNA-editing mechanisms on their RNA, error-prone RdRPs are acceptable.
    (D) The high mutation rate leads to increased variability and rapid evolution.

30. RNA is well known as an information molecule. However, RNA can also have enzymatic functions. Ribozymes are enzymes made of RNA. The RNA world hypothesis states that early life used RNA as both information molecules and catalysts. The hypothesis also says that DNA and protein evolved later. DNA is a more durable information molecule than RNA. Protein is a more versatile structural and catalytic molecule than RNA. Except for some RNA viruses, DNA is the molecule of heredity. It is replicated and passed to offspring. The RNA viruses that use RNA as their genomes use RNA-dependent RNA polymerases (RdRPs) made of protein to copy RNA from RNA. In 2013, Antonio C. Ferretti and Gerald F. Joyce reported on a pair of RNA enzymes they had designed that were capable of copying themselves. The doubling rate of these RNA-based RdRPs was about 20 minutes (about the same as that of many bacteria). What is the significance of this finding?

(A) It proved beyond any doubt that RNA evolved first.
(B) It showed that neither DNA nor protein were required for the replication of heritable information.
(C) It demonstrated that spontaneous generation was possible.
(D) It disproved the RNA world hypothesis because it showed that RNA could be useful only as an intermediate between DNA and protein.

Questions 31–35

Scientists create a large rectangular Petri plate that has separate sections. Each section contains bacterial media with different concentrations of an antibiotic, as shown in the figure below. A clonal culture of *E. coli* bacteria is inoculated on the section of the plate with no antibiotic. The adjacent segment has an antibiotic concentration just high enough to kill the bacteria (3 units). Each subsequent segment has 10 times more antibiotic than the previous segment.

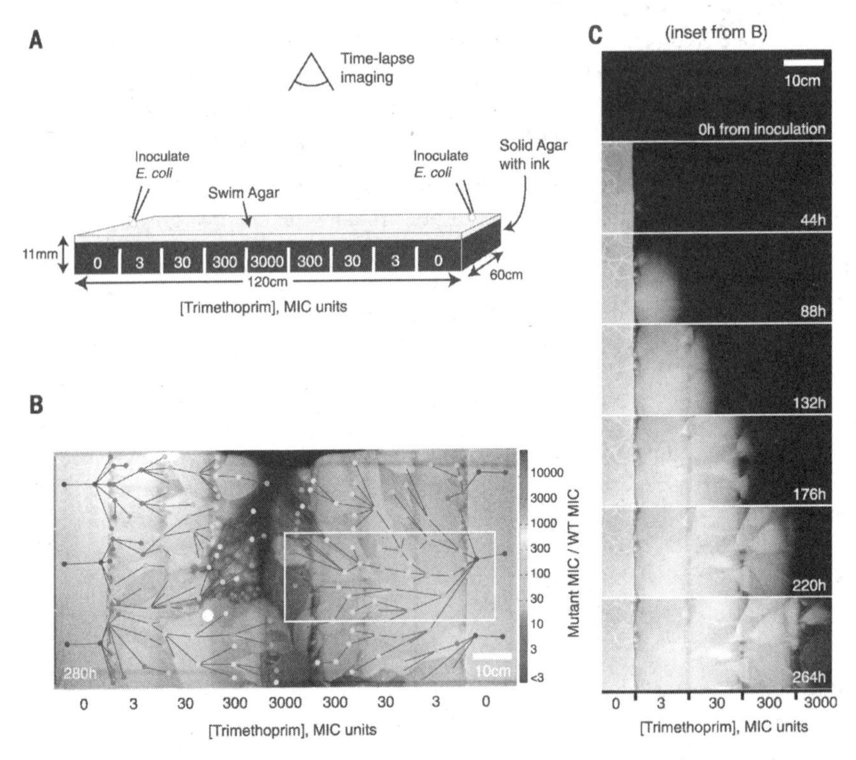

Figure I A. The experimental setup of the Petri plate. **B.** A photograph of the colonized plate. The dots show the locations where new mutations arose. The branches depict the spread of the mutant bacteria. **C.** A time course series showing colonization progress over 264 hours.

The bacteria colonize the plate until the 0 unit segment is covered. After a short period of no expansion, single spots form in the 3 unit segment and then expand to cover the segment. This is repeated on each subsequent segment.

31. What is the source of variation in the bacterial population?

(A) mutations that occur randomly as the bacteria multiply
(B) transformation that occurs due to the bacteria taking up plasmids from the environment
(C) transduction that occurs when viruses mistakenly package host DNA
(D) conjugation between two bacteria that are exchanging DNA

32. What is the reason for the pause before colonizing each subsequent section?

(A) The bacteria stop growing and begin mutating. Once a mutation has occurred that would make the bacteria resistant to the antibiotic, the bacteria start growing again.
(B) As the bacteria multiply in the current section, individuals are being pushed over the line into the next segment. Susceptible ones die and are not seen. Eventually, a randomly produced mutant that can survive is pushed into the next section. It multiplies and covers the section.
(C) Since some of the antibiotic diffuses over the line, bacteria closest to the line gradually acquire resistance and grow once they have complete resistance to that concentration of antibiotic.
(D) Bacteria near the line upregulate genes involved in detoxifying substances. Once they are expressing enough detoxifiers, they spread into the next segment.

33. What would be the result if the scientists inoculated the initial bacteria on the end of the plate with 3,000 units of antibiotic?

(A) The bacteria would mutate to be able to grow in the presence of 3,000 units of antibiotic.
(B) Bacteria that were already resistant to 3,000 units of antibiotic would be selected for.
(C) There would be no growth on the plate.
(D) The bacteria would take up plasmids from the environment that would permit them to grow in the presence of 3,000 units of antibiotic.

34. **A scientist collected resistant bacteria growing on the segment containing the highest antibiotic concentration. That scientist also collected susceptible bacteria growing on the segment with no antibiotic. The scientist plated each strain on 10 agar plates containing no antibiotic and then measured the amount of growth over a period of 7 days. The antibiotic-susceptible strain grew twice as fast as the resistant strain. Which of the following is the best explanation for this?**

    (A) The antibiotic-resistant strain devoted energy to processes that made it able to survive in the presence of antibiotics. In the absence of antibiotics, bacteria that did not spend energy on resistance (the susceptible strain) were able to use that energy for growth.
    (B) Growing on antibiotics stunted the growth of the antibiotic-resistant strain.
    (C) When exposed to the mutant strain, the original strain increased its metabolic efficiency to outcompete the mutant.
    (D) Since mutations had built up in the resistant strain, that strain was more susceptible to viruses in the environment.

35. **Another group of researchers used a similar setup as the one described, but they used different antibiotics in each segment instead of different concentrations of the same antibiotic. Which of the following was the likely result?**

    (A) The bacteria died because they had resistance to the wrong antibiotic.
    (B) A similar pattern emerged. As the bacteria encountered a new environmental challenge, individuals with mutations that allowed them to survive in the new environment were selected.
    (C) The bacteria detected the new drug and chose genes from the environment to overcome the challenge.
    (D) The bacteria spread faster than in the original setup because adaptation to a new drug is a simpler process than adaptation to a higher concentration of the same drug.

Questions 36–38

All of life as we know it needs water. Water's ability to form up to 4 hydrogen bonds gives it many unique properties. Hydrogen bonds form when a hydrogen bound to a nitrogen (N), oxygen (O), or fluorine (F) interacts with a lone pair of electrons on another nitrogen, oxygen, or fluorine. The oxygen atoms shown on both a fatty acid (Figure I) and a phospholipid (Figure II) have lone pair electrons that are available for hydrogen bonding. The long chains of carbon and hydrogen shown toward the bottom of both figures do not contain such lone pairs, nor do they contain any N, O, or F. Phospholipids form phospholipid bilayers and are the primary components of membranes.

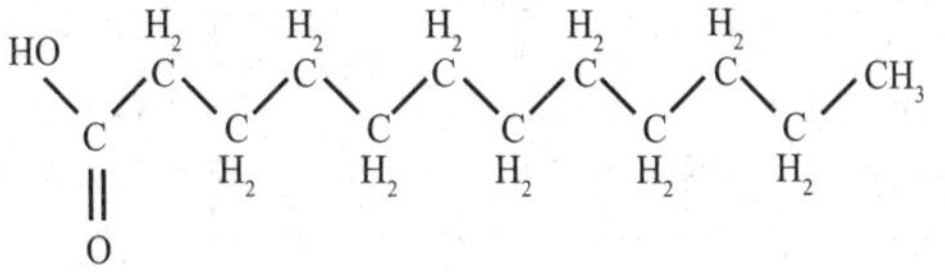

Figure I The structure of a fatty acid

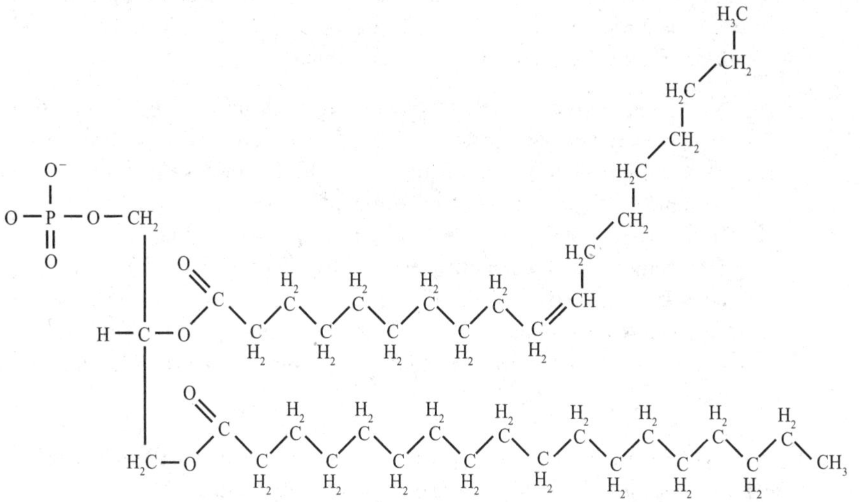

Figure II The structure of a phospholipid, which is a major component of membranes

36. Phospholipids have a polar end that can form hydrogen bonds and is hydrophilic, and a nonpolar end that is hydrophobic. Based on this information and the information given in the preceding material, which of the following is true about the fatty acid shown in Figure I?

(A) The fatty acid is unlike the phospholipid because the fatty acid does not have a hydrophilic end that interacts with water. (The fatty acid is completely hydrophobic.) Therefore, phospholipids can form bilayers in water while fatty acids cannot.
(B) Like the phospholipid, the fatty acid has both a hydrophilic end and a hydrophobic end. Therefore, fatty acids can also form bilayers in water.
(C) The fatty acid is unlike the phospholipid because the fatty acid does not have a hydrophobic end that will not interact with water. (The fatty acid is completely hydrophilic.) Therefore, phospholipids can form bilayers in water while fatty acids cannot.
(D) Translation of mRNA by ribosomes produces fatty acids. The sequence of carbon atoms incorporated in the polymer chain determines whether or not the fatty acid will be able to form a bilayer.

37. One hypothesis for an evolutionary pathway that would lead to modern membranes includes a step in which early membranes were composed of fatty acids. Which of the following best explains why fatty acids are a good candidate for a component of early membranes?

(A) Since fatty acids are capable of catalyzing reactions and are simpler than protein-based enzymes, they are likely to have evolved first.
(B) Phospholipids likely evolved into fatty acids because the complex structure of the phospholipid was unnecessary.
(C) Both fatty acids and phospholipids are capable of storing genetic information. Fatty acids likely evolved first because they contain hereditary material.
(D) Fatty acids are simpler in structure than phospholipids, but they still retain the ability to form a barrier that defines the inside and outside of a cell.

38. Which of the following would provide the best support for the idea that early protocells had membranes composed of fatty acids?

(A) the discovery of an organism that incorporates fatty acids in its modern membranes
(B) the discovery of an abiotic mechanism by which fatty acids could be spontaneously produced from inorganic carbon
(C) the elucidation of the metabolic pathway that builds fatty acids from acetyl CoA
(D) the discovery of a mechanism by which fatty acids interact with nucleic acids to produce protein

Questions 39–40

Fatty acids will spontaneously form spherical bilayers called vesicles when placed into water. When researchers added fatty acids to solutions containing spherical vesicles (Step 1), the free fatty acids were incorporated into the vesicles. Since this process happened faster than the vesicles could absorb water, the surface area of the vesicles grew faster than the volume of the vesicles. Subsequent agitation (Step 2) fragmented the structure into multiple vesicles.

39. **Which of the following diagrams best depicts the scenario described?**

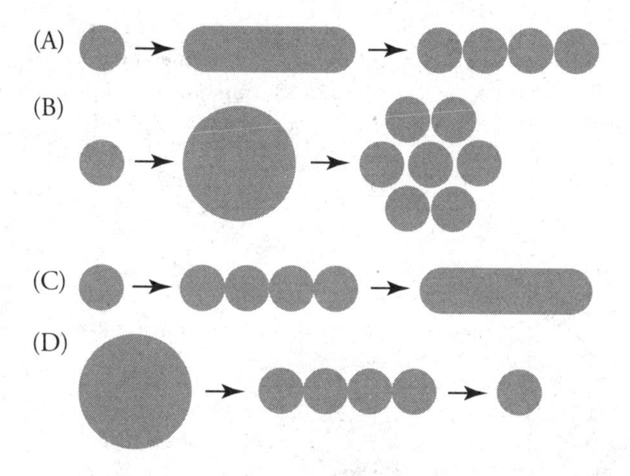

40. **The processes described in the passage are analogous to which of the following cellular processes?**

 (A) replication and death by apoptosis
 (B) growth and competition for resources
 (C) variation and natural selection
 (D) growth and cell division

Questions 41–45

The American chestnut (*Castanea dentata*) was the dominant tree species in the Appalachian Mountains before the 1900s. The nuts were prized as a special food item, and the wood was sought for its strength and beauty. A fungal parasite (*Chryphonectria parasitica*) was accidentally introduced to the United States around 1904 on Japanese chestnut (*Castanea crenata*) trees. By 1940, American chestnut trees were nearly absent from the forest. In response to the rapid spread of the disease, many foresters performed salvage harvests, cutting trees down before they got infected. American chestnut trees still resprout from roots in the Appalachian Mountains but succumb to the disease before producing seeds.

Since 1983, the American Chestnut Foundation has been working on breeding efforts to introduce resistance genes from Asian species of chestnut into American chestnut. To do this, the foundation first hybridized Asian species of chestnut with American chestnut and then backcrossed to American chestnut. At each generation, the foundation screened offspring for blight resistance.

Recently, researchers have turned to genetic engineering to develop blight resistant trees. As the chestnut blight fungus grows out from its infection site, it produces the toxin oxalate. Oxalate kills plant cells, which the fungus then consumes. Wheat plants have a gene that codes for the enzyme oxalate oxidase, which breaks down oxalate. Researchers spliced oxalate oxidase out of wheat and inserted it into the genome of the American chestnut. Resistance was measured by inoculating the fungus on the leaves and measuring how much plant tissue the fungus was able to kill (necrosis length) after 3 days. The results are presented below.

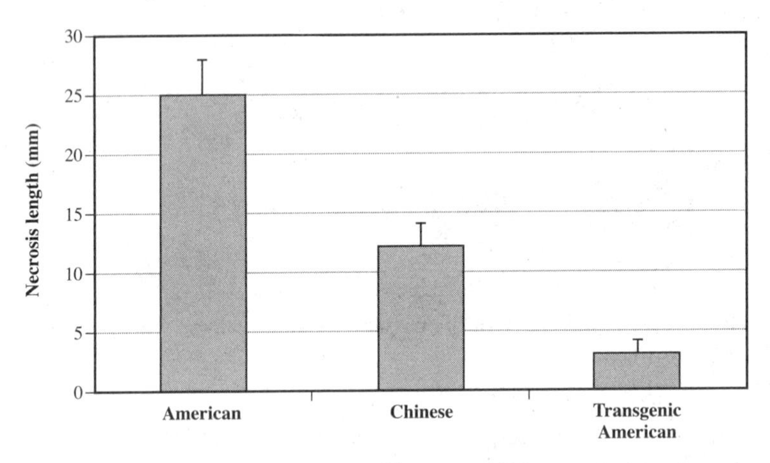

Figure I Fungus resistance in American, Chinese, and Transgenic American chestnut

41. **Which of the following is NOT true from an evolutionary point of view?**

   (A) Salvage harvesting in advance of the disease spread prevented any observations of preexisting resistance to the fungus.
   (B) Trees resprouting from stumps cannot naturally contribute to evolution because they do not live to produce offspring.
   (C) Japanese chestnut trees would be a good source of resistance genes because they probably coevolved with the pathogen.
   (D) Trees that resprout will eventually acquire resistance by repeated exposure to the pathogen.

42. Which of the following best explains why the fungus was able to spread so rapidly?

(A) The American chestnut tree had no genetic resistance to the disease.
(B) The fungus had no natural pathogens in North America.
(C) There were no large geographic gaps between populations of American chestnut trees.
(D) All of these likely contributed to the spread of the pathogen.

43. Which of the following best explains why the American Chestnut Foundation's breeding efforts include backcrossing the trees to American chestnut?

(A) to increase the amount of American chestnut genes
(B) to increase the amount of disease resistance
(C) to test for sexual compatibility
(D) to improve the taste of the fruit

44. Which of the following is an accurate interpretation of the data shown in Figure I?

(A) Transgenic American chestnut was significantly less resistant than either the American species or the Chinese species.
(B) There was no significant difference in resistance among the three different types of trees.
(C) Transgenic American chestnut showed levels of resistance that were no different than those shown by the Chinese chestnut.
(D) Transgenic American chestnut showed significantly better resistance to the fungus than did the Chinese chestnut.

45. Which of the following would be a valid evolutionary argument against widespread distribution of this strain of Transgenic American chestnut?

(A) Transgenes cause allergies in humans. It would be impossible for many people to enjoy the forest if this strain were planted in the wild.
(B) Transgenic American chestnut would hybridize with existing blight-susceptible individuals, diluting the resistance gene in the population.
(C) The American chestnut population would have essentially no genetic variability, making it highly likely that some other pathogen would become widespread.
(D) It is unethical to disrupt the natural course of evolution toward more highly evolved species.

46. A popular criticism of the theory of evolution is the assertion that adaptations seen in nature could not have evolved by random chance. Which of the following best explains the misunderstanding demonstrated by this criticism?

(A) Evolution involves organisms striving to better themselves and their species. It is a nonrandom process.
(B) The criticism contains no misunderstandings.
(C) If given enough time, random events can make anything happen.
(D) Evolution requires both the random variability of mutation and the nonrandom process of natural selection.

Questions 47–49

The diagram below shows phylogenetic trees for different host species of animals (capital letters) and the viruses that infect them (lowercase letters). All species listed as the leaves of the phylogenetic trees represent currently existing species.

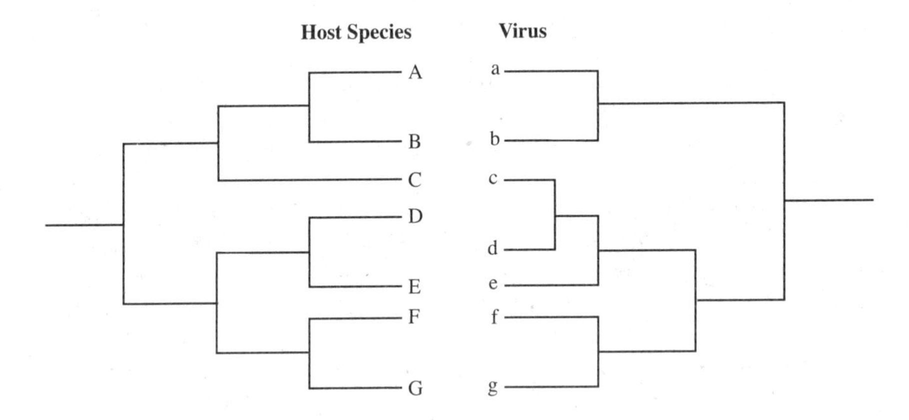

47. Which of the following is a reasonable conclusion based on the data in the phylogenetic trees?

(A) The branching patterns are highly dissimilar because the organisms underwent disruptive selection.
(B) The branching patterns are very similar because the hosts and their viruses underwent speciation at the same time.
(C) The patterns of speciation are similar because the viruses chose hosts that are most closely related to them.
(D) The branching patterns are dissimilar because evolution is a completely random process.

48. **Which of the following is true regarding host species C and virus c?**

(A) Host species C is most closely related to host species A and host species B, while virus c is most closely related to viruses d, e, f, and g.
(B) Host species C is most closely related to host species B and host species D, while virus c is most closely related to viruses a and b.
(C) Host species C is most closely related to host species D, host species E, host species F, and host species G, while virus c is most closely related to viruses a and b.
(D) Host species C is most closely related to virus c.

49. **The pair host species C and virus c is an exception to the pattern of evolution shown. Which of the following is the most likely explanation for virus c being able to infect host species C?**

(A) Virus c and host species C are the ancestral species from which all other species in these trees evolved.
(B) The host species that virus c relied on went extinct, forcing the virus to mutate so that it could choose a different host species.
(C) There is no exception shown. Virus c and host species C follow the pattern of the other pairs.
(D) Virus c was able to "jump the species barrier" to infect a host species with which it did not coevolve.

Questions 50–52

The diagram below depicts the distribution of the greenish warbler, a ring species of birds that inhabit the foothills of the Himalayas. These mountains are an impassable region that birds can fly around but not across. Each ellipse in the diagram represents a different population of birds. Where the populations overlap, birds from different populations can mate and produce viable hybrid offspring. Populations A + B, B + C, C + D, and D + E all produce viable offspring. The exception is populations A + E. When birds from these two different populations mate in the region of habitat overlap, they produce eggs that never hatch.

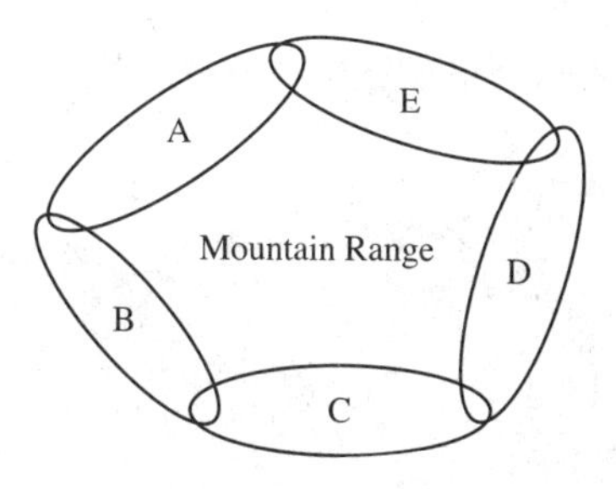

50. **Which of the following predictions about the genetic distance between the different populations is most likely?**

    (A) Populations A and B are more closely related to each other than populations D and E are to each other.
    (B) Populations A and D are more distantly related to each other than populations B and C are to each other.
    (C) Populations B and C are more distantly related to each other than populations D and E are to each other.
    (D) Populations A and E are more closely related to each other than populations A and D are to each other.

51. **Compared to populations B, C, and D, populations A and E have longer and more complex songs. Which of the following is the most likely explanation for this?**

    (A) Birds learn the calls of their parents. Since populations A and E are geographically isolated, there is no reason for their calls to be similar.
    (B) Populations A and E are exposed to different predators that respond differently to the calls.
    (C) Since there is no evolutionary significance associated with bird songs, individuals are able to sing any song they like.
    (D) Since there is a significant cost associated with hybridizing between populations A and E, there is a selection pressure for longer and more complex songs that help individuals select appropriate mates.

52. **What would be the result if populations B, C, and D were destroyed?**

    (A) Populations A and E would hybridize to preserve the species as a single genetic unit.
    (B) Populations A and E would no longer be able to exchange genetic information and would be considered two different species.
    (C) Without the support of populations B, C, and D, the remaining populations would also go extinct.
    (D) Populations A and E would recolonize the habitat vacated by populations B, C, and D and would be able to interbreed successfully in region C.

---

53. **The figure below shows the underlying bone structure of three different animals.**

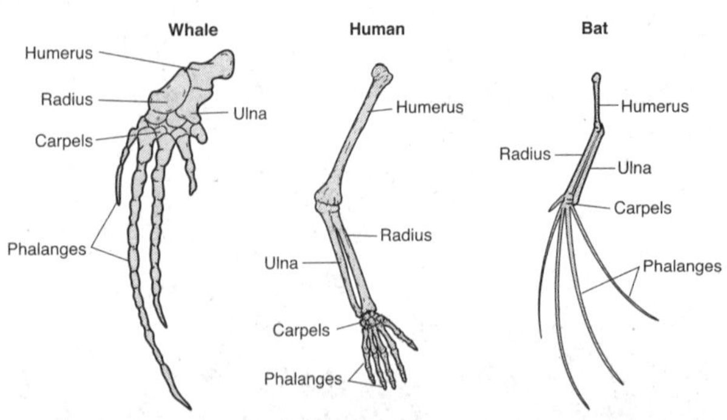

Which of the following conclusions can be reasonably drawn from this figure?

(A) All of these animals had a common ancestor that had the same underlying bone pattern.
(B) The similarities of the bone structures are the result of convergent evolution in which similar environmental forces caused different organisms to evolve similarly.
(C) The underlying bone structures are so different that we can conclude that each line evolved independently from different single-celled ancestors.
(D) The similarities in the bone structures suggest that all of these animals evolved from the whale.

Questions 54–55

Hox genes are a group of genes that code for transcription factors and act as switches to control organ identity in segmented animals. An AP Biology student performed a phylogenetic analysis of amino acid sequences for the proteins Hox-A1 (HA1) and Hox-B1 (HB1) for five mammal species. The protein Hox-D8 (HD8) from a mouse (*Mus musculus*) was used as an outgroup. The results are presented in the phylogenetic tree below.

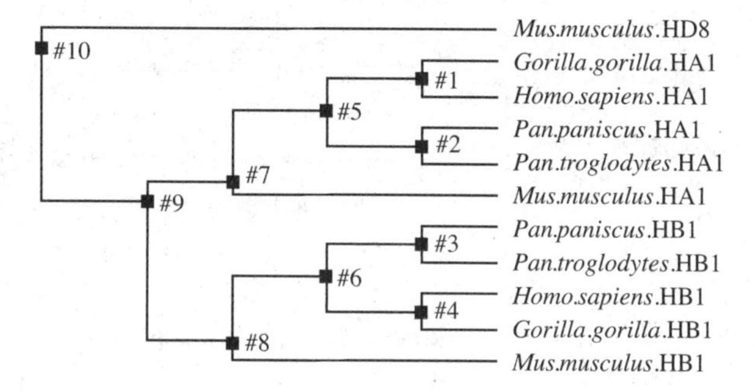

54. **The variety of *Hox* genes are the result of gene duplications. Which of the following conclusions about the timing of the gene duplication event that led to HA1 and HB1 is supported by the information in the passage and by the phylogenetic tree?**

(A) The duplication event occurred before the divergence of mammals from one another.
(B) The duplication event occurred more than once after these mammal species diverged.
(C) The duplication event occurred after mice diverged from the primates (*Gorilla, Homo*, and *Pan*) but before the primates diverged from one another.
(D) The data do not support the idea that *Hox* genes arose by gene duplication.

55. **Which of the following statements is a reasonable conclusion to draw from the phylogenetic tree?**

    (A) Humans (*Homo sapiens*) evolved from gorillas (*Gorilla gorilla*).
    (B) Gorilla HA1 amino acid sequences share more similarity with HA1 amino acid sequences from chimpanzees (*Pan troglodytes*) and bonobos (*Pan paniscus*) than with HA1 amino acid sequences from humans (*Homo sapiens*).
    (C) The HA1 gene in the mouse (*Mus musculus*) shares more similarities with the HA1 genes in other organisms than it does with the HB1 gene in mice.
    (D) The nodes labeled 7 and 8 suggest that the *Gorilla, Homo*, and *Pan* species all evolved from mice.

Questions 56–57

The question "Which came first, the chicken or the egg?" is presented by many people as an unsolvable conundrum. The phylogenetic tree shown below is supported by molecular analysis of many different proteins.

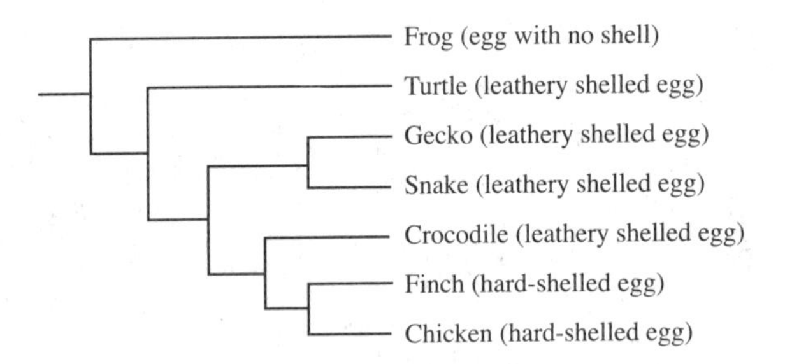

56. **Which of the following provides the best scientific answer to the question "Which came first, the chicken or the egg"?**

    (A) Since all of the animals shown on the tree make eggs, the egg must have evolved after the animals diverged from a common ancestor.
    (B) The chicken and the egg evolved at the same time because you cannot have one without the other.
    (C) Since all of the existing species on the tree make eggs, egg production must be an ancestral trait and must have existed in the common ancestor. Therefore, the egg came before the chicken.
    (D) It is impossible to draw a conclusion based on the information presented.

57. **The crocodile, finch, and chicken share a more recent common ancestor than do any of the other species shown on the tree. Which of the following is the most likely description of the type of egg their most recent common ancestor would have made?**

(A) The most recent common ancestor of the crocodile, finch, and chicken likely made an egg without a shell because frogs are ancestors to all of them and made that type of egg.
(B) The ancestor of the crocodile, finch, and chicken likely made a hard-shelled egg because two of the three make hard-shelled eggs.
(C) The ancestor of the crocodile, finch, and chicken likely made an egg with a leathery covering because species more distantly related to all of them also produced leathery shelled eggs.
(D) It is impossible to draw that kind of conclusion from the data presented.

---

58. **The diagram below shows two phylogenetic trees that represent the hypothetical evolutionary histories of fireflies. The tree on the left was based on molecular analysis of DNA sequences. The tree on the right was based on morphological characteristics. Lines between the two trees connect individual species.**

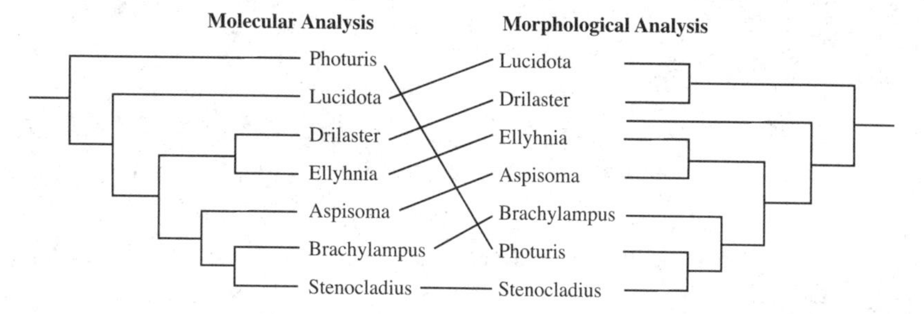

Which of the following is a reasonable conclusion from the phylogenetic trees shown?

(A) No significant differences exist between the phylogenetic relatedness predicted by the molecular analysis of DNA sequences and the morphological analysis.
(B) Molecular analysis of DNA sequences incorrectly predicted the phylogenetic relatedness of firefly species.
(C) Morphological analysis incorrectly predicted the phylogenetic relatedness of firefly species.
(D) Significant differences in the hypothesized phylogenies exist. Additional morphological and molecular studies should be done to resolve the positions of different species.

Questions 59–62

Students performed an experiment on Wisconsin Fast Plants. The students started by counting the number of trichomes (hairy structures) on the underside of the first true leaf of each plant. Then they selected seeds from the 5% baldest or hairiest plants in each generation to plant for the next generation. The graph below shows data from the original population (bars with diagonal lines), the bald-selected population (bars with no shading), and the hairy-selected population (bars with solid black shading) after 4 generations.

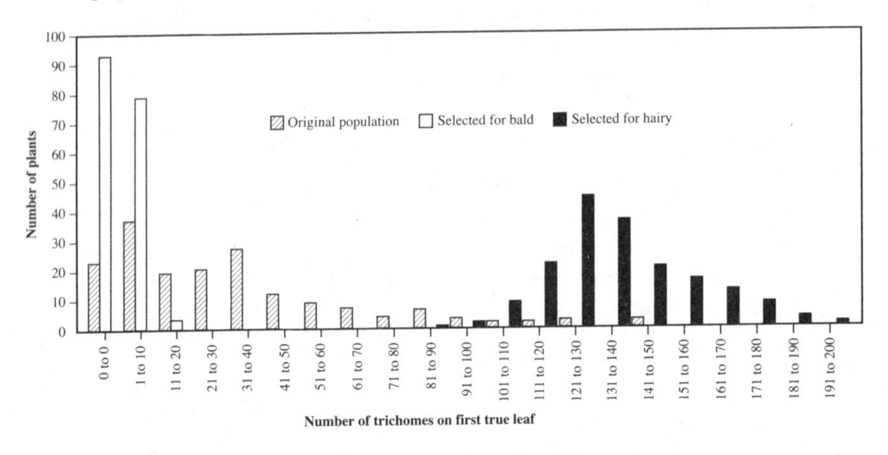

59. **Which of the following were the students investigating?**

   (A) sexual selection
   (B) natural or artificial selection
   (C) genetic drift
   (D) immigration and emigration

60. **Which of the following would be an appropriate control group?**

   (A) a group of plants that were grown under the same conditions as the selected plants but with no selection pressure for 4 generations
   (B) a group of plants created by hybridizing the bald-selected and hairy-selected plants
   (C) ensuring that all plants were exposed to the same amount of light
   (D) using the same microscope at the same magnification to examine the number of trichomes on each plant leaf

61. **Which of the following is an important controlled variable mentioned in the description of the experiment?**

(A) selecting only bald plants
(B) comparing the first true leaf of each plant
(C) growing the plants under the same amount of light
(D) using the same microscope at the same magnification to examine the number of trichomes on each plant leaf

62. **The average trichome number for the bald plants was 2.5, while it was 31.3 for the hairy plants. Which of the following accurately interprets these data?**

(A) These data are statistically different because 2.5 is less than 10% of 31.3.
(B) These data are not statistically different because 2.5 is more than 5% of 31.3.
(C) The graph shows significant overlap of the ranges, so the average trichome numbers are not significantly different.
(D) Since the ranges of the two average trichome numbers don't overlap, the two average trichome numbers are clearly different.

Questions 63–66

Students compared transpiration rates in Wisconsin Fast Plants with no trichomes (bald) and those with many trichomes (hairy). They planted individual plants in containers. The students measured the mass of each plant and container when fully watered and after 4 days without water. Half of the plants were subjected to a fan, and the other half were placed nearby without a fan. All plants were located 15 cm from fluorescent lights at room temperature, which fluctuated from 20°C to 22°C. The data are presented below.

Fan or No Fan	Bald or Hairy	Average % Change in Mass	Standard Error
Fan	Bald	39	2.1
Fan	Hairy	21	4.1
No fan	Bald	24	3.9
No fan	Hairy	7	2.5

63. Did the fan have a statistically significant effect?

(A) The fan did not have a statistically significant effect because the average standard error was too high.
(B) The fan did not have a statistically significant effect because 21 (for fan, hairy) and 24 (for no fan, bald) were closer together than the average standard error for these two groups.
(C) Since the differences between the averages are greater than the standard errors for the hairy plants, the results are likely to be statistically significant. That is not the case for the bald plants.
(D) Since the differences between the averages are greater than the standard errors, the fan likely had a statistically significant effect for both bald and hairy plants.

64. Was there a statistically significant effect of the trichomes in the presence of and in the absence of the fan?

(A) The trichomes did not have a statistically significant effect because the standard error was too high.
(B) The trichomes did not have a statistically significant effect because 21 (for fan, hairy) and 24 (for no fan, bald) were closer together than the standard error for these two groups.
(C) Since the standard errors are smaller than the difference between the averages, the trichomes likely had a statistically significant effect for plants in the wind and in still air.
(D) The trichomes had a statistically significant effect for plants that were exposed to the fan but not for plants that were not exposed to the fan.

65. Which of the following best explains the effect of the fan on transpiration?

(A) The fan causes evaporated air to move away from the leaf surface, creating lower water potential in the air and making it more likely for water molecules to evaporate from the leaf.
(B) The fan makes the temperature drop, leading to higher water potential in the air and making it more likely for water molecules to evaporate from the leaf.
(C) The fan provides energy for the plant to perform active transport of water out of the xylem.
(D) The fan has no effect on transpiration.

66. Which of the following best explains the effect of the trichomes on transpiration?

(A) Hairy leaves with trichomes trap water vapor next to the leaf surface. As a result, the water potential near the leaf surface is higher than it would be in a bald leaf.
(B) Trichomes increase the surface area of the leaf and therefore lead to more transpiration.
(C) Trichomes prevent water from settling on the leaf, which leads to increased transpiration rates.
(D) Trichomes have no effect on transpiration.

Questions 67–71

Figures I through IV are graphs that show the frequency of allele *A1* in simulated populations over time. The table shows both forward and backward mutation rates of allele *A1*.

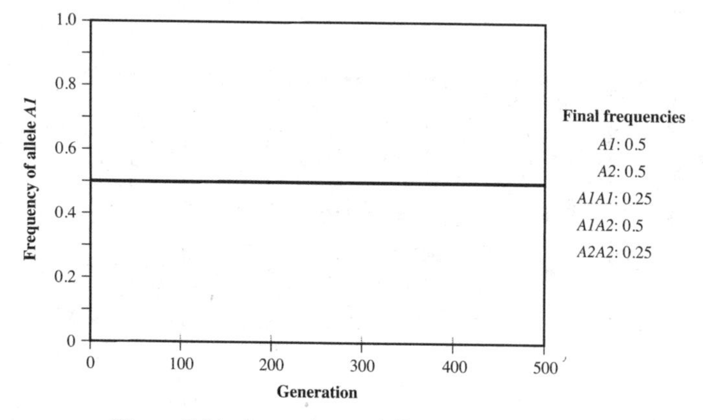

Figure I Ideal population (all assumptions met)

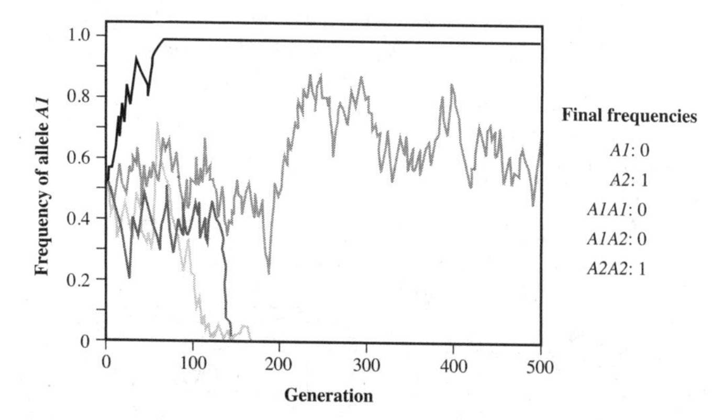

Figure II Small population (100 individuals)—four runs of the simulation are shown on the same graph

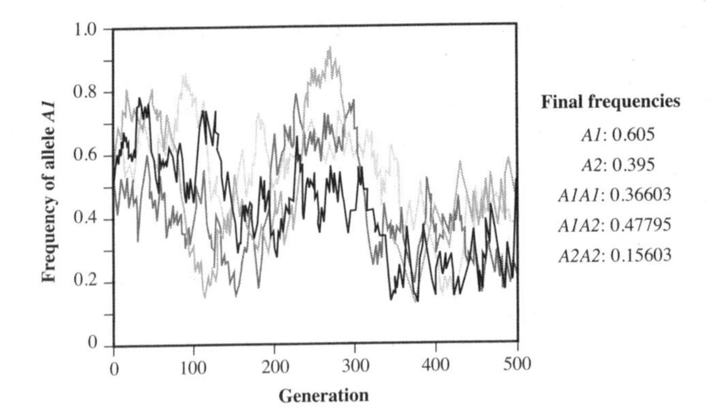

Figure III Small population (100 individuals) with immigration from a larger population—four runs of the simulation are shown on the same graph

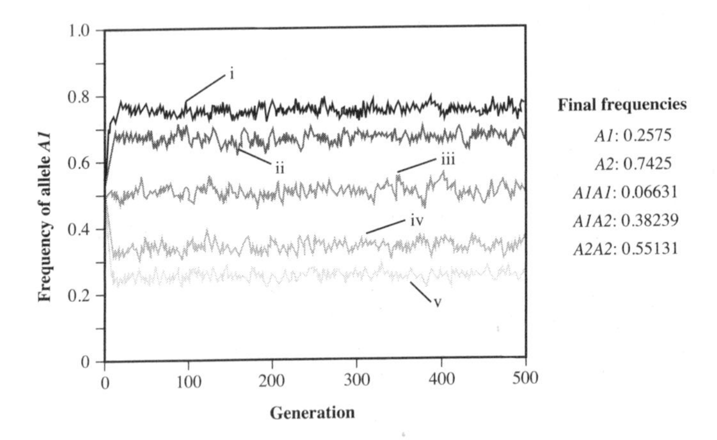

Figure IV The effect of different mutation rates on a population of 1,000—five runs of the simulation are shown on the same graph with different mutation rates as shown in the table that follows

Line	Forward Mutation Rate	Backward Mutation Rate
i	0.3	0.1
ii	0.2	0.1
iii	0.1	0.1
iv	0.1	0.2
v	0.1	0.3

67. **Which of the following is the most likely explanation for the variability in the small population shown in Figure II in comparison to the infinitely large population in Figure I?**

(A) Sexual selection is much more likely in a small population as each individual knows all of the others. This leads to more variability.
(B) Inbreeding is more common in a small population. This leads to a higher mutation rate and more variability.
(C) Since individuals in a small population are a larger portion of the whole than in an infinitely large population, random fluctuation is more pronounced in the small population. This is known as genetic drift.
(D) The infinitely large population could have been more variable. It simply wasn't in this particular simulation.

68. **Which of the following is a consequence of small populations that has an impact on wildlife conservation and is demonstrated in the comparison of Figures I and II?**

(A) Small populations are more likely to lose genetic variability due to genetic drift.
(B) Large populations are more likely to lose genetic variability because outliers are selected against.
(C) Small populations are more likely to lead to higher variability due to increased mutation rates.
(D) Sexual selection is more intense in small populations. This leads to increased genetic variability.

69. **Compare Figures I, II, and III. The simulations in Figure III use the same conditions as in Figure II. However, Figure III includes immigration from a large population on a regular basis. Which of the following is a consequence of immigration?**

(A) Immigration exacerbates genetic drift because the alleles from the larger population distort the natural allele frequencies of the smaller population.
(B) Since immigration introduces a variety of alleles from the larger population, the effects of genetic drift are mitigated.
(C) Immigration has no effect on the smaller population's diversity.
(D) Immigration decreases genetic drift because immigration removes deleterious alleles from the gene pool.

70. **Figure IV shows the effect of different mutation rates. In all cases, the population changes and then reaches a relatively stable level. Which of the following is true about these populations?**

(A) Since mutations are occurring, the populations never reach Hardy-Weinberg equilibrium.
(B) The populations are in Hardy-Weinberg equilibrium at all times.
(C) Evolution is occurring when the allele frequencies are changing but then approaches an equilibrium state with no net mutation.
(D) Evolution is occurring when the allele frequencies stay the same.

71. **Which of the following best explains why the population represented in the top line of Figure IV reaches a relatively stable allele frequency?**

(A) All the deleterious alleles were removed from the population.
(B) Natural selection stopped.
(C) There are more copies of allele *A2* in the population. Since there is also a higher mutation rate of allele *A2* as compared to allele *A1*, the copies of *both alleles* reach equilibrium.
(D) There are more copies of allele *A1* in the population. Even though each one of these copies is less likely to mutate to *A2*, there are enough copies to balance the copies of *A2* mutating back to *A1*.

---

72. **Adelgids are insects that feed on the sap of trees. A particular species of adelgid was accidentally imported into North America from Japan. It was able to feed on native hemlock and spruce trees, which are abundant in the forests of eastern North America. Which of the following likely had the greatest impact on native forests?**

(A) Since the adelgid did not have any native predators or pathogens in the new environment, it likely spread uncontrollably.
(B) North American birds ate this new resource and kept the adelgid population small.
(C) A virus that infects native adelgids evolved to attack the foreign species.
(D) The adelgid species interbred with native adelgids, eventually demonstrating a founder effect.

Questions 73–77

Maple syrup urine disease (MSUD) is an autosomal recessive genetic disorder caused by a mutated form of an enzyme that is responsible for breaking down the amino acids valine, leucine, and isoleucine. The disorder causes a buildup of these amino acids, resulting in urine that smells like maple syrup. If left untreated, the disorder causes damage to the central nervous system, coma, and death. In the world population, 1 in 185,000 babies have this condition. In the Old Order Mennonite (OOM) population (total population approximately 27,000), however, 1 in 380 children are born with this disorder.

A man, with no symptoms of MSUD, whose mother had the disorder, married a woman who had no ancestors in the OOM population. Since they were planning on having children, they consulted a genetic counselor about their concerns about MSUD.

73. **Which of the following is the most likely explanation for the difference in frequency of MSUD between the world population and the OOM population?**

    (A) This is an example of the founder effect. The original OOM population had, by random chance, a higher frequency of the allele for MSUD.
    (B) This is an example of natural selection. Since the OOM population generally eats a diet rich in valine, leucine, and isoleucine, there is selective pressure for the MSUD allele.
    (C) This is an example of heterozygote advantage. Heterozygotes are more likely to survive mosquito-borne illnesses.
    (D) OOM families tend to have more children than the average family, which leads to a higher rate of mutation.

74. **Which of the following is the best advice that the genetic counselor could give concerning the man's genotype?**

    (A) Since your mother had the disorder and you received your X chromosome from her but do not have symptoms, you have no chance of passing on the allele to your child.
    (B) Since your mother had the disorder, she was homozygous recessive for the MSUD allele. Therefore, you have a 50% chance of passing the allele to your child.
    (C) Since your mother had the disorder, you are part of the OOM population and therefore you have a 1 in 380 chance of passing on the allele to your child.
    (D) Since your mother had the disorder and you received your X chromosome from her, you have a 100% chance of passing on the allele to your child.

75. **Which of the following is the best advice that the genetic counselor could give concerning the woman's genotype?**

(A) Since you do not come from the OOM population, there is absolutely no chance that you have the allele or that you can pass on the allele to your children.
(B) Since you have no ancestors in the OOM population, you have a 1 in 185,000 chance of having a baby with the disease if you have a child with someone with no family history of the disease. The odds that you are a carrier are therefore approximately 0.4%.
(C) Since you are marrying into the OOM population, you have a 1 in 380 chance of having a child with the disorder.
(D) Since your mother-in-law had the disorder, she was homozygous recessive for the MSUD allele. Therefore, you have a 50% chance of having the allele.

76. **The couple's first child does not have the disease. Which of the following is true concerning the risk for future children?**

(A) Since the first child did not have the disease, the second child has 0% chance of inheriting the disease.
(B) The odds for the second child having the disease remain the same as the odds were for the first child having the disease.
(C) Since the first child did not have the disease, the second child has a higher chance of inheriting the disease.
(D) Although the odds are not zero for the second child, they are less than they were before the first child was born.

77. **The couple's second child is born with MSUD. What are the odds that their third child will have the disorder?**

(A) 100%
(B) 50%
(C) 25%
(D) 0%

Questions 78–79

The *Pax-6* gene is involved in eye development in mice. A mutated version of the gene produces an eyeless phenotype in mice. Fruit flies, squid, and flatworm strains exist that also have eyeless phenotypes. Researchers transformed these eyeless strains of fruit flies, squid, and flatworms with the wild-type *Pax-6* gene from mice. The fruit flies, squid, and flatworms that resulted all developed eyes typical of their species. For example, the fruit fly developed a compound eye, not a mouse eye.

78. **Which of the following is most likely to be true of the *Pax-6* gene?**

    (A) *Pax-6* codes for a transcription factor protein involved in turning on the genes that make different types of eyes in the different species.
    (B) *Pax-6* has completely different sequences in fruit flies, squid, and flatworms and produces different kinds of eyes as a result.
    (C) *Pax-6* contains all of the information for making the different kinds of eyes, but only part of the gene is used in each organism.
    (D) *Pax-6* codes for ribosomes involved in producing mRNA.

79. **Based on the information provided, which of the following is an accurate statement that is consistent with a modern understanding of evolution?**

    (A) This experiment suggests that eyes evolved multiple times in different lineages.
    (B) Since the same DNA sequence can turn on the production of different types of eyes, the different types of eyes likely shared a common origin.
    (C) This experiment reinforces the idea that a structure as complex as the eye couldn't have evolved.
    (D) Since the same DNA sequence can produce different types of eyes, it must contain all the information to make compound as well as lens-type eyes.

Questions 80–84

Students in an AP Biology class obtained 10 fruit fly colonies that had been kept for 50 generations. Half of the colonies (M1–M5) were fed maltose. The other half (S1–S5) were fed starch.

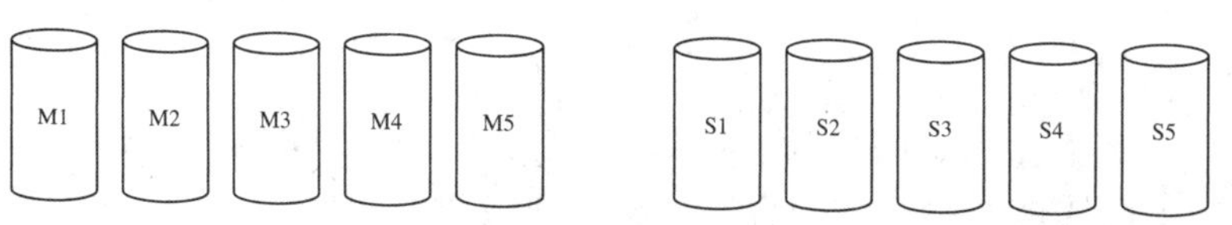

The students conducted mate choice experiments. They introduced 12 virgin males and 12 virgin females into a chamber and video recorded each chamber for 90 minutes. The students replayed the videos on high speed and recorded the number of mating bouts for each trial. They conducted all possible combinations of matings, some of which are shown in Tables I, II, and III.

Table I Matings of M1 and M2 Males and Females

	Females	
Males	**M1**	**M2**
M1	22	21
M2	23	20

Table II Matings of S1 and S2 Males and Females

	Females	
Males	**S1**	**S2**
S1	23	25
S2	21	22

Table III Matings of M1 and S1 Males and Females

	Females	
Males	**M1**	**S1**
M1	22	8
S1	7	21

80. **Which of the following is the most appropriate null hypothesis for the data shown in Table I?**

    (A) The observed values do not differ significantly from a 9:3:3:1 ratio (48, 16, 16, and 5).
    (B) The observed values do not differ significantly from a 1:1:1:1 ratio (21.5 for each square).
    (C) The observed values differ significantly from a 1:1:1:1 ratio (21.5 for each square).
    (D) The observed values differ significantly from a 3:1 ratio (64.5 and 21.5).

81. **The chi-square (χ^2) value for the results shown in Table I is 0.23. Which of the following is a correct interpretation of the χ^2 statistic?**

    (A) There are 3 degrees of freedom. Using the 0.01 cutoff, we reject the null hypothesis because the χ^2 value is less than 11.34.
    (B) There is 1 degree of freedom. Using the 0.05 cutoff, we reject the null hypothesis because the χ^2 value is less than 3.84.
    (C) There are 3 degrees of freedom. Using the 0.01 cutoff, we accept the null hypothesis because the χ^2 value is less than 11.34.
    (D) There are 4 degrees of freedom. Using the 0.05 cutoff, we reject the null hypothesis because the χ^2 value is less than 9.49.

82. Which of the following is the most reasonable conclusion to draw from Tables I and II?

(A) Flies grown on the same media preferentially hybridize with individuals from a different population rather than mate within their own population.
(B) Flies grown on the same media show no preference for mates from either within or outside of their own population.
(C) Males and females grown on different media were less likely to mate than males and females grown on the same media.
(D) Males and females grown on starch were more likely to hybridize with individuals from another population whereas flies grown on maltose were more likely to select mates grown on maltose.

83. Which of the following is the most reasonable conclusion to draw from Table III?

(A) Flies grown on different media preferentially hybridized rather than mate within their own population.
(B) All males preferred females grown on starch, while all females preferred males grown on maltose.
(C) Males and females grown on different media were less likely to mate than males and females grown on the same media.
(D) Fruit flies do not have brains capable of mate selection. Therefore, random mating always takes place.

84. Which of the following is the most reasonable evolutionary interpretation of the data shown in all three tables?

(A) Genetic drift alone is sufficient to create reproductive isolation in fruit flies.
(B) Immigration and emigration explain the variation seen in the results.
(C) Selection for growth on different media had a greater impact on mate choice than isolation alone.
(D) Mate choice has little effect in creating new species.

Questions 85–88

Students performed an experiment on a species of *Tetrahymena*. They placed the organisms into solutions with different concentrations of salt and then measured the rate of contraction of the contractile vacuole. Contractile vacuoles collect and expel water from the cell. Each of the 32 students measured the number of contractions for 1 minute at each molarity. The class data are presented in the table below as averages with standard errors of the mean (SEM).

Osmolarity (mM NaCl)	Number of Contractions per Minute	SEM
500	0	0.1
400	4.2	1.9
300	11.4	0.9
200	14.7	1.4
100	16.8	1.1
0	21.1	1.3

85. **Which of the following best explains why the contractile vacuole did not contract at 500 mM NaCl?**

    (A) *Tetrahymena* must be isotonic to the environment at 500 mM NaCl. Therefore, no contraction is necessary.
    (B) *Tetrahymena* die at 500 mM NaCl because the solution is hypertonic to their environment.
    (C) *Tetrahymena* are hypertonic to the environment at 500 mM NaCl. Therefore, no contraction is necessary.
    (D) ATP production is impossible at 500 mM NaCl. Therefore, no contraction is possible.

86. **Which of the following would most likely be true if *Tetrahymena* were exposed to 600 mM NaCl solution?**

    (A) The rate of contraction would remain zero but the cells would stay the same size due to increased aquaporins in the membrane.
    (B) The contraction rate would increase to expel the excess NaCl.
    (C) The cells would shrink due to water moving from the cells into the hypertonic environment.
    (D) The cells would burst as a result of water moving into them due to the hypertonic environment.

87. **Which of the following describes a possible feedback cycle that maintains osmotic homeostasis in *Tetrahymena*?**

    (A) When the internal osmolarity falls, passive transport mechanisms are activated. These transport pure water into the contractile vacuole, restoring internal solute concentrations. This is a positive feedback loop.
    (B) When the internal osmolarity falls, passive transport mechanisms are activated. These transport pure water into the contractile vacuole, restoring internal solute concentrations. This is a negative feedback loop.
    (C) When the internal osmolarity falls, active transport mechanisms are activated. These transport pure water into the contractile vacuole, restoring internal solute concentrations. This is a positive feedback loop.
    (D) When the internal osmolarity falls, active transport mechanisms are activated. These transport pure water into the contractile vacuole, restoring internal solute concentrations. This is a negative feedback loop.

88. **Which of the following best represents the data graphically?**

    (A) 

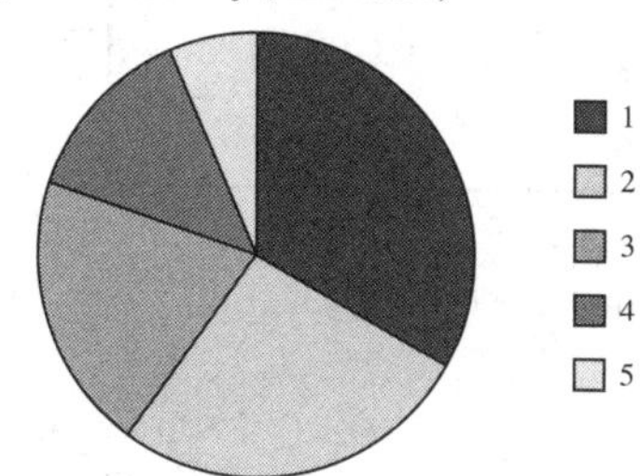

(B) 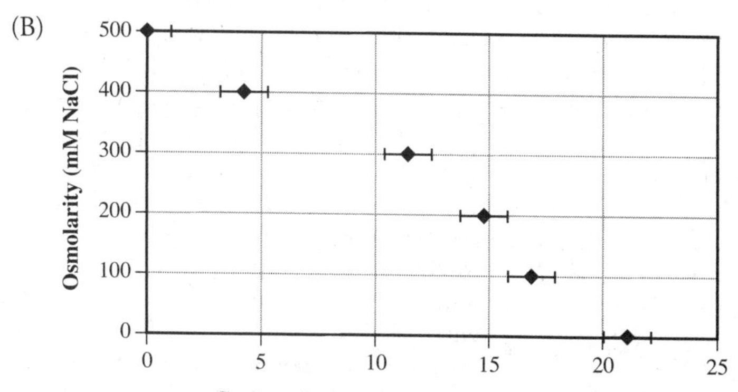

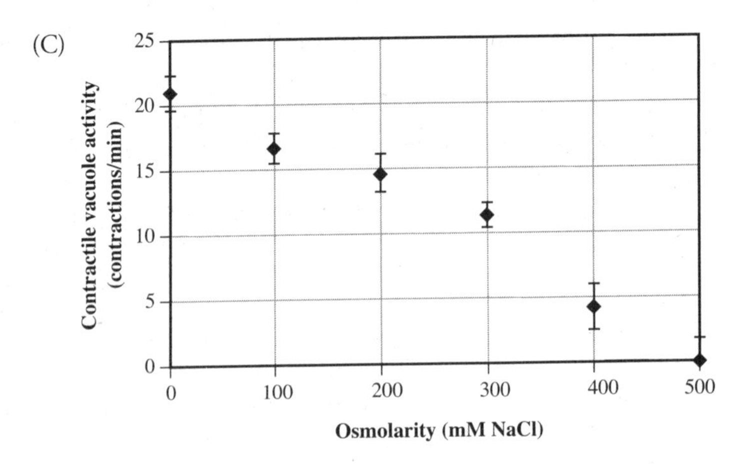

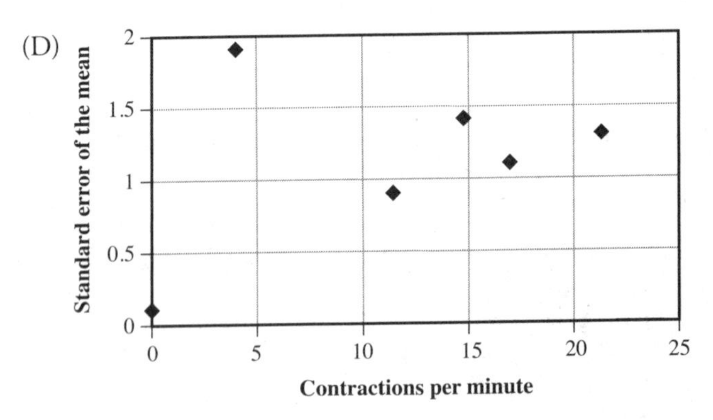

Questions 89–92

Students performed serial dilutions. They added 1 mL of a culture of yeast to 9 mL of culture medium in a test tube and mixed the contents of the tube. The students then took 1 mL of this 1/10 dilution, added it to another 9 mL of culture medium in a test tube, and mixed the contents of the tube. They continued to repeat this dilution process. The students then took 100 μL from each test tube, including the undiluted culture, and plated each sample onto a separate agar plate. After incubating the plates in identical conditions for the same amount of time, the students counted the number of yeast colonies on each plate.

Dilution	Number of Yeast Colonies
0	Lawn
1	Lawn
2	Lawn
3	Too numerous to count
4	230
5	19

89. Which of the following is closest to the number of colonies you would expect on dilution 3?

(A) 200
(B) 2,000
(C) 20,000
(D) 200,000

90. Which of the following is closest to the density of yeast in the dilution 4 tube?

(A) 23 yeast/mL
(B) 230 yeast/mL
(C) 2,300 yeast/mL
(D) 23,000 yeast/mL

91. Which of the following is closest to the density of yeast in the original tube?

(A) 2.3×10^{-7} yeast/mL
(B) 2.3×10^{7} yeast/mL
(C) 2.3×10^{-3} yeast/mL
(D) 2.3×10^{3} yeast/mL

92. Which of the following is an accurate explanation of "lawn" growth?

(A) Lawn growth is caused when bacteria fuse membranes to create one large cell.
(B) At the concentrations on each agar plate, there are too few yeast to make individual colonies.
(C) Viruses behave like blades of grass, causing the bacteria to lyse and appear as a smear on the plate.
(D) At the concentrations on each agar plate, there are enough yeast initially so that the colonies overlap and the plate becomes completely covered.

Questions 93–96

The diagram below represents the hypothetical phylogenetic history of a group of organisms. The most recent time is labeled IV, and the most ancient time is labeled I. Each branch on the tree represents groups of species at the order level. (An order is a group of families that contains groups of genera that contain groups of species.)

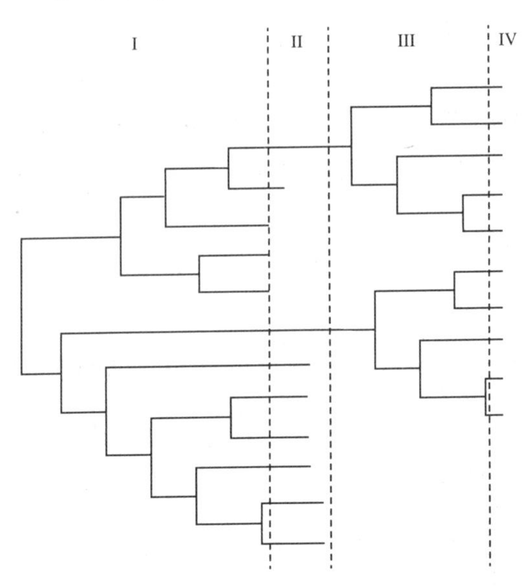

93. **Which of the following is true concerning time period II?**

    (A) Large numbers of species emerged during this time.
    (B) Large numbers of species went extinct during this time.
    (C) Existing species interbred.
    (D) Species in period III competed and only a few remained in period II.

94. **Which of the following is NOT a plausible explanation for the effect seen during time period II?**

    (A) the impact of a large comet, which rapidly lead to changes in global temperature and ocean chemistry
    (B) a change from an oxygen-free atmosphere to one containing a large percentage of oxygen over a small geological time period
    (C) a long period of volcanic activity that slowly reshaped the landscape
    (D) a geologically rapid period of change in atmospheric and oceanic oxygen concentrations due to the colonization of land by plants

95. Which of the following is an accurate description of the events in time period III?

(A) The number of species decreased due to the effect of a disaster such as a comet impact.
(B) An adaptive radiation took place in which there was a rapid diversification of life.
(C) The species present entered a period of stasis, in which stabilizing selection prevented changes in phenotypes.
(D) Genetic diversity increased to a level not seen before in the phylogenetic history shown in this diagram.

96. What might have happened in time period II to cause the results in time period III?

(A) Inbreeding among existing species led to their extinction. This allowed species that hybridized to diversify.
(B) Most of the existing species accumulated mutations and died out. Those that survived had mutations that were selected for.
(C) Massive overpopulation of the environment by a few orders caused species in other orders to evolve faster in order to survive.
(D) A disaster such as a comet impact caused species to go extinct. This opened ecological niches to which surviving organisms adapted.

Questions 97–99

The diagram below shows two models of evolution.

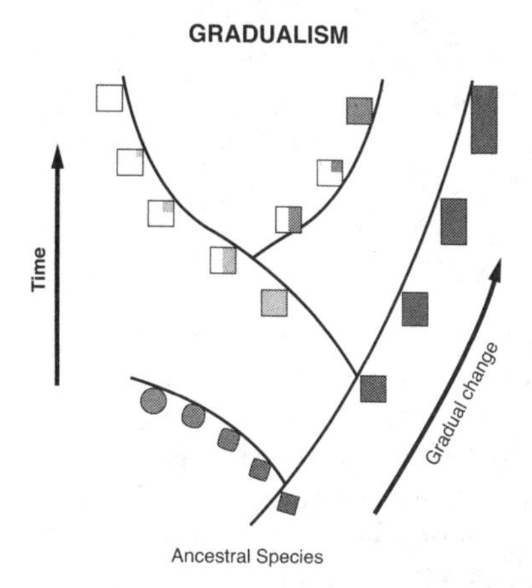

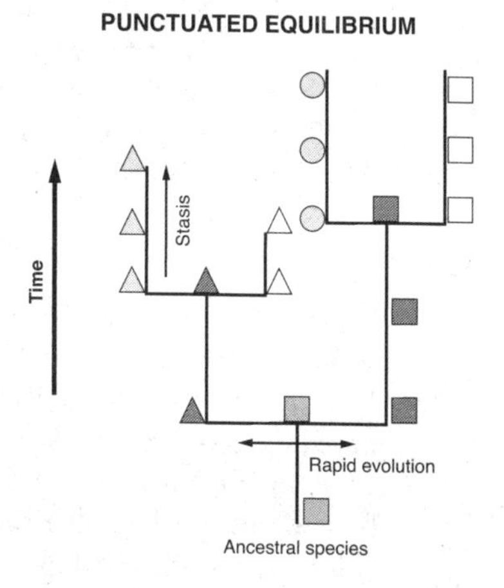

97. **Which of the following best describes the main difference between the two models?**

(A) One model includes evolution by natural selection, while the other doesn't.
(B) One model assumes that the rate of evolution is relatively constant, while the other proposes that the rate of change is variable.
(C) One model leads to a greater number of species in a shorter period of time.
(D) There is no substantive difference between the two models.

98. **Which of the following would be a prediction made by punctuated equilibrium that would not be made by gradualism?**

(A) It should always be possible to find intermediate forms between known fossil species.
(B) Natural selection over long periods of time will lead to a diverse number of forms.
(C) Given a complete fossil record, there would still be changes in organisms that appear to occur rapidly.
(D) Organisms will eventually give rise to sentient beings.

99. **According to the punctuated equilibrium model, which of the following explains why there are periods of stasis?**

(A) Genetic drift causes periods of stasis because random fluctuations are inherently stabilizing.
(B) Periods of stasis represent times when organisms are content in their environment and are not striving to change.
(C) During periods of stasis, species are under stabilizing selection in which the extreme phenotypes are selected against.
(D) During periods of stasis, populations are under diversifying selection in which the extreme phenotypes are selected for.

Questions 100–103

The goldenrod gall fly, *Eurosta solidaginis,* lays eggs on tall goldenrod (*Solidago altissima*). The larvae that hatch burrow into the stem of the plant and make spherical swellings in the stem. These larvae are preyed upon by woodpeckers. Woodpeckers tend to select larger galls. A species of parasitoid wasp lays eggs in the larvae inside of galls. These wasps attack smaller galls, probably because the larva must be within reach of their ovipositor. The data that follow, which were collected by students in AP Biology classes, were from the same field of goldenrod over a period of 3 years.

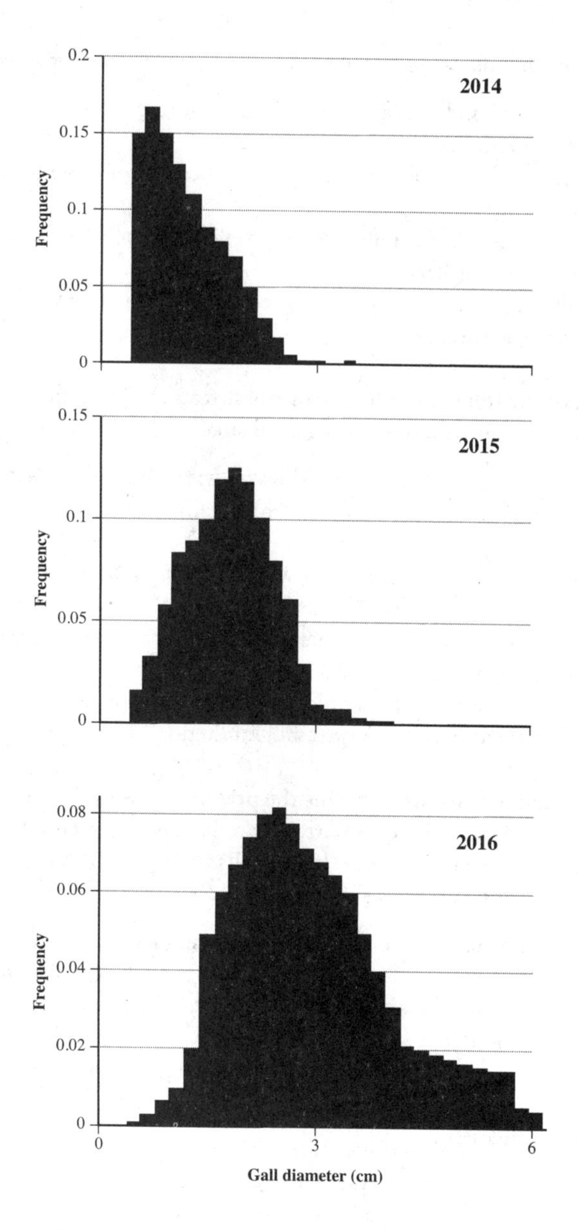

100. **Which of the following is an accurate description of the data?**

(A) The frequency distribution shifted toward smaller galls over the period of 3 years.
(B) The gall size stayed about the same over 3 years.
(C) The gall size fluctuated widely over 3 years.
(D) The frequency distribution shifted toward larger galls over the period of 3 years.

101. Which of the following statements is consistent with the data shown?

(A) Gall size is under directional selection because of fewer parasitoids or more woodpeckers.
(B) Gall size is under directional selection because of more parasitoids or fewer woodpeckers.
(C) Gall size is under stabilizing selection because of more parasitoids and more woodpeckers.
(D) Gall size is under diversifying selection because of fewer parasitoids and fewer woodpeckers.

102. Which of the following experiments best tests the hypothesis that parasitoid wasps choose to lay eggs on smaller galls?

(A) Present newly matured, mated wasps with a choice of various-sized galls in the laboratory. Record which galls the females lay eggs on more frequently.
(B) Examine a number of smaller galls in the field for wasp damage. Record what percentage of small galls have damage.
(C) Present woodpeckers with galls containing parasitoids to see if the woodpeckers eat them.
(D) Measure the average length of each wasp's ovipositor and compare it with the radius of the largest gall infested.

103. Another hypothesis that explains the presence of wasp larvae on only smaller galls is that the wasps attempt to lay on larger galls, but their ovipositor does not reach into larger galls. The result is failure. Which of the following best tests this hypothesis?

(A) Present mated female wasps with one gall each. Examine which size galls were used by the wasps and which galls contained viable larvae.
(B) Present mated female wasps with only large galls to see if any attempt to lay eggs on them.
(C) Collect large galls and dissect them, looking for wasp damage.
(D) Present woodpeckers with wasp larvae, and count how many times the larvae are eaten.

Questions 104–106

The table below compares the genomes of a plant nucleus to those of the chloroplast and cyanobacteria.

Genome	Number of Base Pairs	Number of Chromosomes	Chromosome Shape	Ribosome Size (S)
Plant nucleus	1×10^9	Many	Linear	80
Chloroplast	1×10^6	One	Circular	70
Cyanobacteria	1×10^6	One	Circular	70

104. Which of the following is a reasonable conclusion based on the data presented?

(A) Chloroplast DNA and cyanobacteria DNA should share little to no sequence similarities.
(B) Plant cells regularly incorporate modern cyanobacteria in their cells.
(C) When evolving compartmentation, convergent evolution caused plant cells to develop something similar to cyanobacteria.
(D) Chloroplasts are more closely related to cyanobacteria than they are to the nucleus of the cell in which the chloroplasts exist.

105. Which of the following is a reasonable hypothesis for the evolution of the chloroplast?

(A) Chloroplasts evolved as intracellular parasites after plants colonized land.
(B) Chloroplasts evolved as starch storage organelles from the mitochondria, an organelle that produces large amounts of starch from respiration.
(C) Chloroplasts originated long before land plants when ancestors of cyanobacteria became engulfed in heterotrophic cells.
(D) Chloroplasts evolved as outgrowths of the endoplasmic reticulum.

106. If you were to compare the ribosomal DNA of the bacterium *E. coli* to the ribosomal DNA from the genomes in the table, which of the following would be the most accurate phylogenetic tree?

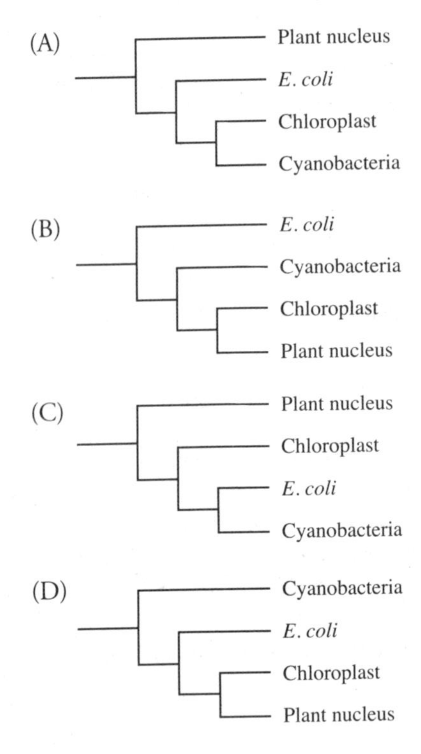

GRID-IN QUESTIONS

Questions 107–108

Maple syrup urine disease (MSUD) is an autosomal recessive genetic disorder caused by a mutated form of an enzyme that is responsible for breaking down the amino acids valine, leucine, and isoleucine. The disorder causes a buildup of these amino acids, resulting in urine that smells like maple syrup. If left untreated, the disorder causes damage to the central nervous system, coma, and death. The table below summarizes data about the frequency of MSUD.

Population	Frequency of MSUD	Population Size
Old Order Mennonite (OOM)	1 in 380	27,000
World (all humans)	1 in 185,000	7,500,000,000

Note: When working through the grid-in, long free-response, and short free-response questions throughout this book, be sure to keep scratch paper handy to work out any calculations.

107. Calculate the frequency of the recessive MSUD allele in the world population, assuming that the population is in Hardy-Weinberg equilibrium. Report your answer to the nearest ten thousandth.

108. Calculate the frequency of the recessive MSUD allele in the OOM population, assuming that the population is in Hardy-Weinberg equilibrium. Report your answer to the nearest ten thousandth.

---

109. The graph below shows the frequency of the homozygous recessive phenotype white fur in a population of mice.

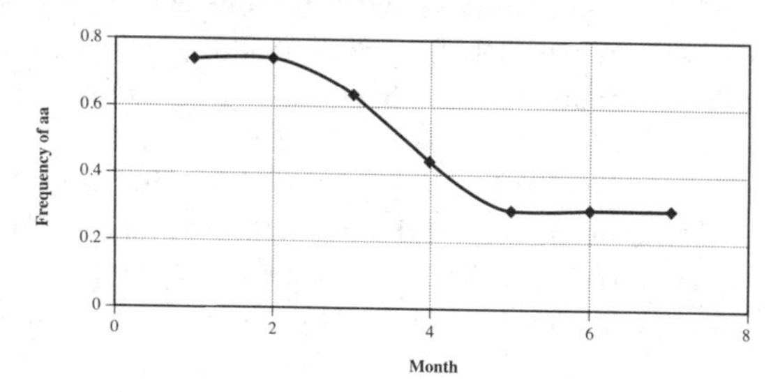

Using the data from the graph, calculate the frequency of the dominant allele during June (month 6), assuming that the population is in Hardy-Weinberg equilibrium. Express your answer to the nearest hundredth.

Questions 110–115

Students in 9 schools located at different latitudes in North America collected wild samples of the fruit fly *Drosophila melanogaster* in August. Approximately 100 flies were collected by each school. Each fly was screened for mutations in a gene for diapause using DNA sequencing. (Diapause is a period of slowed development associated with unfavorable conditions.) Flies that are homozygous for the diapause minus mutation do not enter diapause in response to exposure to a short day (long night).

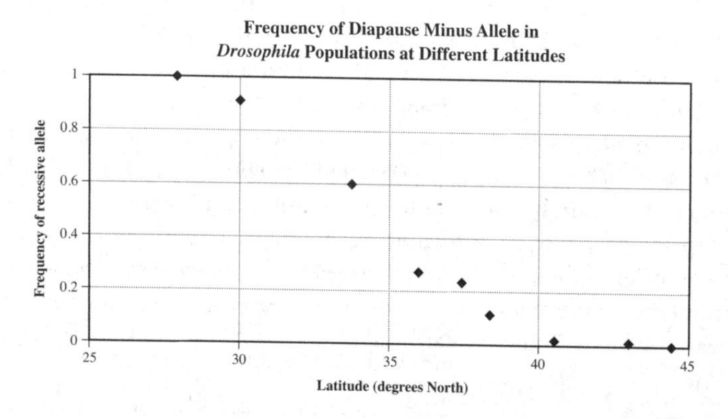

110. Assuming that the population at 30° N latitude is in Hardy-Weinberg equilibrium, what percentage of the population is expected to be unable to enter diapause when exposed to a short day (long night)? Express your answer as a number between 0 and 100 to the nearest whole number.

111. Assuming that the population at 30° N latitude is in Hardy-Weinberg equilibrium, what percentage of the population is expected to be heterozygous at this gene locus? Express your answer as a number between 0 and 100 to the nearest whole number.

112. Assuming that the population at 30° N latitude is in Hardy-Weinberg equilibrium, what is the allele frequency for the dominant allele? Express your answer as a number between 0 and 1 to the nearest tenth.

113. Assuming that the population at 34° N latitude is in Hardy-Weinberg equilibrium, what percentage of the population is expected to be unable to enter diapause when exposed to a short day (long night)? Express your answer as a number between 0 and 100 to the nearest whole number.

114. Assuming that the population at 34° N latitude is in Hardy-Weinberg equilibrium, what percentage of the population is expected to be heterozygous at this gene locus? Express your answer as a number between 0 and 100 to the nearest whole number.

115. Assuming that the population at 34° N latitude is in Hardy-Weinberg equilibrium, what is the allele frequency for the dominant allele? Express your answer as a number between 0 and 1 to the nearest tenth.

---

116. An AP Biology class performed an experiment on natural selection in a population of fruit flies exposed to red-backed jumping spiders. The fruit flies varied only in their body color. One form was black, and the other form was light brown. Body color is controlled by a single gene locus in which the black allele is dominant. A pure breeding black male fly was mated with a pure breeding light brown female fly. The offspring were permitted to mate to produce an F2 generation. The F2 fruit flies were introduced into chambers with dark colored walls. Three of the chambers contained 10 red-backed jumping spiders, and the other three chambers did not contain spiders. After 24 hours, the surviving flies were removed from the chambers and were permitted to mate and lay eggs. The students counted the number of black offspring and the number of light brown offspring of 200 of the offspring that resulted. Assuming that all of the assumptions of the Hardy-Weinberg equilibrium were met, how many fruit flies should show the recessive phenotype? Express your answer to the nearest whole number of flies.

Questions 117–118

Students in an AP Biology class obtained 10 fruit fly colonies that had been kept for 50 generations. Half of the colonies (M1–M5) were fed maltose. The other half (S1–S5) were fed starch.

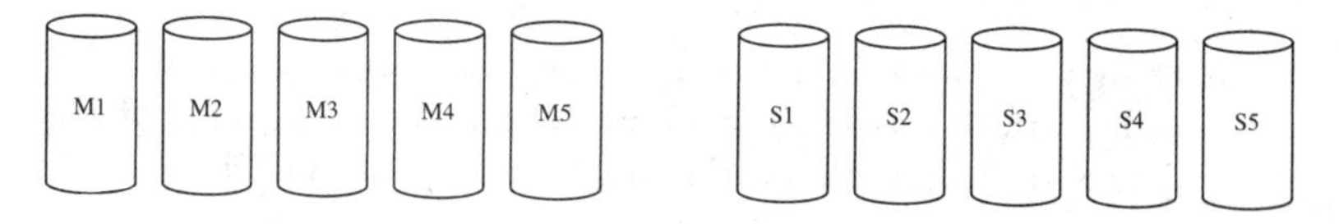

The students conducted mate choice experiments. They introduced 12 virgin males and 12 virgin females into a chamber and video recorded each chamber for 90 minutes. The students replayed the videos on high speed and recorded the number of mating bouts for each trial. They conducted all possible combinations of matings, some of which are shown in Tables I, II, and III.

Table I Matings of M1 and M2 Males and Females

	Females	
Males	**M1**	**M2**
M1	22	21
M2	23	20

Table II Matings of S1 and S2 Males and Females

	Females	
Males	**S1**	**S2**
S1	23	25
S2	21	22

Table III Matings of M1 and S1 Males and Females

	Females	
Males	**M1**	**S1**
M1	22	8
S1	7	21

117. **Calculate the chi-square (χ^2) value for the data in Table II. Record your answer to the nearest hundredth.**

118. **Calculate the chi-square (χ^2) value for the data in Table III. Record your answer to the nearest hundredth.**

LONG FREE-RESPONSE QUESTIONS

119. Microevolution (the change in allele frequencies in a population) and speciation (the generation of new species) have been observed in both the wild and in the laboratory.

    (a) **Design** a controlled experiment to cause microevolutionary change and speciation artificially in a sexually reproducing organism of your choice. **Include** in the description of your experimental design:

    - A rationale for your choice of organism
    - The conditions under which the organism will be grown
    - How you will assess microevolutionary change
    - The criteria you will use to determine whether you have created a new species
    - How you will use mathematics to analyze the data

    (b) **Identify** which of the five assumptions of the Hardy-Weinberg equilibrium will NOT be met in your experiment. **Discuss** the importance of TWO of these unmet assumptions with respect to microevolutionary change and speciation.

120. A veterinarian collected data on birth weights for lambs born on farms in his region over 25 years. The table below shows the percentage of lambs born in different weight categories for two different years.

Weight (kg)	Percentage of Births in 1990	Percentage of Births in 2015
0.9	2	0
1.4	6	0
1.8	11	0
2.3	17	2
2.7	20	4
3.2	18	8
3.6	12	16
4.1	7	23
4.5	5	24
5.0	2	15
5.4	0	5
5.9	0	2
6.4	0	1

Note: For the purposes of distinguishing the bold words in the long free-response and short free-response questions in this book, the editors have taken the extra measure of also underlining these action words to help readers recognize them more easily.

(a) **Graph** the birth weights for 1990 and 2015 on the grid provided.

(b) **Describe** the selective forces that act on high birth weights and on low birth weights.

(c) **Compare** the distributions in 1990 and 2015. **Explain** the differences in terms of the selective forces you identified in part (b).

121. Workers on a factory ship fishing in the North Pacific Ocean noticed a high proportion of fish with abnormally small eggs. They reported this to fishery scientists, who tracked the problem to a recessive mutation in a gene involved in the development of sperm and eggs. Homozygous recessive fish are sterile. Researchers sampled 1,000 fish and determined whether the fish were fertile or sterile. Using a polymerase chain reaction (PCR) and electrophoresis-based technique, they also determined the genotypes of each fish. The data are presented in the table below.

	Fertile		**Sterile**
Phenotype	915		85
Genotype	**AA**	**Aa**	**aa**
Observed (o)	468	447	85
Expected (e)			
$\frac{(o-e)^2}{e}$			

(a) Based on the phenotype data, **calculate** the expected number of each genotype. Record your answers in the table above. Assume that the population is in Hardy-Weinberg equilibrium.

(b) **State** the null hypothesis, **calculate** the chi-square statistic, and **interpret** the test in terms of the null hypothesis.

(c) **Identify** the selective forces that may be affecting the population. **Explain** how these forces could lead to the data shown.

SHORT FREE-RESPONSE QUESTIONS

122. Students in 9 schools located at different latitudes in North America collected wild samples of the fruit fly *Drosophila melanogaster* in August. Approximately 100 flies were collected by each school. Each fly was screened for mutations in a gene for diapause using DNA sequencing. (Diapause is a period of slowed development associated with unfavorable conditions.) Flies that are homozygous for the diapause minus mutation do not enter diapause in response to exposure to a short day (long night).

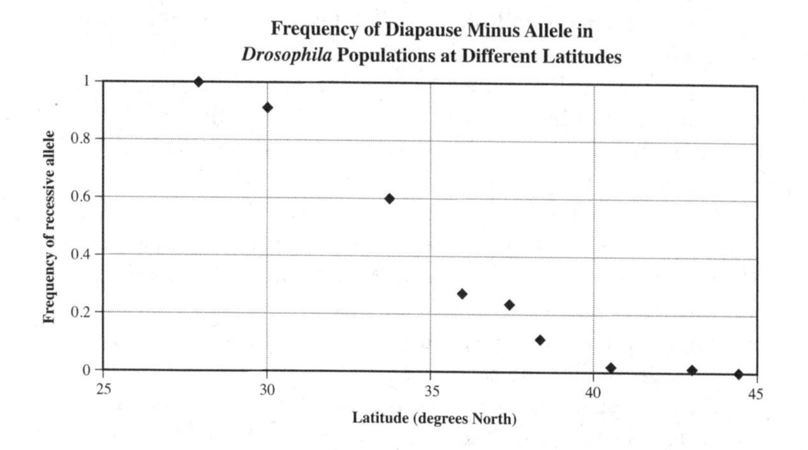

Describe the relationship between the frequency of the recessive allele and the latitude at which the samples were collected. **Identify** ONE evolutionary cost and ONE benefit to diapause. **Explain** the pattern shown in the graph.

123. Use the information below to **construct** a phylogenetic tree for the selected organisms using the tree provided on page 62. **Indicate** at which point each of the derived characters arose. **Describe** the most recent common ancestor of the cat and the lizard in terms of the characters listed, and **explain** your description.

Organism	Derived Character		
	Backbone	**Legs**	**Hair**
Earthworm	Absent	Absent	Absent
Salmon	Present	Absent	Absent
Lizard	Present	Present	Absent
Cat	Present	Present	Present

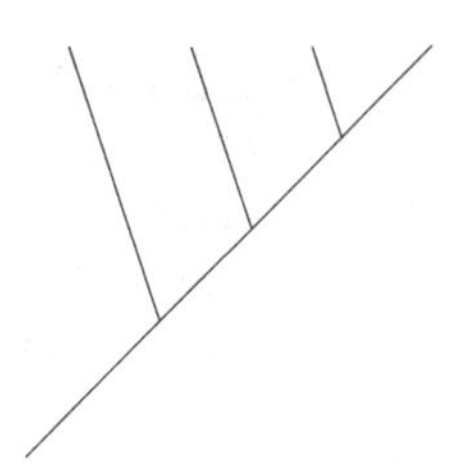

124. Pitcher plants and bromeliads are two groups of plants that are capable of collecting rainwater. Bromeliads grow as epiphytes on the branches of trees where there is very little soil to hold water. These epiphytic plants catch and store rainwater in depressions in their leaves. Bromeliads absorb this water during dry weather. Some of these bromeliad species are also carnivorous, absorbing nutrients from insects and other debris that fall into the water. Pitcher plants are one group of plants that have leaves that are modified to be pitfall traps that are partially filled with water. Animals walking on the edge of the pitcher often slip on the edge, fall in, and cannot get out. This triggers the plant to secrete digestive enzymes into the pitcher. The plant then absorbs the nutrients and water that were in the pitcher. *Nepenthes spathulata* is a species of pitcher plant that has large pitfalls and can eat animals as large as mice. Other species in the genus have smaller pitfalls and eat insects.

The diagram below shows the steps by which the leaf structure of plants might have become modified over generations from a flat leaf to a pitcher shape.

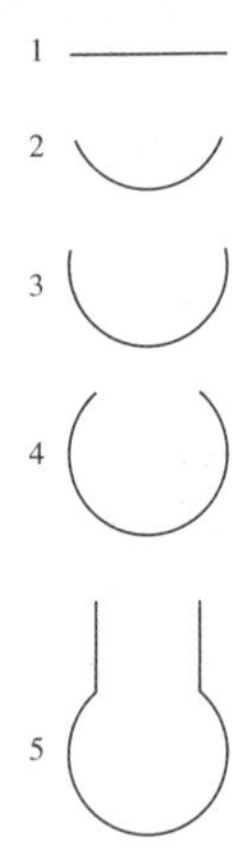

(a) **Identify** TWO possible adaptive functions that might lead stepwise from shape 1 to shape 5.

(b) **Describe** how genetic variability and natural selection may have worked together in this case to cause adaptation.

Answers and explanations for all of the questions in this chapter can be found on pages 275–308.

Big Idea 2—Energy

Answers and explanations can be found on page 309.

CHAPTER 2

BIG IDEA 2
Biological systems utilize free **energy** and molecular building blocks to grow, to reproduce, and to maintain dynamic homeostasis.

For the complete list of big ideas, learning objectives, enduring understandings, essential knowledge, and science practices, refer to the "AP Biology Course and Exam Description" from the College Board:

https://secure-media.collegeboard.org/digitalServices/pdf/ap/ap-biology-course-and-exam-description.pdf

MULTIPLE-CHOICE QUESTIONS

Questions 125–131

Many species of salmonid fish hatch in freshwater streams and then swim downstream to the ocean, where they live for anywhere between 3 and 5 years. When they are sexually mature, the fish swim upstream to spawn in freshwater streams. The switch from freshwater to ocean water (saltwater) and back provides significant osmoregulatory challenges.

Researchers conducted a series of experiments on osmoregulation in salmonid fish. Sixty fish each from three different species were reared for 8 months in freshwater (FW) and then 30 fish from each species were transferred to saltwater while the other 30 fish from each species were kept in freshwater. The researchers measured plasma osmolarity (Figure I), sodium-potassium pump activity (Figure II), and mRNA levels for two different forms (Form A and Form B) of the sodium-potassium pump (Figure III).

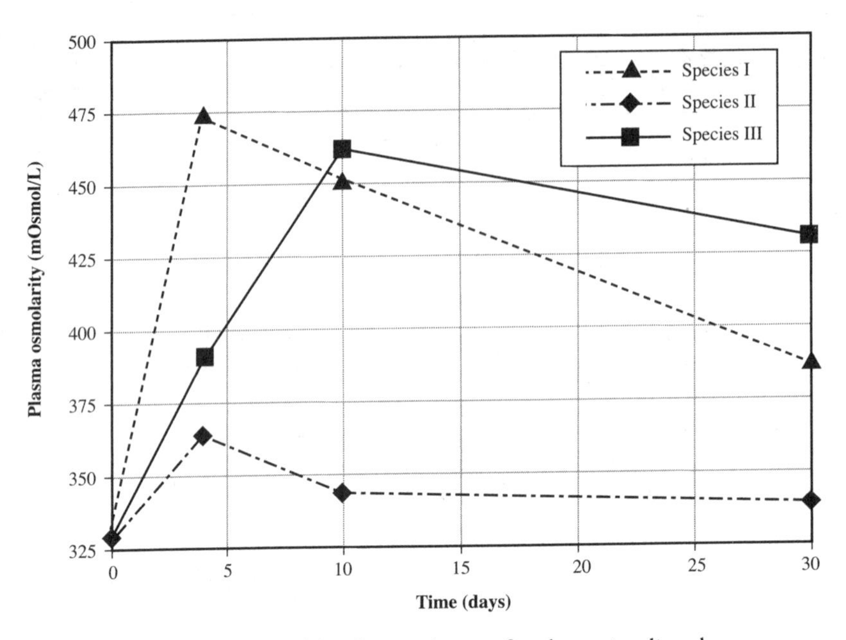

Figure I Plasma (blood) osmolarity of each species directly following the transfer to saltwater (fish kept in freshwater maintained a plasma osmolarity of 325 mOsmol/L and are not included on this graph)

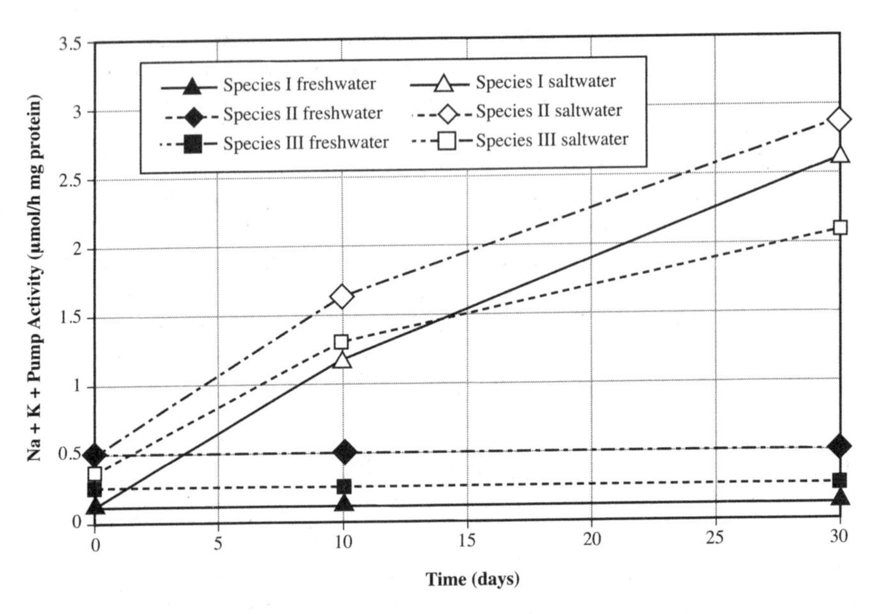

Figure II The activity of the Form B sodium-potassium pump in fish kept in freshwater and in those transferred to saltwater

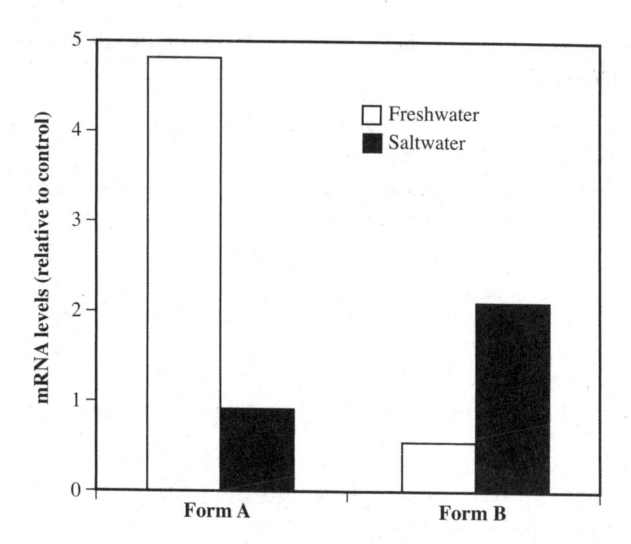

Figure III mRNA data for Form A protein and Form B protein in Species II only

125. **According to Figure I, which species had the most rapid change in blood plasma osmolarity between days 4 and 10?**

(A) Species I
(B) Species II
(C) Species III
(D) There were no differences in the rate of change in plasma osmolarity for this time period.

126. **According to Figure II, which species had the highest sodium-potassium pump activity in freshwater?**

(A) Species I
(B) Species II
(C) Species III
(D) There were no differences in the sodium-potassium pump activity in freshwater.

127. **Which species of fish adapted most rapidly to the change in salinity according to Figure I?**

(A) Species I
(B) Species II
(C) Species III
(D) It is impossible to determine based on the information presented.

128. **What was the purpose of keeping some of each type of fish in freshwater (see Figure II)?**

(A) The researchers did not want to kill all of the fish because they needed some to use in future experiments.
(B) The fish kept in freshwater were the control groups and were used to compare to the fish exposed to saltwater.
(C) There was not enough saltwater to expose all of the fish to it at the same time.
(D) It is not possible to tell what the purpose was from the description presented.

129. **Which of the following is a reasonable conclusion based on Figures I and II?**

(A) Freshwater fish should not be placed into saltwater because their tissues will be hypertonic to the environment and shrink.
(B) Fish exposed to saltwater will pump sodium into their plasma in order to become isotonic to the external environment.
(C) Exposure to saltwater led to an increase in the activity of the sodium-potassium pump in all three species. This pumping caused plasma osmolarity to decrease toward normal levels.
(D) Fish exposed to high salinity levels found in saltwater will choose to lay eggs that are better adapted to living in saltwater.

130. **According to Figure III, which of the following is true about the expression of the sodium-potassium pump gene?**

(A) The mRNA for Form A protein is expressed more in saltwater than it is in freshwater.
(B) The mRNA for Form B protein is expressed more in saltwater than the mRNA for Form A protein is expressed in freshwater.
(C) The mRNA for Form A protein is expressed more in saltwater than the mRNA for Form B protein is expressed in saltwater.
(D) The mRNA for Form B protein is expressed more in saltwater than it is in freshwater.

131. **Which of the following is a reasonable conclusion based on Figure III?**

(A) The mRNA for Form B protein has higher expression levels in saltwater because the protein is responsible for pumping solute from the water into the gills. The mRNA for Form A protein has higher expression levels in freshwater because the protein is primarily responsible for pumping solute out of the gills.
(B) Form B protein has a different ATP binding site than does Form A protein.
(C) The gene for Form A protein is present only in the DNA of freshwater fish. The gene for Form B protein is present only in the DNA of saltwater fish.
(D) The mRNA for Form B protein has higher expression levels in saltwater because the protein is responsible for pumping solute out of the gills. The mRNA for Form A protein has higher expression levels in freshwater because the protein is primarily responsible for pumping solute from the water into the gills.

Questions 132–135

Two species of trees were grown in the same field. Students measured the leaf water potential every hour, starting at 5 A.M. and ending at 8 P.M., for 10 individuals of each species. Each point on the graph below represents the average of 10 individuals.

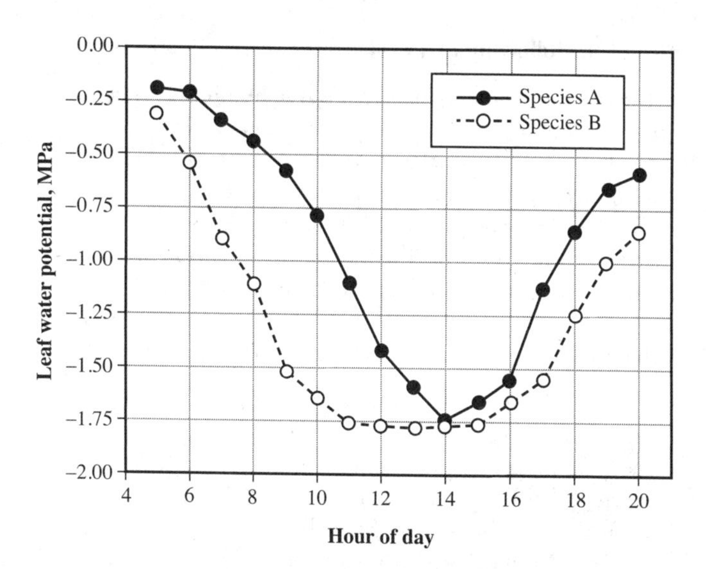

132. Assuming that both trees were in the same soil, which of the following identifies which tree used the most water and why?

(A) Species B used the most water because it had a higher leaf water potential for a longer period of time. A leaf with a higher leaf water potential has a stronger pull on a water column and uses more water.
(B) Species A used the most water because it had a higher leaf water potential for a longer period of time. A leaf with a higher leaf water potential has a stronger pull on a water column and uses more water.
(C) Species B used the most water because it had a lower leaf water potential for a longer period of time. A leaf with a lower leaf water potential has a stronger pull on a water column and uses more water.
(D) There was no difference in the water potentials of the leaves. Therefore, they used the same amount of water.

133. Which species had the faster rate of change in leaf water potential?

(A) Species A had the faster rate of change between hours 12 and 14, while Species B had the faster rate of change between hours 5 and 8.
(B) Species B had the faster rate of change between hours 12 and 14, while Species A had the faster rate of change between hours 5 and 8.
(C) Species B had the fastest rate of change between hours 5 and 12.
(D) There were no differences in the rate of change in leaf water potential for the two species.

134. Which of the following best explains the shape of both curves?

(A) The leaf water potential decreases as light intensity increases because increased photosynthesis requires more gas exchange.
(B) The leaf water potential decreases as the humidity increases because humidity decreases the water potential of the air.
(C) The leaf water potential increases as the soil dries out.
(D) The leaf water potential increases as light intensity increases because the plant closes stomata in response to light.

135. Which of the following best explains the difference between the two species?

(A) Species B more readily opens stomata in response to light than does Species A.
(B) Species B has a thicker cuticle than that in Species A.
(C) Species B has hairy leaves, which slow the air flow over the leaf surface, while Species A has no hair on its leaves.
(D) Species B has fewer stomata than that in Species A.

Questions 136–138

Researchers measured predawn water potential (MPa) in two different clones of hybrid poplar trees growing in a greenhouse. Trees were grown in 30 boxes, each of which contained 6 individuals of each clone. Half of the boxes were watered 3 times per week (watered group). The others were watered only when the predawn water potential fell below –1 MPa (stressed group). The data are presented below. Note that the error bars represent one standard error above and below the mean for 3 observations per treatment combination on days 278, 283, 306, 308, and 334 of the year and for 30 observations per treatment combination on days 285, 309, and 336 of the year.

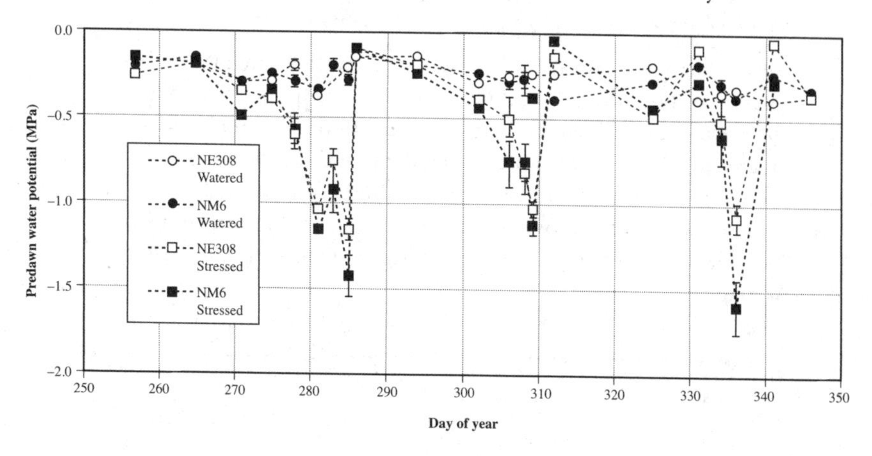

136. **Which of the following is an accurate statement concerning the water potential on day 309?**

   (A) NE308 had a much lower water potential than NM6, regardless of whether the plants were watered or not.
   (B) Stressed NM6 and stressed NE308 had much lower water potentials than watered NM6 and watered NE308.
   (C) There were no differences between any of the treatments.
   (D) None of these statements is accurate.

137. **Which of the following is an accurate statement concerning the water potential on day 336?**

   (A) Both clones experienced the same water potential.
   (B) Clone NE308 experienced significantly lower water potential than clone NM6 in the stressed treatment.
   (C) Clone NM6 experienced significantly lower water potential than clone NE308 in the stressed treatment.
   (D) The result is an outlier and should be disregarded.

138. **The researchers grew the two clones in the same soil in order to subject them to the same level of drought stress. Were the researchers able to accomplish this goal?**

   (A) No. Even though the plants shared the same soil environment, NM6 did not experience as low a water potential as did NE308.
   (B) Yes. Since the clones shared the same soil environment, any measured difference must be due to experimental error.
   (C) No. Even though the plants shared the same soil environment, NE308 did not experience as low a water potential as did NM6.
   (D) It is not possible to tell from the data shown.

Questions 139–141

Researchers measured the mass and the respiration rate of 17 populations of primates, including 16 different species. Two of the populations were of humans: one human population lived in an area in which food is scarce (subsistence diet), and the other human population lived in an industrialized Western country with an abundance of food (Western diet). The data are presented below. Note that the data points for some of the animals with low masses overlap.

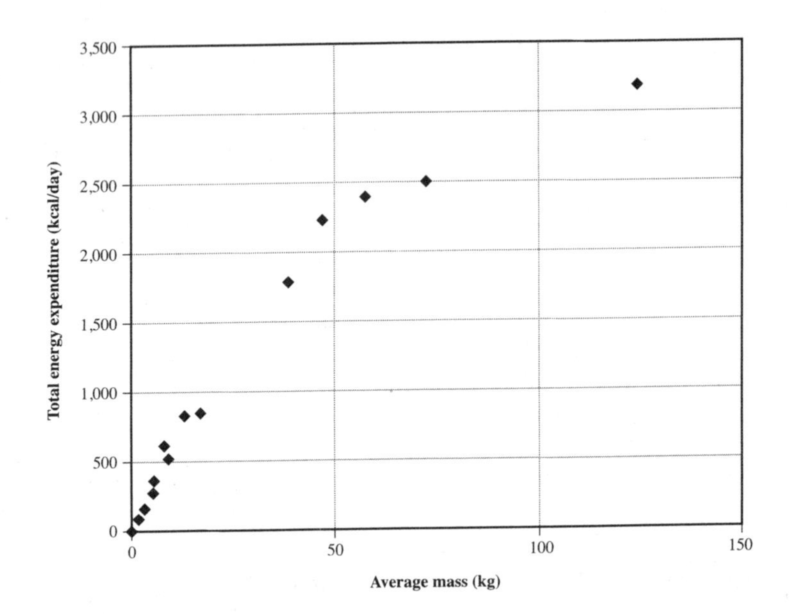

139. Two different populations of humans are represented in the graph. One population has an average mass of 45 kg (subsistence diet), and the other has an average mass of 70 kg (Western diet). Which of the following is true about the total energy expenditure (TEE) of the two populations?

(A) The TEE of both groups is the same.
(B) The subsistence diet population is more sedentary than the Western diet population.
(C) The subsistence diet population had a TEE of about 3,300 kcal/day, while the Western diet population had a TEE of just under 1,000 kcal/day.
(D) The subsistence diet population had a TEE of about 1,700 kcal/day, while the Western diet population had a TEE of about 2,500 kcal/day.

140. Based on the data, which of the following is a reasonable conclusion about the relationship between total energy expenditure (TEE) and average mass?

(A) There is no relationship between TEE and average mass.
(B) The larger an organism is, the higher is its TEE.
(C) The smaller an organism is, the higher is its TEE.
(D) Humans have the highest TEE per unit of average mass.

141. Which of the following is true about the metabolic rate, which can be expressed as the total energy expenditure (TEE) per unit mass (kcal/day • kg)?

(A) There is no relationship between the metabolic rate and the size of an organism.
(B) The smaller an organism is, the higher is its metabolic rate.
(C) The larger an organism is, the higher is its metabolic rate.
(D) Humans have the highest metabolic rate per unit mass.

Questions 142–144

Fishery managers collected data on blue crab population levels in the Chesapeake Bay over 27 years. Samples were taken during December. The data presented below are for crabs larger than 2.4 inches. Males of this size in December will grow to harvestable size (5 inches) during the summer crabbing season. Females of this size in December will grow large enough to bear eggs during the summer crabbing season.

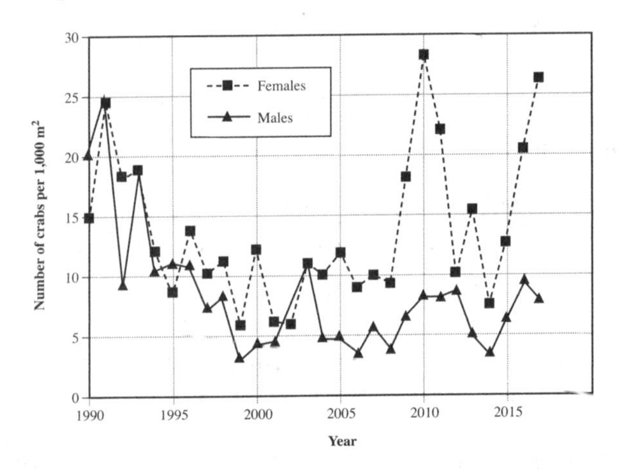

142. **Which of the following is a reasonable hypothesis for the decline in population density from 1990 to 1999?**

   (A) Overfishing led to reduced population sizes.
   (B) Pollution from upstream wastewater treatment plants and runoff from land killed crabs.
   (C) Residential land development on waterfront property involved clearing aquatic vegetation for boating. This led to decreased habitats for immature crabs.
   (D) All of these are reasonable hypotheses.

143. **Compare the average number of males and females over the five-year period between 1990 and 1995 with the average number of males and females over the five-year period between 2005 and 2010. The averages for males and females from 1990 to 1995 are more similar than the averages for males and females from 2005 to 2010. There were fewer males than females between 2005 and 2010 than there were between 1990 and 1995. Which of the following is consistent with this trend?**

   (A) Natural selection selected against males from 1990 to 1995.
   (B) Female crabs were legally protected during the 2005 to 2010 period but not between 1990 and 1995.
   (C) Polluted waters affected males more than females from 1990 to 1995.
   (D) Natural selection selected against females from 2005 to 2010.

144. From the perspective of maintaining a healthy crab population, what is the rationale behind protecting female crabs?

(A) Female crabs are rarer than males because they have a higher natural death rate.
(B) Increasing the number of females decreases the spread of disease in crabs.
(C) Males are more likely to bite humans than females. Reducing the number of males makes the waters more attractive to swimmers.
(D) Since one male can fertilize the eggs of a large number of females, harvesting males has much less effect on population levels than does harvesting females.

Questions 145–146

Foresters attempting to grow trees as biofuels studied the growth of poplar trees over 40 years. The data are presented below.

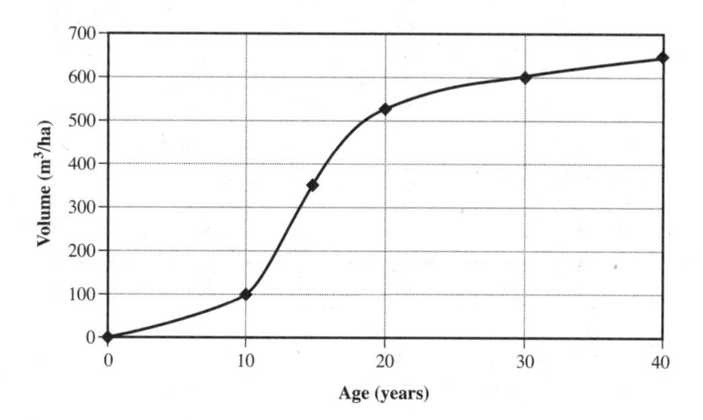

145. At what age are the trees growing the fastest?

(A) 10
(B) 15
(C) 20
(D) 35

146. After harvesting trees, the growers plan to replant. What age trees should the growers harvest in order to maximize the amount of biomass?

(A) 10
(B) 15
(C) 20
(D) 35

Questions 147–151

A class of AP Biology students examined the rate of respiration in germinating peas. They used the volume method, in which potassium hydroxide (KOH) is used to absorb carbon dioxide. The students exposed peas to water to induce germination. The peas were exposed to water for 5, 3, 1, or 0 days. The peas were placed into respirometers, and the change in volume was recorded for 20 minutes. An additional respirometer contained glass beads. The volume change in the glass beads was subtracted from the volume change seen in the other respirometers. The data are presented in the graphs below.

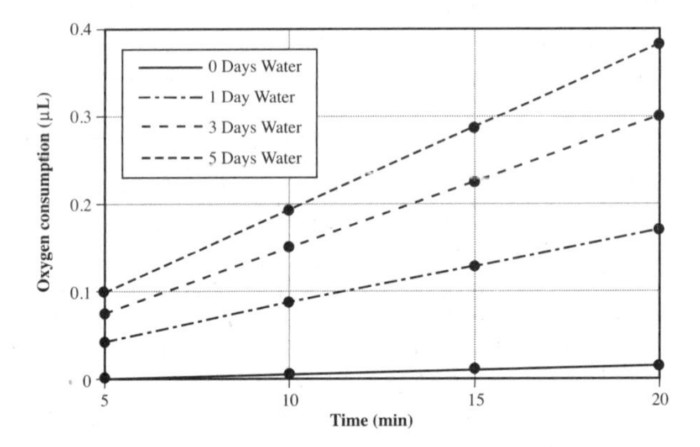

Figure I Respiration as a function of time for peas exposed to water for various durations of time

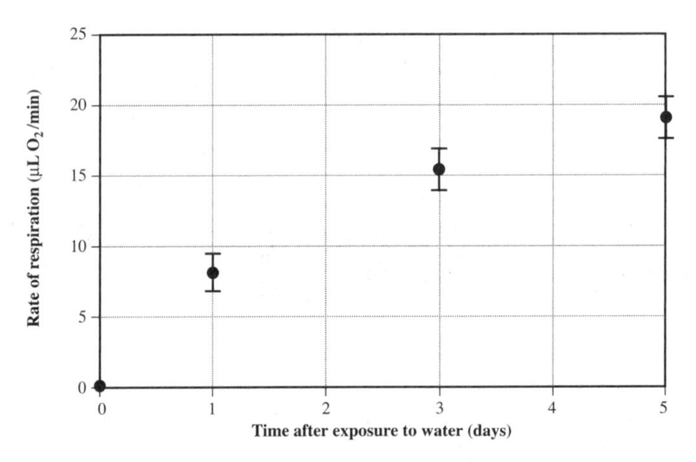

Figure II Rate of respiration as a function of time after exposure to water (days)

147. What experimental question was the class most likely investigating?

(A) Does the rate of photosynthesis change after pea seeds are exposed to water?
(B) Does the rate of respiration change after pea seeds are exposed to water?
(C) Do pea seeds germinate faster when exposed to water?
(D) At what point do germinating pea seeds begin to produce oxygen?

148. What was the source of energy for the germinating pea seeds?

(A) light energy from photosynthesis
(B) energy provided from the breakdown of water into hydrogen and oxygen
(C) chemical energy stored in the soil as molecular nitrogen
(D) chemical energy stored in the seeds as starch

149. Why did the students include peas that were not exposed to water?

(A) They did this to determine the amount of water the peas could draw from the air.
(B) They did this to determine the amount of energy stored in a pea.
(C) The peas exposed to 0 days of water were the control and provided a basis for comparing the respiration rates.
(D) They did this to compare the weights of peas with different durations of water exposure, including no exposure.

150. What was the function of the glass beads?

(A) They were used to determine whether glass beads respired.
(B) They were used to remove carbon dioxide from the air inside the respirometer.
(C) The glass beads were not really necessary, which is why their results were excluded from the graphed data.
(D) They were used to control for changes in volume not due to respiration.

151. What is the relationship between Figure I and Figure II?

(A) There is no relationship between the two figures.
(B) Figure I plots the slopes of the lines in Figure II.
(C) Figure II plots the slopes of the lines in Figure I.
(D) Figure II shows photosynthesis, while Figure I shows respiration.

Questions 152–156

Students in an AP Biology class made separate 0.3 M solutions of maltose, sucrose, lactose, and sucralose. The sugars maltose, sucrose, and lactose all have the molecular formula $C_{12}H_{22}O_{11}$. Sucralose is an artificial sweetener that has the molecular formula $C_{12}H_{19}Cl_3O_8$. The students placed each 0.3 M solution into its own 250 mL flask. The solutions were then sterilized in an autoclave. After the solutions cooled to room temperature, brewer's yeast (*Saccharomyces cerevisiae*) cultures were introduced. An airlock (an S-shaped tube with water in it) was then attached to each flask, and students recorded the number of bubbles produced per minute.

Sugar Source	Number of Bubbles Produced per Minute	Standard Error
Maltose	11	2.1
Sucrose	9	2.4
Lactose	0	0
Sucralose	0	0

152. Which of the following is true about a comparison of maltose and sucrose?

(A) Maltose is a much better food source for yeast than sucrose is because maltose produced more bubbles than did sucrose.
(B) Maltose and sucrose appear to be equivalent energy sources since there was no significant difference between the numbers of bubbles produced.
(C) Although maltose was a better energy source in terms of the number of bubbles produced per minute, sucrose was a better energy source in terms of the standard error.
(D) There are insufficient data to draw a comparison.

153. Which of the following explains why there were no bubbles produced in the lactose flask?

(A) Yeast do not have the gene for lactase, the enzyme responsible for breaking down lactose.
(B) Lactose does not contain carbon.
(C) The lactose solution was hypertonic to the yeast cells, causing them to shrink.
(D) Lactose is toxic to yeast.

154. Which of the following is a reason for using equal molar solutions of each chemical?

(A) One mole of each of these compounds contains the same amount of carbon. Therefore, equal amounts of carbon dioxide can be produced from the complete breakdown of each chemical.
(B) Using equal numbers of molecules of each chemical makes possible the direct comparison of the number of bubbles produced per minute.
(C) Using equal molarities of each chemical ensures that the osmotic conditions are the same for each culture.
(D) All of these are good reasons for using equal molar solutions of each chemical.

155. A group of students vigorously shook half of the flasks containing maltose before adding yeast. The rate of bubble production in the shaken flasks was more than double that in the unshaken flasks over a period of 24 hours. Which of the following is the best explanation for this observation?

(A) Shaking introduced high levels of oxygen into the solution. This led to a higher rate of respiration.
(B) Shaking distributed the yeast more evenly, which led to less competition for nutrients.
(C) Shaking caused more nucleation sites for bubbles to form, so the gas came out of the solution more rapidly.
(D) Shaking introduced higher levels of nitrogen gas into the solution. Yeast use this form of nitrogen to make amino acids.

156. Which of the following is the most appropriate graphical representation of the data?

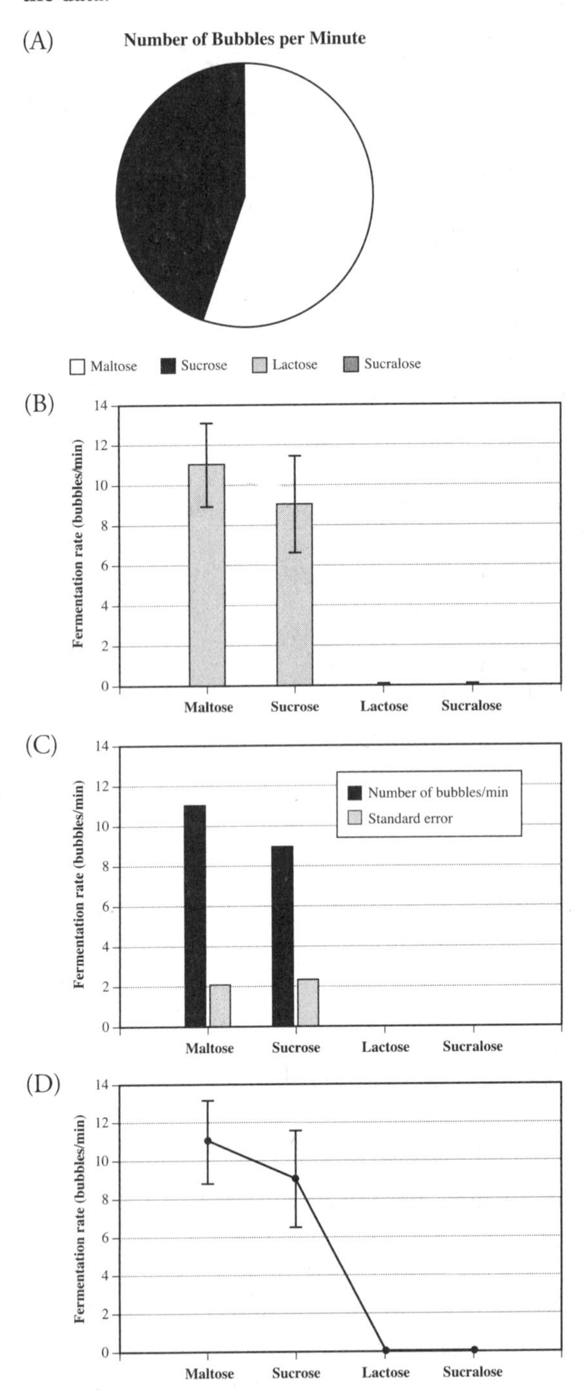

157. Yeast produce the enzyme sucrase, which allows the yeast to metabolize sucrose but not sucralose. The structures of sucrose and sucralose are shown below. The only differences are the replacement of 3 hydroxyl (OH) groups on sucrose with 3 chlorine (Cl) atoms on sucralose. This changes the shape and charge of the molecule.

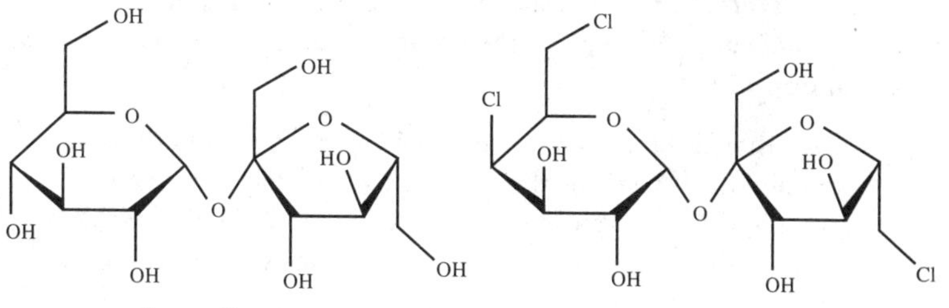

Sucrose (Sugar) **Sucralose (Splenda)**

Which of the following is the most likely explanation for why yeast can metabolize sucrose but not sucralose?

(A) Sucralose has the correct charge and shape to fit into and bind to the active site of sucrase, while sucrose does not.
(B) Sucrose has the correct charge and shape to fit into and bind to the active site of sucrase, while sucralose does not.
(C) Sucralose does not have the correct charge and shape to fit into and bind to the allosteric site of sucrase, but sucrose does.
(D) Sucralose is a much larger molecule than sucrose. Therefore, it fits more tightly into and binds more strongly to the active site of sucrase than does sucrose.

158. A 133 kg (250 lb) bicyclist rode 200 kilometers (124 miles), burning 10,309 kcal. He took in 3,000 kcal in food. Hydrolysis of 1 mole of ATP releases 7.3 kcal, and the molar mass of ATP is 507 g/mol. The bicyclist calculates that he used 716 kg (1,575 lb) of ATP (2.2 lb/kg). How is this possible?

(A) It isn't possible. He made a mathematical error in his calculations.
(B) ATP was recycled and replenished by chemicals that were more energy dense.
(C) He ate enough ATP-containing food while riding to make up the difference.
(D) Physiological respiration accounts for the extra weight because O_2 weighs more than CO_2.

Questions 159–160

Sex hormone–binding globulin (SHBG) is a protein responsible for carrying the steroid hormones estrogen and testosterone through the blood.

159. **Abuse of synthetic testosterone decreases the expression of SHBG. Which of the following is likely to be the immediate result once steroid use is discontinued?**

(A) Even though SHBG levels would be low, the amount of testosterone reaching cells would be the same as it was during steroid abuse.
(B) Since SHBG levels are low, less testosterone would be available to cells.
(C) Since SHBG levels are low, more testosterone would be available to cells.
(D) SHBG levels would increase so that the amount of testosterone reaching cells would be the same as it was prior to steroid abuse.

160. **The diagram below is a schematic representation of SHBG.**

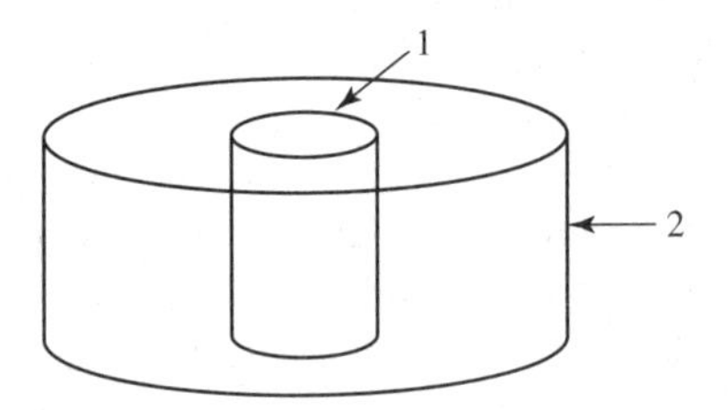

Which of the following would most likely be true about the structure of SHBG?

(A) The structure labeled 1 would be polar and shaped like sex hormones. The structure labeled 2 would be nonpolar, allowing SHGB to dissolve in membranes.
(B) Sex hormones would bind to the structure labeled 2. The structure labeled 1 would attach to hemoglobin.
(C) The structure labeled 1 would be nonpolar and shaped like sex hormones. The structure labeled 2 would be polar so SHBG could dissolve in blood.
(D) The structure labeled 2 would be polar so that SHBG could reside in the lipid portion of the phospholipid bilayer. The structure labeled 1 would create a nonpolar pore through which sex hormones could diffuse.

Questions 161–170

The graph below depicts the Gibbs free energy change (ΔG) in a reversible reaction. The reaction from left to right is called the forward reaction, and the reaction from right to left is called the backward reaction. The equation for Gibbs free energy is

$$\Delta G = \Delta H - T\Delta S$$

where ΔH is change in enthalpy, T is the temperature in Kelvin, and ΔS is the change in entropy.

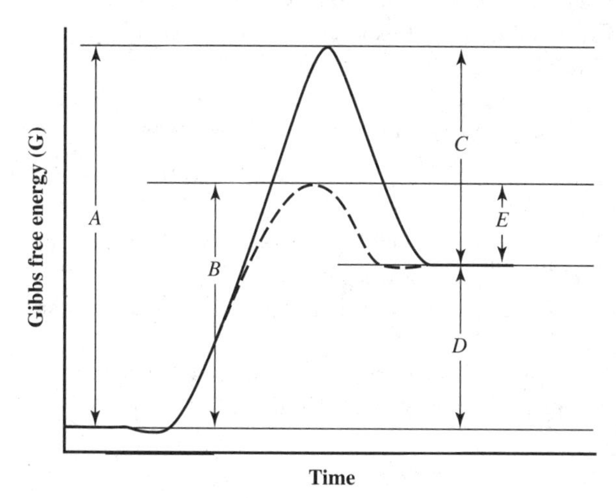

161. **Which of the following statements is true regarding the forward reaction depicted in the graph?**

(A) It could be depicting photosynthesis since that is an endergonic reaction.
(B) It could be depicting respiration since that is an exergonic reaction.
(C) It could be depicting ATP hydrolysis since that is an endergonic reaction.
(D) It could be depicting fermentation since that is an exergonic reaction.

162. **In the reaction depicted, the forward and backward reactions can occur at room temperature. In the diagram, the energy level of the reactants is pictured on the left, and that of the products is pictured on the right. Which of the following would be true for this reaction?**

(A) The forward reaction should occur more rapidly than the backward reaction.
(B) The forward reaction should occur more slowly than the backward reaction.
(C) The forward and backward reactions should occur at the same rate.
(D) The forward reaction would be faster with an enzyme, while the backward reaction would be faster without one.

163. Which of the following is true of ΔG for the forward reaction?

(A) ΔG is zero because entropy is created.
(B) ΔG is negative because the reactants have more energy than the products.
(C) ΔG is positive because the products have more energy than the reactants.
(D) ΔG cannot be determined from the graph as depicted.

164. Which of the following is true of ΔH for the backward reaction?

(A) The reaction could be endothermic as long as enough entropy is created.
(B) The reaction must be endothermic under all conditions.
(C) The reaction must be exothermic under all conditions.
(D) None of these is true.

165. Which of the following must be true for the change in entropy (ΔS)?

(A) Entropy increases in all reactions and must be positive for both the forward and backward reactions.
(B) The changes in entropy for the forward and backward reactions are equal but have opposite signs.
(C) The change in entropy for both the forward and backward reactions is random.
(D) At equilibrium, the entropy of the forward reaction must equal the entropy of the backward reaction.

166. On the graph, line *A* indicates

(A) the activation energy for the catalyzed forward reaction
(B) the activation energy for the uncatalyzed forward reaction
(C) the total energy change for the forward reaction
(D) the activation energy for the uncatalyzed backward reaction

167. On the graph, line *B* indicates

(A) the activation energy for the catalyzed forward reaction
(B) the activation energy for the uncatalyzed forward reaction
(C) the total energy change for the forward reaction
(D) the activation energy for the uncatalyzed backward reaction

168. On the graph, line *C* indicates

(A) the activation energy for the catalyzed forward reaction
(B) the activation energy for the uncatalyzed forward reaction
(C) the total energy change for the forward reaction
(D) the activation energy for the uncatalyzed backward reaction

169. **On the graph, line *D* indicates**

(A) the activation energy for the catalyzed backward reaction
(B) the activation energy for the uncatalyzed forward reaction
(C) the total energy change for the forward reaction
(D) the activation energy for the uncatalyzed backward reaction

170. **On the graph, line *E* indicates**

(A) the activation energy for the catalyzed backward reaction
(B) the activation energy for the uncatalyzed forward reaction
(C) the total energy change for the forward reaction
(D) the activation energy for the uncatalyzed backward reaction

---

171. **Abscisic acid (ABA) is a plant hormone produced by drought-stressed plants. In a controlled experiment, researchers found that the application of ABA to petunias delayed wilting for up to 5 days. Which of the following is the most likely connection between natural ABA production and drought stress?**

(A) ABA is a pheromone that signals parasitoid wasps to attack caterpillars.
(B) ABA is a chemical signal that stimulates processes in plants that decrease water loss by transpiration.
(C) ABA causes fruit ripening.
(D) ABA causes plant cells to divide.

Questions 172–177

Rhizobium radiobacter is a bacterium that causes tumors in plants. It accomplishes this by transferring a portion of the tumor-inducing (Ti) plasmid into a host cell. A diagram of the Ti plasmid is shown below.

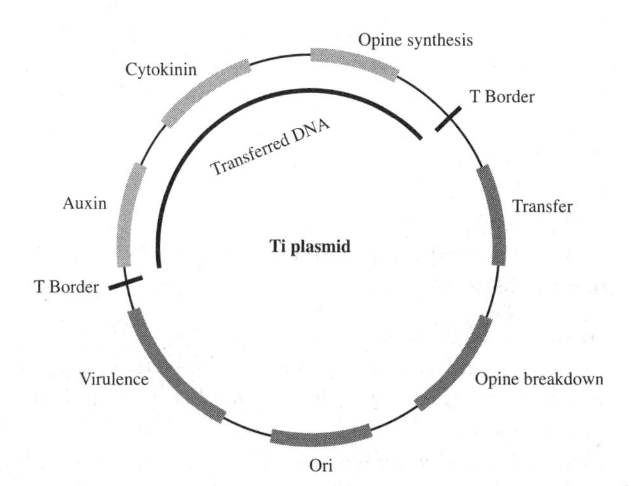

When reviewing this diagram, keep in mind the following:

- Ori = origin of replication
- Virulence = proteins regulating T-DNA transfer
- T Border = the border of the transferred region of DNA
- Auxin = a plant hormone that causes cells to grow larger
- Cytokinin = a plant hormone that causes cell division
- Opine = an unusual compound that contains carbon and nitrogen
- Transfer = makes it possible to transfer the DNA between the borders into the plant

The Ti plasmid can be used in genetic engineering to transfer genes into plants. When using the Ti plasmid to make transgenic plants, antibiotic resistance genes are also added. One is a gene for resistance to antibacterial compounds, and the other is a gene for resistance to compounds that are toxic to plants.

172. **According to the diagram of the Ti plasmid, 3 genes are transferred into the plant. Which of these genes is/are important in causing tumor formation?**

 (A) auxin and cytokinin only
 (B) opine synthesis only
 (C) transfer and opine synthesis only
 (D) ori only

173. **The bacterium transfers the opine synthesis gene but retains the opine breakdown gene. What is the advantage of creating a compound that will immediately be broken down?**

 (A) It is energetically wasteful for the bacterium, which is a hallmark of disease.
 (B) It leads to the proliferation of vascular tissue in the tumor.
 (C) Since other organisms do not have the ability to use opine, the bacterium causes the plant to make a compound from which only the bacterium can benefit.
 (D) There is no benefit to this mechanism.

174. **Imagine that you replaced the auxin, cytokinin, and opine synthesis genes with a gene that encoded a green fluorescent protein (GFP). Which of the following would be the result?**

 (A) The tumors would glow.
 (B) The DNA in the transfer region would not be transferred because auxin, cytokinin, and opine synthesis genes produce proteins involved in transferring DNA.
 (C) No tumors would form, but any cells that were transformed would glow.
 (D) The plasmid would not replicate.

175. Imagine that you could delete the origin of replication from the plasmid. Which of the following would be the result?

(A) The tumors would glow.
(B) The DNA in the transfer region would not be transferred because the ori gene produces a protein involved in transferring the plasmid into the plant cell.
(C) No tumors would form, but any cells that were transformed would glow.
(D) The plasmid would not replicate.

176. The Ti plasmid has been used by genetic engineers to introduce novel genes into plants. Which of the following genes would be necessary to use the Ti plasmid for this purpose?

(A) auxin, cytokinin, opine synthesis, and opine breakdown
(B) transfer, virulence, and ori
(C) auxin, cytokinin, and transfer
(D) All of these genes are necessary.

177. When the Ti plasmid is used to transform plants, a gene is included for resistance to kanamycin (a compound that is toxic to both bacteria and plant cells). On which part of the plasmid is the kanamycin resistance gene placed and why is it located at that particular spot?

(A) The kanamycin resistance gene is located on the transfer DNA so that any cells that are transformed are resistant to kanamycin. This makes it possible to grow only those cells that were transformed by the bacteria.
(B) The kanamycin resistance gene is located on the portion of DNA retained by the bacterium so that the bacterium can continue to grow in the presence of kanamycin. This makes it possible to grow only those cells that were transformed by the bacteria.
(C) As long as the kanamycin resistance gene is present, it does not matter which portion of the plasmid contains it.
(D) The kanamycin resistance gene is located in such a way as to straddle the T Border. This way, when the DNA is transferred, the bacteria are no longer resistant to kanamycin.

Questions 178–181

Childbirth involves the processes shown below.

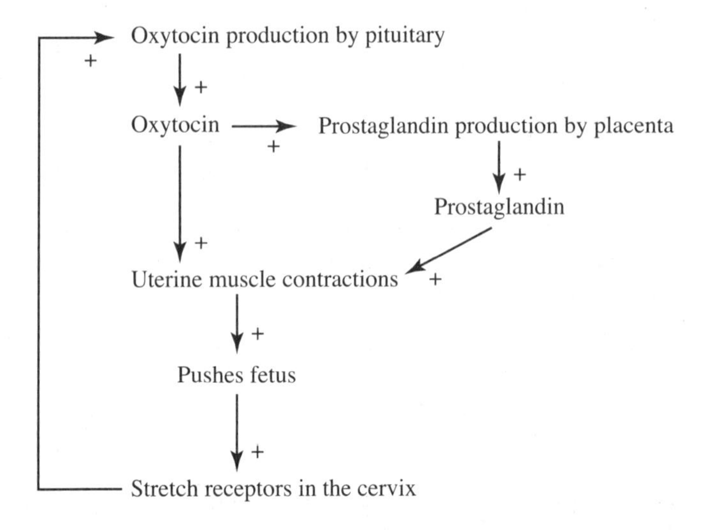

178. **Once the baby is delivered, what happens to the levels of prostaglandin and oxytocin according to the diagram?**

(A) Since the fetus is no longer activating the stretch receptors, oxytocin and prostaglandin levels no longer rise.
(B) Oxytocin levels no longer rise, but prostaglandin levels continue to rise.
(C) Oxytocin levels continue to rise, but prostaglandin levels fall.
(D) Oxytocin is broken down by the placenta, and the uterine muscles break down the prostaglandin, decreasing the levels of both.

179. **Intravenous oxytocin is often administered to women with overdue babies. Which of the following best matches the reasoning for this practice?**

(A) Oxytocin directly stimulates stretch receptors in the cervix to cause dilation so that the fetus can more easily pass through the birth canal.
(B) Oxytocin is given intravenously to stimulate the production of milk.
(C) Intravenous oxytocin decreases the production of oxytocin by the pituitary and helps to return the woman to a homeostatic steady state.
(D) Intravenous oxytocin stimulates the production of prostaglandins and induces muscle contractions. The pressure on the cervix increases, producing more oxytocin and leading to an amplification of the response.

180. In some cases, a fetus may be positioned so that the head does not push against the cervix. Which of the following is a likely consequence of this?

(A) If the stretch receptors are not activated, less oxytocin will be produced by the pituitary. This will lead to an increase in prostaglandin production, which will return the body to a homeostatic steady state.
(B) The sensory mechanisms in the uterus will cause uneven contractions to position the fetus correctly.
(C) Even though the fetus is not stimulating the stretch receptors, muscle contractions will continue to become stronger until the baby is delivered.
(D) Since the fetus is not activating the stretch receptors in the cervix, positive feedback amplification of muscle contractions will not occur.

181. An amniotomy is a procedure in which the amniotic sac is ruptured, releasing the amniotic fluid into the uterus. Amniotic fluid contains high levels of prostaglandin. What is the effect of this procedure?

(A) The released prostaglandin decreases the production of natural prostaglandin by the placenta.
(B) The released prostaglandin stimulates muscle contractions.
(C) The released prostaglandin leads to less oxytocin being produced by the pituitary.
(D) Sufficient information is not presented. The correct answer depends on when in the birth process the procedure was performed.

Questions 182–183

Nectar is a dilute aqueous solution composed primarily of the sugars sucrose, fructose, and glucose. All three of these sugars have the empirical formula CH_2O.

182. Which of the following is true of nectar?

(A) It is an excellent source of energy because the sugars it contains are readily metabolized by glycolysis.
(B) It is an excellent source of energy for long-term storage since carbohydrates are the most energy dense macromolecules.
(C) It is an excellent source of nucleotides because it is rich in nitrogenous bases.
(D) It is a poor source of water since the concentration of sugars is high enough to cause osmotic shrinking in cells.

183. Honeybees (*Apis mellifera*) feed primarily on nectar but also collect and eat pollen. Which of the following statements is true concerning a honeybee's need to eat pollen?

(A) Pollen contains nitrogen, which is necessary for building carbohydrates.
(B) Pollen contains amino acids, which are necessary for building proteins.
(C) Pollen contains sulfur, which is necessary for building nucleic acids.
(D) All of the above are true.

---

184. A bicyclist claimed that commuting 15 miles by bicycle was a "zero-emission activity" and that it did not release any carbon dioxide. The cyclist's heart rate monitor indicated that he burned 935 kcal during his commute. Which of the following is true regarding his claim?

(A) The bicyclist's claim is correct because he did not release any more carbon dioxide than what he would have released if he didn't ride the bike.
(B) The bicyclist's claim is technically incorrect because respiration is used to harvest energy from food and release carbon dioxide as a waste product. However, driving a car the same distance would release far more carbon dioxide.
(C) The bicyclist's claim is correct because the carbon dioxide he released while riding the bike was first fixed by photosynthesis.
(D) The bicyclist's claim is incorrect because the processes that allow him to make ATP to drive his muscles release far more carbon dioxide as a waste product than does driving a car the same distance.

Questions 185–187

An ecologist determined the sex ratio of spring peeper frogs (*Pseudacris crucifer*) in lakes surrounded by suburban homes and in lakes surrounded by forests. The percentage of each sex in each habitat is presented in the table below. Chi-square analysis showed no significant difference in the sex ratio from 1:1 in forested lakes but did show a highly significant difference in suburban lakes.

	Percentage of Males	**Percentage of Females**
Lakes Surrounded by Forests	49	51
Lakes Surrounded by Suburban Homes	23	77

185. Which of the following hypotheses is NOT consistent with the data presented?

(A) Natural selection by predators who eat males more frequently favors females in suburban environments.
(B) In spring peeper frogs, gender is determined entirely by the sex chromosomes (X and Y).
(C) Environmental chemicals in the suburban environment skew the gender ratio toward females in the suburban environment.
(D) Increased competition for habitats among males in the suburban habitat leads to high male mortality.

186. Which of the following experiments would best test the hypothesis that chemicals in the environment skewed the sex ratio in the suburban environment?

(A) Raise frogs from eggs in the laboratory using water samples collected from the suburban environment and from the forested lakes. Determine the sex ratio in each water source.
(B) Observe the sex ratios of frogs in the suburban environment and in the forested lakes over a period of 5 years.
(C) Add estrogen to the forested lakes to change the sex ratio.
(D) Convert forested lakes to suburban environments by building houses, and observe the sex ratio.

187. Which of the following experiments would best test the hypothesis that predators consumed more males in suburban environments?

(A) Observe frogs in suburban environments and in forested lakes to see what time of day they are most active.
(B) Recover the stomach contents of frog predators in both environments, and determine the sex ratio of consumed frogs.
(C) Collect predators from suburban environments, and transfer them to forested lakes.
(D) Collect frogs from forested lakes, and transfer them to suburban environments.

Questions 188–191

The diagram below shows the energy pyramid for a fishery that supports tuna fish as the top-level carnivore.

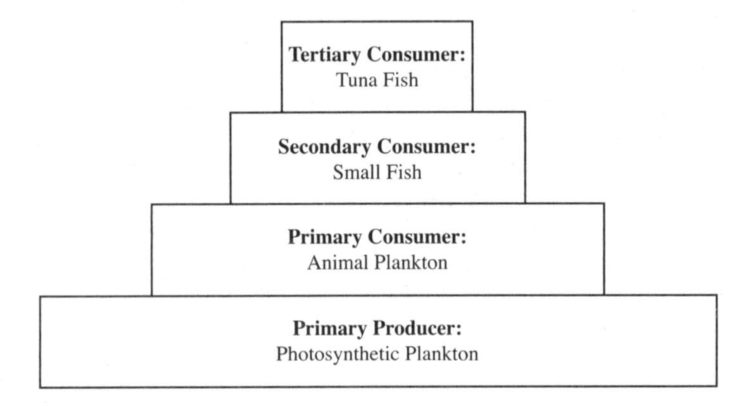

188. Assuming that there are 2,000 J of energy available in the form of small fish, which of the following best matches modern theories of energy flow in ecosystems?

(A) There are 10,000 J of energy available in photosynthetic plankton.
(B) There are 2 J of energy in tuna fish.
(C) There are 2,000 J of energy in animal plankton.
(D) There are 200 J of energy in tuna fish.

189. Humans can choose to feed at different trophic levels. Eating which of the following would support the maximum number of people?

(A) photosynthetic plankton
(B) animal plankton
(C) small fish
(D) tuna fish

190. Which of the following activities could increase the number and/or size of tuna fish, and therefore the amount of energy, that could be harvested?

(A) removing mercury from the environment
(B) harvesting 10% of the small fish
(C) harvesting 10% of the animal plankton
(D) fertilizing patches of ocean to increase the number of photosynthetic plankton

191. Scientists measured mercury levels in tissue samples of organisms from different trophic levels in this ecosystem. Which of the following is true?

(A) Tuna fish have the lowest concentrations of mercury because there are fewer tuna fish than other organisms.
(B) Small fish have higher concentrations of mercury than photosynthetic plankton because the toxin concentration gets magnified at each trophic level.
(C) Animal plankton have the highest concentrations of mercury because most of them are classified as invertebrates.
(D) Photosynthetic plankton have the highest concentrations of mercury because they actively take up mercury as protection against animal plankton.

Questions 192–193

The diagram below depicts the thylakoid membrane of the chloroplast.

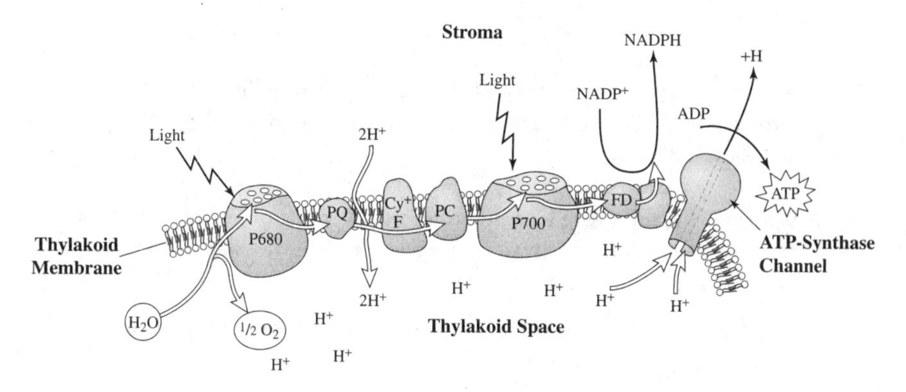

Imagine that you could treat chloroplasts with an indicator dye that turns red in acidic conditions, yellow in neutral conditions, and blue in basic conditions.

192. If the light was on, which of the following would be an accurate description of the chloroplast?

(A) The stroma would appear red, and the thylakoid space would appear blue.
(B) The stroma would appear yellow, and the thylakoid space would appear red.
(C) Both the stroma and the thylakoid space would appear yellow.
(D) The thylakoid space would appear red, and the stroma would appear blue.

193. If the light was off, which of the following would be an accurate description of the chloroplast?

(A) The stroma would appear red, and the thylakoid space would appear blue.
(B) The stroma would appear yellow, and the thylakoid space would appear red.
(C) Both the stroma and the thylakoid space would appear yellow.
(D) The thylakoid space would appear red, and the stroma would appear blue.

Questions 194–195

The diagram below represents the mitochondrion.

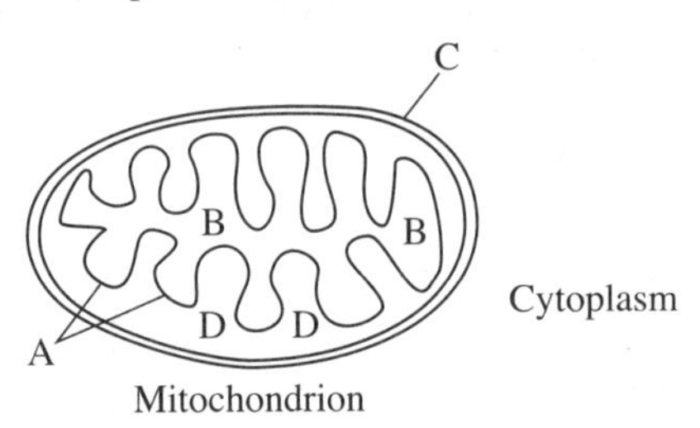

Imagine that you could treat mitochondria with an indicator dye that turns red in acidic conditions, yellow in neutral conditions, and blue in basic conditions.

194. Which of the following best describes the distribution of color that you would expect if both pyruvate and oxygen were in the solution that contains the mitochondria?

(A) The structure labeled B would appear red, and the structure labeled D would appear blue.
(B) The structure labeled A would appear red, and the structures labeled B, C, and D would appear yellow.
(C) All of the structures would appear yellow.
(D) The structure labeled D would appear red, and the structure labeled B would appear blue.

195. Which of the following best describes the distribution of color that you would expect if pyruvate was in the solution that contains the mitochondria but oxygen was not?

(A) The structure labeled B would appear red, and the structure labeled D would appear blue.
(B) The structure labeled A would appear red, and the structures labeled B, C, and D would appear yellow.
(C) All of the structures would appear yellow.
(D) The structure labeled D would appear red, and the structure labeled B would appear blue.

Questions 196–197

The Krebs cycle (also known as the citric acid cycle or tricarboxylic acid cycle) is a metabolic pathway that occurs in the mitochondria. The Krebs cycle is shown in the figure below.

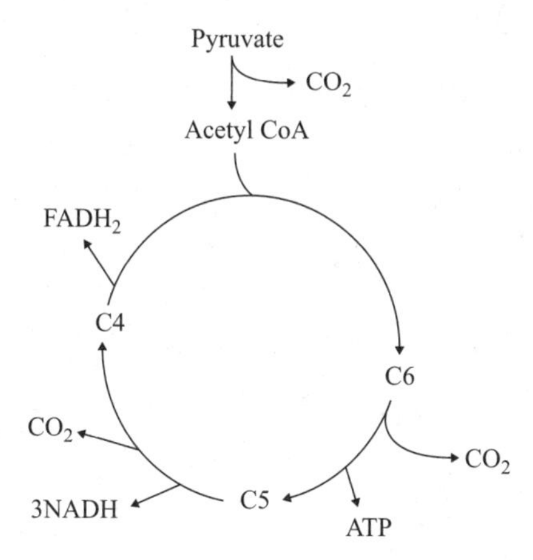

196. Some organisms are capable of running the Krebs cycle backward. Which of the following would be an accurate description of these organisms?

(A) They would be heterotrophic because they use the Krebs cycle.
(B) They would be methanogens because the Krebs cycle uses methane (CH_4).
(C) They would be denitrifiers because they convert nitrates into atmospheric nitrogen.
(D) They would be autotrophic because they would make organic compounds from inorganic compounds.

197. According to the diagram, which of the following is necessary for organisms to run the Krebs cycle backward?

(A) an energy source in the form of ATP and electrons provided by NADH and $FADH_2$
(B) a source of electrons from sunlight and carbon from acetyl CoA
(C) a source of energy from stored fat to convert into acetyl CoA
(D) ADP, inorganic phosphate, NAD^+, and $FADH^+$

Questions 198–199

The diagram below shows the Calvin cycle, which occurs in the chloroplast.

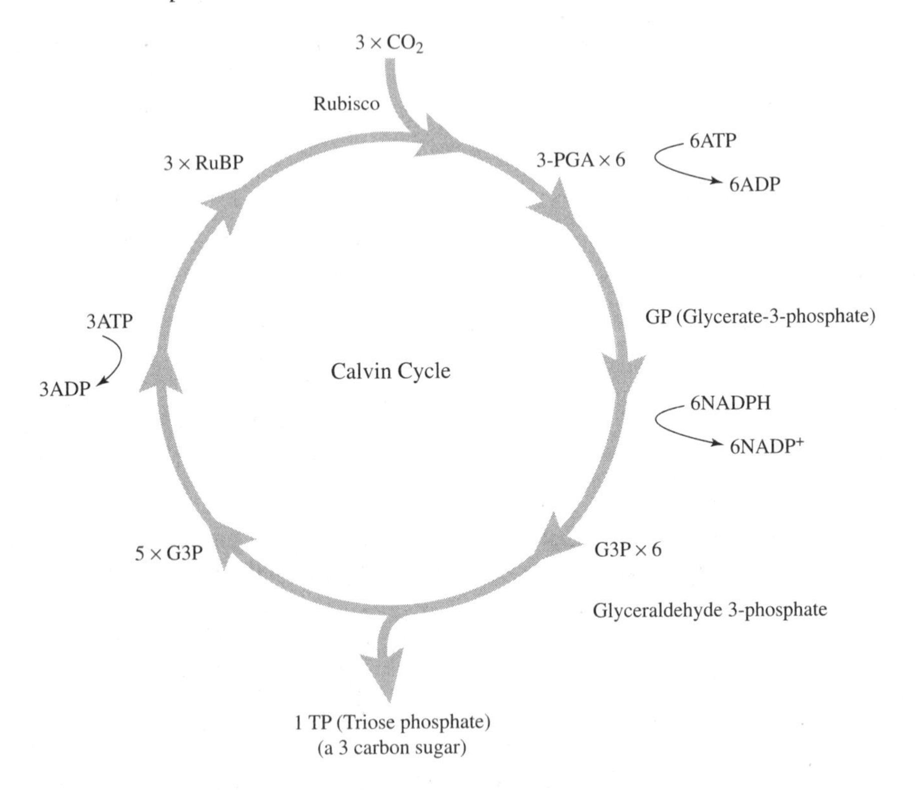

198. The reactions shown in this diagram of the Calvin cycle are often referred to as the "light-independent reactions." They are sometimes called the "dark reactions" in some publications. These names are misleading because the Calvin cycle does not occur in the dark. Rubisco, a key enzyme in the Calvin cycle, is inactivated in dark conditions. Which of the following would be a consequence if rubisco was not inactivated in the dark and was able to catalyze reactions when there was no light present?

(A) Without the light reactions to produce ATP and NADPH, the ADP and $NADP^+$ levels would be high. Therefore, the cycle would release carbon dioxide because the cycle would run backward.
(B) Rubisco would break down carbon dioxide into carbon and oxygen.
(C) Rubisco would continue to fix carbon, producing an excess of 3-PGA. Without the ATP from the light reactions, 3-PGA would build up to toxic levels and cause the stomata to close permanently.
(D) No reactions would occur without something to provide activation energy. This mechanism for turning rubisco off is unnecessary.

199. When rubisco combines RuBP (a 5-carbon compound) with carbon dioxide, the reaction produces 2 molecules of 3-PGA. Each molecule of 3-PGA contains 3-carbon atoms. Sometimes, rubisco uses oxygen instead of carbon dioxide to react with the RuBP. Which of the following would be the result of this alternate reaction?

(A) One molecule of 3-PGA and one molecule with 2 carbon atoms would be produced.
(B) ATP would be produced.
(C) Two molecules of carbon dioxide would be produced.
(D) Two molecules of 3-PGA would be produced.

Questions 200–202

The diagram below shows oxidative phosphorylation, a process that occurs on the inner membrane of the mitochondrion. It uses a series of proteins called the electron transport chain to pump hydrogen ions across the membrane. In this process, electrons are donated to the electron transport chain by NADH and $FADH_2$. These electrons are ultimately passed to oxygen, which acts as the final electron acceptor. The hydrogen ion gradient created across the membrane is used to power the ATP synthase, which drives the synthesis of ATP from ADP and inorganic phosphate.

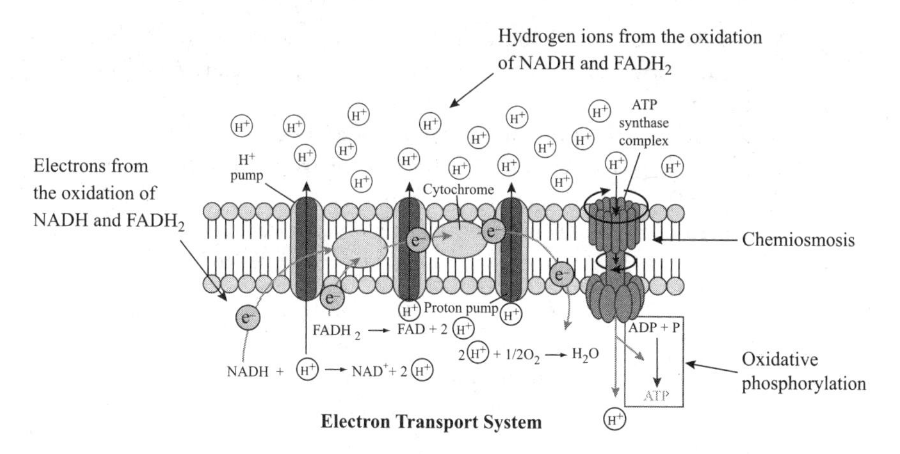

200. According to the diagram, which of the following is true concerning the ability of NADH and $FADH_2$ to drive the production of ATP?

(A) NADH and $FADH_2$ provide exactly the same amount of energy to drive the production of ATP. This is because they both donate electrons to the electron transport chain.
(B) NADH provides less energy than $FADH_2$. This is because NADH has to move through more proteins to reach the ATPase than does $FADH_2$.
(C) $FADH_2$ leads to the production of less ATP because $FADH_2$ donates electrons to hydrogen ions to make them neutrally charged so that they can pass through the membrane. NADH leads to the production of more ATP because NADH donates electrons to the electron transport chain, which pumps hydrogen ions across the membrane.
(D) $FADH_2$ leads to the production of less ATP whereas NADH leads to the production of more ATP. This is because NADH donates electrons to the first of three hydrogen pumps in the electron transport chain whereas $FADH_2$ donates electrons to the second of the three hydrogen ion pumps.

201. The reaction that makes water occurs on the matrix side of the membrane rather than in the intermembrane space. Which of the following would be a consequence if this reaction was performed in the intermembrane space instead of on the matrix side?

(A) Since the reaction produces hydrogen ions, it leads to a higher pH. If the pH was higher in the intermembrane space, more ATP would be made.
(B) Since the reaction produces hydrogen ions, it leads to a lower pH. If the pH was lower in the intermembrane space, more ATP would be made.
(C) Since the reaction consumes hydrogen ions, it leads to a higher pH. If the pH was higher in the intermembrane space, less ATP would be made.
(D) The reaction would have no effect on the amount of ATP made because the reaction doesn't affect the rate of hydrogen pumping.

202. **Which of the following would be the most direct consequence of a lack of oxygen?**

(A) The organism would switch to using fermentation.
(B) Electron transport through the electron transport chain would stop because there would be no final electron acceptor.
(C) The electron transport chain would speed up to bypass the need for oxygen.
(D) The organism would die.

---

203. **The diagram below shows the molecular mechanism by which muscle contraction occurs.**

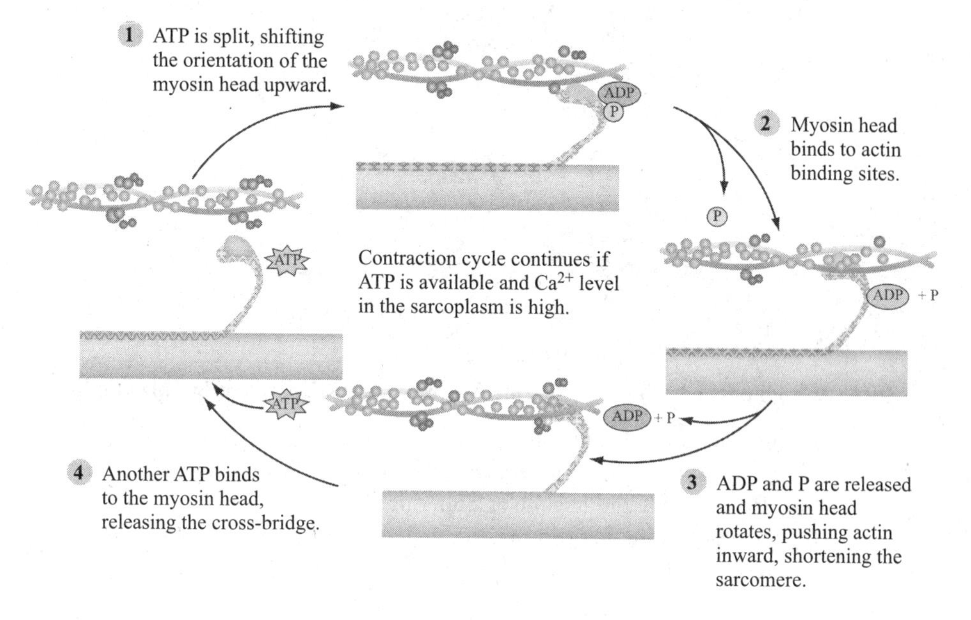

According to the diagram, which of the following would be a consequence of a lack of ATP in the muscle cells?

(A) The myosin head would remain bound to the actin fiber, leading to stiffened muscles typical of rigor mortis.
(B) Myosin heads would fire sporadically, leading to the uncontrollable contraction and relaxation of muscle tissue.
(C) The myosin heads would be unable to bind to the actin fiber, leading to completely relaxed muscles.
(D) The efficiency of the muscles' oxygen usage would increase to compensate for the lack of ATP.

Questions 204–207

Hemoglobin and myoglobin are oxygen-binding molecules. Hemoglobin is found in red blood cells, while myoglobin is found in muscle cells. Humans express different forms of hemoglobin at different times in their lives. Fetal hemoglobin, for example, is expressed only when the fetus is in the womb. The graph below shows the affinity of three different proteins (adult hemoglobin, fetal hemoglobin, and myoglobin) for oxygen.

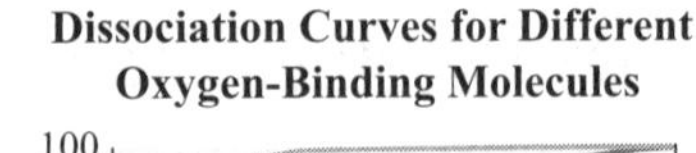

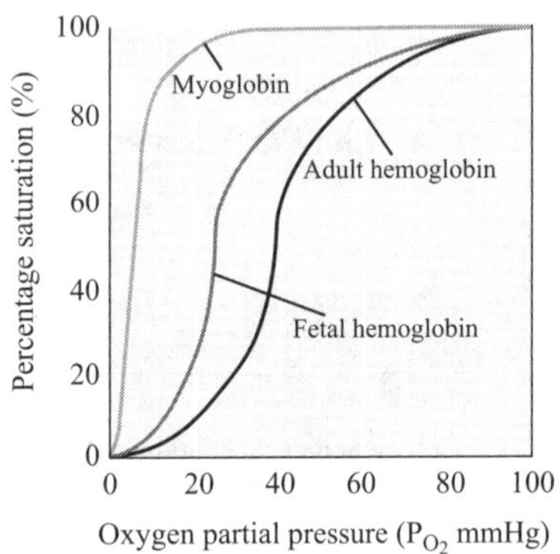

204. Which of the following scientific questions is best answered by this graph?

(A) Which type of hemoglobin is a better source of energy?
(B) How do the amino acid sequences of adult and fetal hemoglobin differ?
(C) How do adult and fetal myoglobin differ?
(D) How tightly do different oxygen carriers bind to oxygen at different oxygen concentrations?

205. At which oxygen partial pressure would oxygen most rapidly move from adult hemoglobin to myoglobin?

(A) 1 mmHg
(B) 10 mmHg
(C) 40 mmHg
(D) 80 mmHg

206. Which of the following best explains the significance of the difference in oxygen binding between fetal hemoglobin and adult hemoglobin?

(A) Fetal hemoglobin is less mature than adult hemoglobin. Therefore, fetal hemoglobin is less efficient at binding oxygen.
(B) The oxygen requirements of a fetus are greater than that of an adult. Therefore, the fetal hemoglobin must bind oxygen with greater affinity.
(C) Fetal hemoglobin binds oxygen more tightly than does adult hemoglobin because the fetus obtains oxygen from adult hemoglobin in the placenta.
(D) Adult hemoglobin is degraded from being used for a long time. It therefore needs more time to bind to oxygen than does fetal hemoglobin.

207. Which of the following best explains the significance of myoglobin's high affinity for oxygen?

(A) Myoglobin's high affinity for oxygen prevents the transfer of oxygen from hemoglobin to myoglobin to ensure an adequate supply of oxygen throughout the body.
(B) Myoglobin's high affinity for oxygen facilitates the direct transfer of oxygen from the alveoli in the lungs to the muscle, bypassing the hemoglobin.
(C) Myoglobin's high affinity for oxygen prevents oxygen from dissolving in the blood plasma.
(D) Myoglobin's high affinity for oxygen facilitates the transfer of oxygen bound by hemoglobin into the muscle cells, where the oxygen will be used for cellular respiration.

Questions 208–209

The diagram below shows the affinity of hemoglobin for oxygen at high and low blood pH. Since carbon dioxide reacts in water to form carbonic acid, oxygenated blood has a higher pH than does deoxygenated blood.

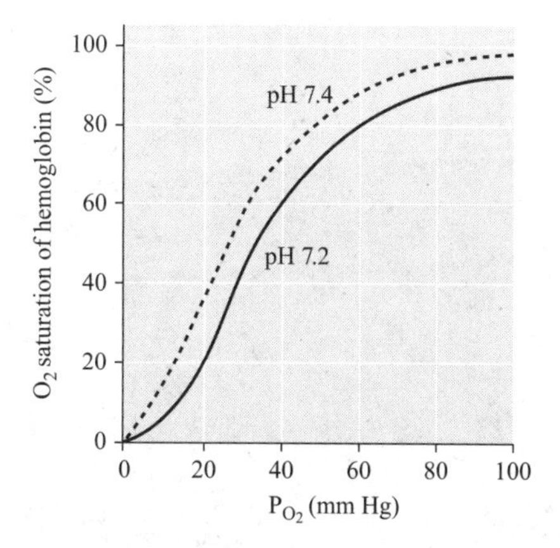

208. Blood returning to the lungs is typically depleted in oxygen and has high concentrations of carbon dioxide. As a result, the blood pH is initially low but rises as carbon dioxide diffuses out of the lungs. Which of the following best explains the significance of the differences between the oxygen affinity of hemoglobin at pH 7.2 and at pH 7.4?

(A) As the pH rises, hemoglobin has a lower affinity for oxygen in the heart where oxygen uptake occurs.
(B) Hemoglobin changes the affinity for oxygen so that the pH will rise.
(C) Hemoglobin has a different affinity for oxygen at a different pH in order to act as a buffer, resisting pH change.
(D) As the pH rises, hemoglobin has a higher affinity for oxygen in the lungs where oxygen uptake occurs.

209. Blood entering body tissues where oxygen is used, such as muscle, has a high concentration of oxygen and a low concentration of carbon dioxide. As a result, the pH of the blood is initially high. However, the pH decreases as oxygen diffuses out of the bloodstream and carbon dioxide diffuses in. Which of the following best explains the significance of the differences between the oxygen affinity of hemoglobin at pH 7.4 and at pH 7.2 in the muscle?

(A) As the pH decreases, hemoglobin has a lower affinity for carbon dioxide in order to prevent lactic acid buildup.
(B) The pH decreases in order to force the oxygen to move into the red blood cells.
(C) As the pH decreases, hemoglobin has a lower affinity for oxygen in the muscles where oxygen is used.
(D) As the pH decreases, myoglobin has a lower affinity for oxygen so that the muscle can store more lactose.

Questions 210–212

Body mass index (BMI) is frequently used by medical professionals to determine an appropriate weight for patients. One alternative to BMI is to measure percent body fat. Figure I and Table I compare these two methods of determining body health.

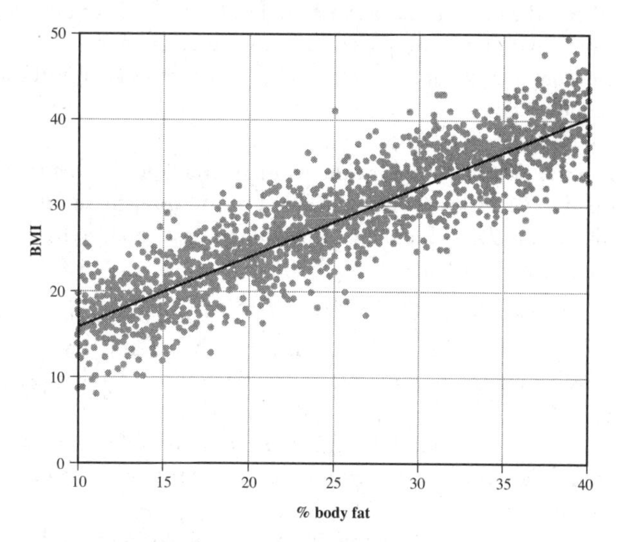

Figure I Body mass index (BMI) as a function of percent body fat (% BF) for women

Table I Classifications for Body Mass Index (BMI) and Percent Body Fat (% BF) for Women

Classification by BMI	Body Mass Index (BMI)	Classification by % BF	Percent Body Fat (% BF)
Underweight	<18.5	Essential	10–13.9
Normal	18.5–24.9	Athletic	14–20.9
Overweight	25.0–29.9	Fit	21–24.9
Obese	30–40	Acceptable	25–31
Extremely obese	>40	Obese	>31

210. Based on Figure I, which of the following is an accurate statement comparing body mass index (BMI) and percent body fat (% BF)?

(A) There is a positive correlation between BMI and % BF.
(B) BMI and % BF are unrelated.
(C) It is possible to calculate the % BF directly from the BMI.
(D) BMI is a much better indicator of physical health than % BF.

211. A woman who completed a triathlon, in which she swam 2.4 miles (3.86 km), rode her bicycle for 112 miles (180.25 km), and ran for 26.22 miles (42.2 km), had a body mass index of 30 and 12% body fat. How should she be classified according to Table I?

(A) BMI indicates that she is underweight. The % BF shows her as obese.
(B) BMI indicates that she is obese. The % BF shows her in the essential class.
(C) Both BMI and % BF place her in the obese category.
(D) BMI indicates that she is obese. The % BF shows her in the acceptable range.

212. Refer to the data collected for the woman who completed a triathlon as described in Question 211. Use Table I to interpret these values. How is it possible for an elite athlete, like the woman described, to have a BMI of 30 and a % BF of 12%?

(A) The athlete's body consumed most of her adipose tissue during the race.
(B) Since % BF does not take into account the weight of adipose tissue, it is possible for an obese person to compete in high-level athletics, such as a triathlon.
(C) Since BMI does not differentiate between mass due to fat and mass due to muscle, an elite athlete could have a BMI that erroneously indicates obesity.
(D) It is not possible. There was an error in measuring the athlete's weight, height, or amount of adipose tissue.

Questions 213–214

Mealworms (the larvae of *Tenebrio molitor*) have been observed to eat Styrofoam* (polystyrene) as their only food source. Mealworms fed a diet containing bactericidal antibiotics were subsequently unable to live on polystyrene as their only food source. When fed fecal matter from the control group (which had not been treated with antibiotics), mealworms regained the ability to use polystyrene as their food source.

*Styrofoam is a trademark of The Dow Chemical Company.

213. Which of the following is the most useful experiment to conduct next?

(A) Feed fruit fly larvae fecal matter from polystyrene-eating mealworms and test the resulting flies for the ability to use polystyrene as an energy source.
(B) Sequence the genome of *Tenebrio molitor*.
(C) Isolate mealworm intestinal bacteria, and see if any of the intestinal bacteria will grow on polystyrene agar on a Petri plate.
(D) Perform a choice experiment in which mealworms are presented different food sources in addition to polystyrene.

214. Which of the following is a likely original source of the gene for polystyrene use?

(A) plasmid transfer by conjugation from other bacteria
(B) random mutation
(C) crossing-over during meiosis
(D) contamination from genetically engineered plants

---

215. Neurons at rest have a membrane potential of –70 mV, meaning that the inside of the neuron is more negatively charged than the surrounding environment. Which of the following is an accurately labeled diagram of the mechanism that creates this resting potential?

(A)

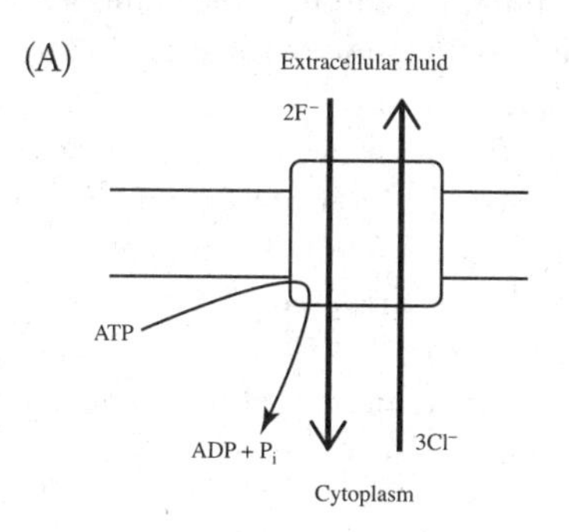

(B)

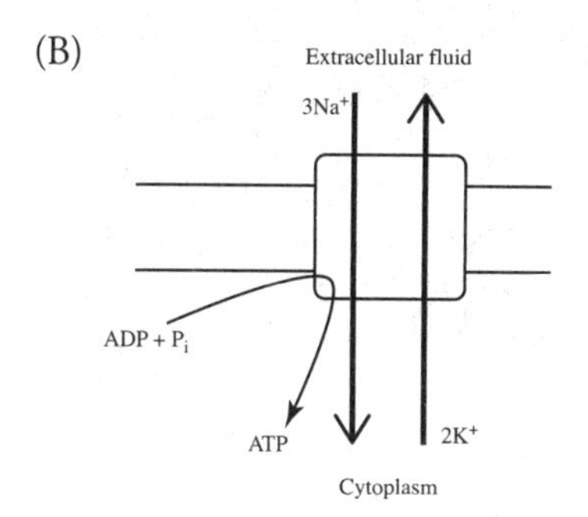

(C)

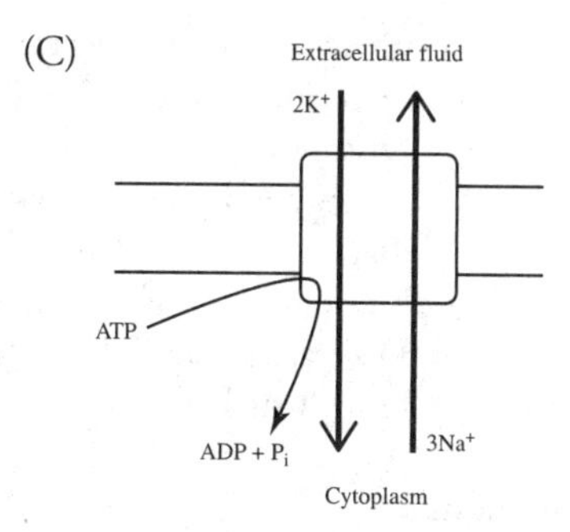

(D)

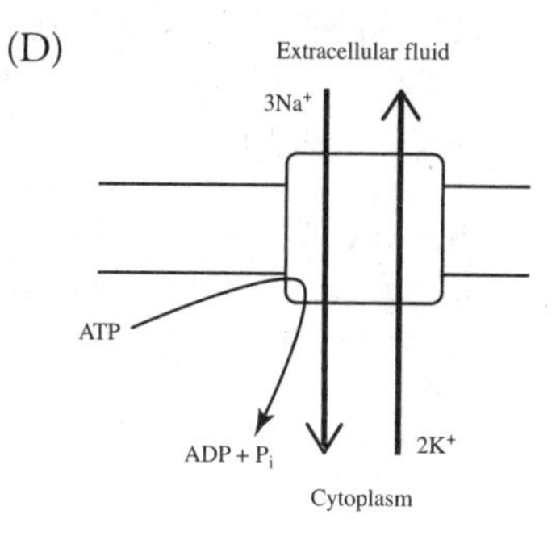

216. During anaerobic exercise, carbon dioxide levels in the blood rise faster than the carbon dioxide can be cleared from the blood. In aqueous solutions, like blood, the following reversible reactions occur:

$$CO_2 + H_2O \rightleftarrows H_2CO_3 \rightleftarrows H^+ + HCO_3^- \rightleftarrows 2H^+ + CO_3^{2-}$$

Which of the following best describes a negative feedback loop that would restore homeostasis?

(A) Dissolved CO_2 increases the blood pH. Receptors in the medulla oblongata detect the change in pH and trigger nerve impulses to the diaphragm to contract less frequently. Decreased ventilation in the lungs increases gas exchange and lowers dissolved CO_2.
(B) Dissolved CO_2 lowers the blood pH. Receptors in the kidneys increase filtration rates in Bowman's capsule. Secretion of carbonate (CO_3^{2-}) in the urine restores blood pH to a steady-state level.
(C) Dissolved CO_2 lowers the blood pH. Receptors in the medulla oblongata detect the change in pH and trigger nerve impulses to the diaphragm to contract more frequently. Increased ventilation in the lungs increases gas exchange and lowers dissolved CO_2.
(D) Decreased blood pH mobilizes carbonate (CO_3^{2-}) deposition in the bone, which absorbs hydrogen ions and returns the pH to neutral.

Questions 217–218

AP Biology students cut russet potatoes and sweet potatoes into cubes 1 cm on a side. They exposed 34 cubes of each variety to 6 different concentrations of sucrose solution. The solutions varied in concentration from 0 M to 1 M. Each cube was measured before and after 24 hours, and the percent change in mass was calculated. The experiment was conducted at 20°C. The results are shown in the graph below.

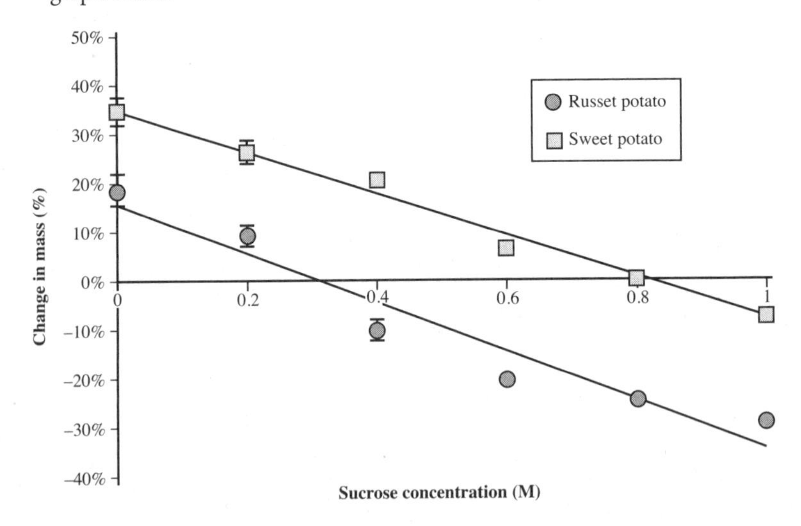

217. Which type of potato has a lower (more negative) solute water potential?

(A) sweet potato
(B) russet potato
(C) The answer cannot be determined from this graph.
(D) The solute water potentials are equal.

218. A student cut five 1 cm cubes of sweet potato and five 1 cm cubes of russet potato so that none of the pieces had any skin on them. She weighed each cube. Next, she placed each cube of sweet potato in direct contact with a cube of russet potato. The student then wrapped each pair of cubes together in plastic tape. This kept the potato cubes in contact with each other and prevented water loss due to evaporation. The figure below shows one of the five pairs of potato cubes.

Russet potato	Sweet potato

After a week, she removed the plastic tape and weighed each cube a second time. Which of the following is the most likely result?

(A) The potato cubes will exchange genetic material. Chimeric cells will form at the interface.
(B) Solutes will flow from the sweet potato into the russet potato. This will cause the russet potato to gain mass and the sweet potato to lose mass.
(C) There will be no net flow of water in either direction. Therefore, the mass of each potato will stay the same.
(D) There will be a net flow of water from the russet potato into the sweet potato. This will cause the sweet potato to gain mass and the russet potato to lose mass.

Questions 219–221

A group of AP Biology students measured dissolved oxygen in samples of pond water incubated at different intensities of light over a period of two days. The data are presented in the graph below.

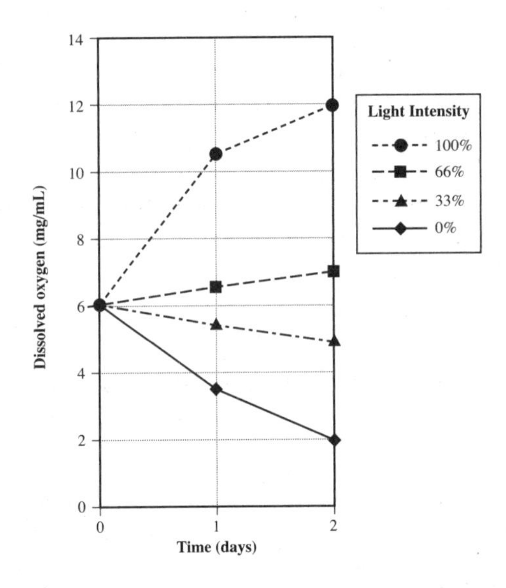

219. Which of the following is most accurate concerning the sample kept in darkness (0% light)?

(A) No photosynthesis took place.
(B) No respiration took place.
(C) Both photosynthesis and respiration took place, but respiration consumed more oxygen than what was generated by photosynthesis.
(D) Both photosynthesis and respiration took place, but photosynthesis generated more oxygen than what was consumed by respiration.

220. Which of the following is most accurate concerning the sample kept in 33% light?

(A) No photosynthesis took place.
(B) No respiration took place.
(C) Both photosynthesis and respiration took place, but respiration consumed more oxygen than what was generated by photosynthesis.
(D) Both photosynthesis and respiration took place, but photosynthesis generated more oxygen than what was consumed by respiration.

221. Which of the following is most accurate concerning the sample kept in 66% light?

(A) No photosynthesis took place.
(B) No respiration took place.
(C) Both photosynthesis and respiration took place, but respiration consumed more oxygen than what was generated by photosynthesis.
(D) Both photosynthesis and respiration took place, but photosynthesis generated more oxygen than what was consumed by respiration.

---

222. Macrophages are white blood cells that recognize non-self cells such as bacteria and perform a process called phagocytosis. The macrophages then engulf bacteria, forming a vesicle called a phagosome. The phagosome then fuses with a lysosome, creating a phagolysosome. The bacteria in the phagolysosome are subjected to digestive enzymes, low pH, and peroxides. These chemicals digest the bacteria. Portions of the bacteria are then excreted or presented on the surface of the macrophage for T cells to recognize. The function of a macrophage is shown in the figure below.

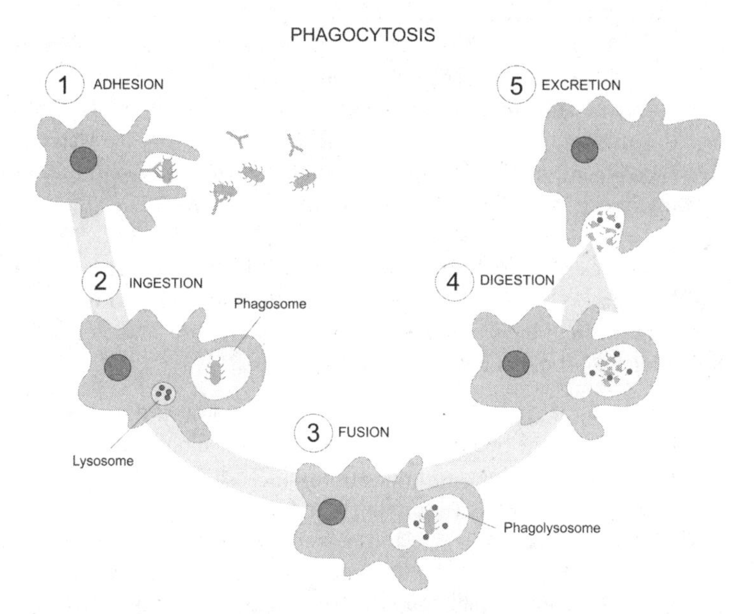

Mycobacterium tuberculosis is an intracellular parasite that survives and divides inside macrophages. Which of the following would NOT contribute to the bacterium's ability to live inside a macrophage?

(A) budding off of the phagosome to avoid fusion with the lysosome
(B) the production of ammonia (a base) to neutralize acids in the lysosome
(C) the production of catalase to detoxify peroxides
(D) avoiding recognition as a non-self cell

Questions 223–224

Manatees are mammals that inhabit shallow, tropical estuaries and rivers. They are herbivorous and have a low metabolic rate, although they are endothermic and regulate their internal body temperature at around 36.4°C. In line with their low metabolic rate, manatees typically swim very slowly. In response to a potential predator, such as an alligator, crocodile, or shark, manatees can swim up to 15–20 mph for short periods of time. The blood returning from a manatee's tail muscles can either flow through the fluke epidermal capillary bed (FECB) in the surface tissues or bypass this structure and return directly via a vein deep in the interior of the fluke.

223. **One hypothesis for the function of the FECB is that it represents a second mechanism by which manatees obtain oxygen for their muscles. If this were true, which of the following would be expected of the anatomy of the tail?**

(A) The fluke should be covered with a thick layer of mucus to prevent pathogens from infecting the bloodstream.
(B) Blood from the FECB should return directly to the lungs, where it can absorb additional oxygen before returning to the fluke muscles.
(C) Oxygenated blood from the lungs should flow to the FECB, bypassing the muscles so that the blood can absorb oxygen from the water.
(D) Blood from the FECB should return directly to the fluke muscles so that it can immediately supply absorbed oxygen.

224. **One hypothesis for the function of the FECB is that it is a mechanism by which manatees dissipate heat produced by the muscles during swimming before returning blood to the well-insulated body core. Which of the following experiments best tests this hypothesis?**

(A) Using 20 manatees, measure the blood flow speed and temperature of the blood in the body core and measure as the blood enters and exits the FECB under resting and fast-swimming conditions.
(B) Using 20 manatees, measure the blood flow speed and temperature of the blood in the body core and measure as the blood enters and exits the FECB in various water temperatures.
(C) Using 20 manatees, measure the blood flow speed and oxygen content of the blood in the body core and measure as the blood enters and exits the FECB in various water temperatures.
(D) Using 20 manatees, measure the blood flow speed and oxygen content of the blood in the body core and measure as the blood enters and exits the FECB under resting and fast-swimming conditions.

Questions 225–227

The diagram below shows portions of a nephron. Filtrate from the blood flows through the proximal convoluted tubule into the descending limb of the Loop of Henle. As filtrate flows down the descending limb, water passively diffuses into the medulla. Filtrate then moves up the ascending limb, which is not permeable to water. The ascending limb has ion channels that allow passive transport of sodium ions at its bottom and active transport of sodium ions toward the top of the limb. Active transport of sodium ions is primarily responsible for the solute gradient in the medulla. Filtrate from the ascending limb of the Loop of Henle flows through the distal convoluted tubule (not shown) and then into the collecting duct (or tubule).

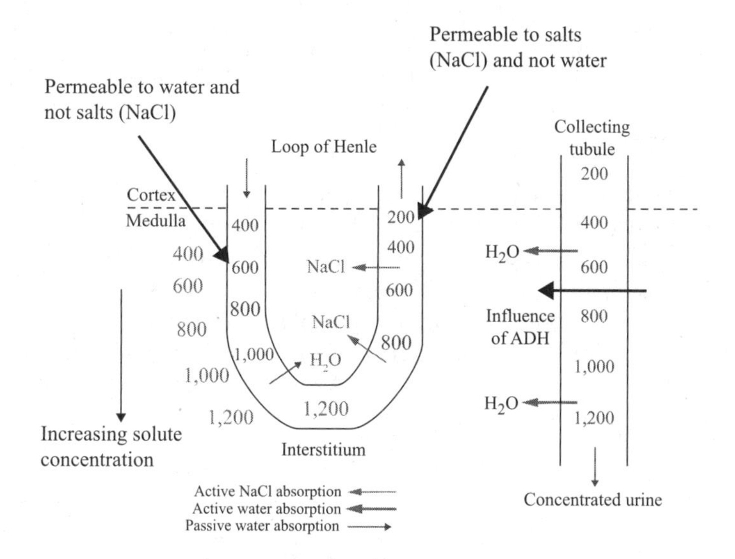

225. One effect of caffeine is to decrease the production of antidiuretic hormone (ADH). ADH causes the insertion of aquaporins in the collecting tubule. Which of the following is the result of the consumption of caffeine?

(A) the production of more concentrated urine as less water is pumped into the collecting duct
(B) the production of more dilute urine as less water is reabsorbed by the kidney
(C) leakage of salt through the aquaporins into the urine, leading to more concentrated urine
(D) a breakdown of the concentration gradient in the Loop of Henle as salt is permitted to leak into the collecting duct

226. A person has just completed a vigorous workout without drinking water. Which of the following is likely true of the individual's collecting duct?

(A) There are many aquaporins in the walls of the collecting duct to allow for the reabsorption of water.
(B) There are very few aquaporins in the walls of the collecting duct so that water will not diffuse out of the duct.
(C) There are more sodium pumps to absorb sodium from the urine.
(D) Urea pumps become more active to secrete waste.

227. Desert animals produce extremely concentrated urine. Which of the following leads to the ability to produce highly concentrated urine?

(A) a short Loop of Henle
(B) a long Loop of Henle
(C) a collecting duct with numerous aquaporins
(D) an ascending limb of the Loop of Henle with numerous aquaporins

Questions 228–231

Dialysis is a procedure that is used to remove toxins from the blood of patients with kidney failure. The blood is pumped through a tube that is immersed in dialysis solution and then returned to the body. Dialysis solution contains molecules that the body needs, such as glucose, at the same concentration as in blood so that there is no net loss of these compounds from the blood. The dialysis solution is also free of toxins, such as urea, so that toxins will diffuse out of the patient's blood into the dialysis solution.

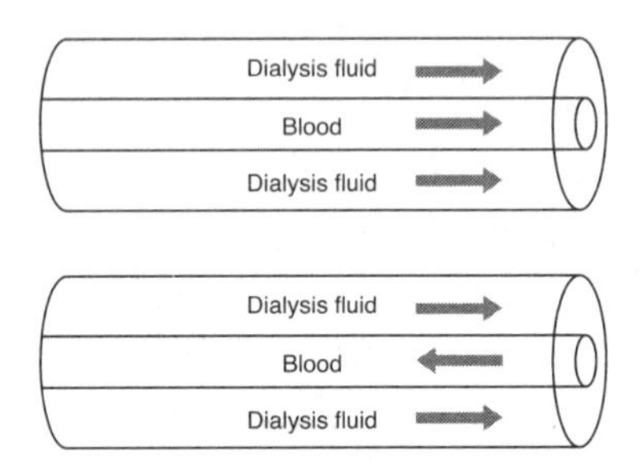

Figure I Two possible ways in which dialysis fluid and blood could flow through the dialysis apparatus

In the top half of Figure I, the blood and the dialysis fluid flow in the same direction. This is known as concurrent flow. In the bottom half of Figure I, the blood and the dialysis fluid flow in opposite directions. This is known as countercurrent flow.

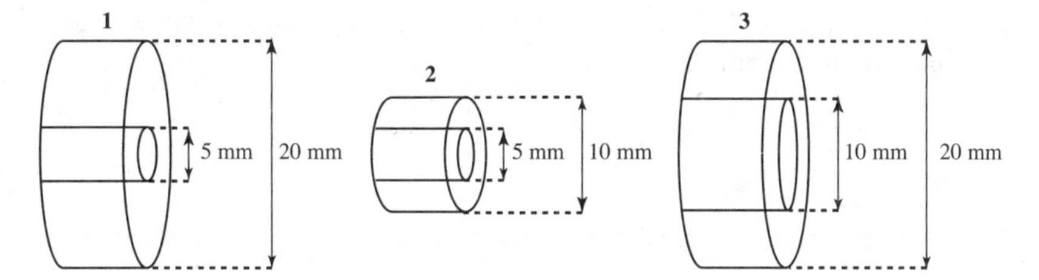

Figure II Different-sized dialysis tubes

In Figure II, you'll see that the diameter of the tubes in the dialysis machine may vary. In all three tubes depicted in Figure II, blood flows through the interior tube, and dialysis fluid flows through the exterior tube.

228. Refer to Figure I. Assuming the flow speed and the tube length are the same, which of the following best explains the most efficient way to remove toxins in the blood using this apparatus?

(A) Concurrent flow is more efficient because there is less friction between the fluids moving in the same direction.
(B) Countercurrent flow is more efficient because the cleanest blood encounters the cleanest dialysis fluid, leading to a concentration gradient along the entire length of the tube.
(C) Concurrent flow is more efficient because the blood and the dialysis fluid have the most time for diffusion to occur.
(D) Neither type of flow is more efficient. As long as the dialysis fluid contains no urea, it will stay clean.

229. Refer to Figure II. Which set of tubes has the most surface area per unit volume for exchange, assuming that the lengths of the tubes are the same?

(A) Set 1
(B) Set 2
(C) Set 3
(D) Set 1 and Set 2 have the same amount of surface area per unit volume and that value is higher than that of Set 3.

230. Refer to Figure II. Which set of tubes has the most total surface area for exchange, assuming that the lengths of the tubes are the same?

(A) Set 1
(B) Set 2
(C) Set 3
(D) Set 1 and Set 2 have the same total surface area and that value is higher than that of Set 3.

231. Refer to Figure II. Which set of tubes has the least efficient diffusion per unit volume of blood?

(A) Set 1
(B) Set 2
(C) Set 3
(D) Set 1 and Set 2 are equally efficient and are less efficient than Set 3.

Questions 232–233

The diagrams below represent different cell shapes. Assume that the three shapes continue into a third dimension (for example, assume that shape I is spherical).

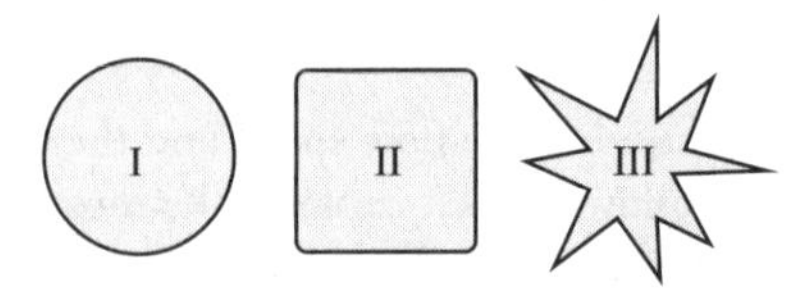

232. Assuming that each cell contains the same volume of cytoplasm, which of these cells would have the highest ratio of surface area to volume?

(A) I
(B) II
(C) III
(D) They would all have the same ratio of surface area to volume.

233. Assuming that each cell contains the same volume of cytoplasm, which of these cells would have the lowest ratio of surface area to volume?

(A) I
(B) II
(C) III
(D) They would all have the same ratio of surface area to volume.

Questions 234–242

A class of AP Biology students performed an experiment to examine the activity of the enzyme peroxidase, which catalyzes the breakdown of hydrogen peroxide into water and oxygen. Students dipped filter paper disks into solutions of peroxidase and then placed each disk at the bottom of a beaker of hydrogen peroxide. The reaction of peroxidase and hydrogen peroxide generates oxygen that collects on the underside of the disk and causes the disk to float. The time it takes the disk to float is proportional to the reaction rate. Plotted points on Figures I, II, III, and IV indicate the average of 15 trials. Where shown, error bars represent the standard error of the mean.

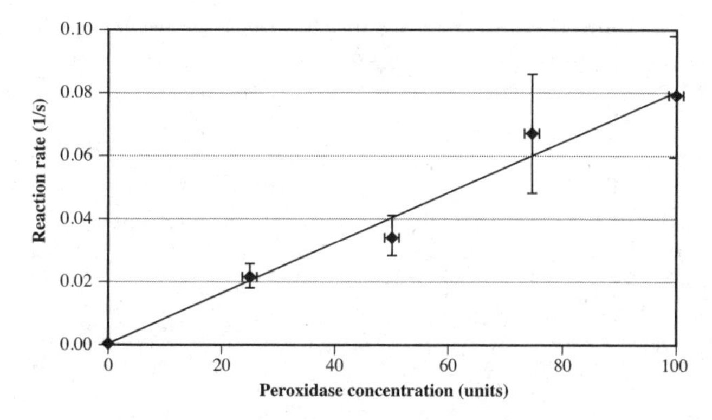

Figure I Altering the concentration of the enzyme (peroxidase)

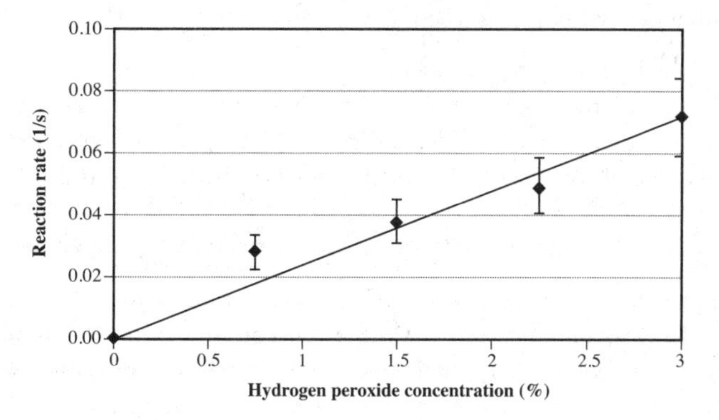

Figure II Altering the concentration of the substrate (hydrogen peroxide)

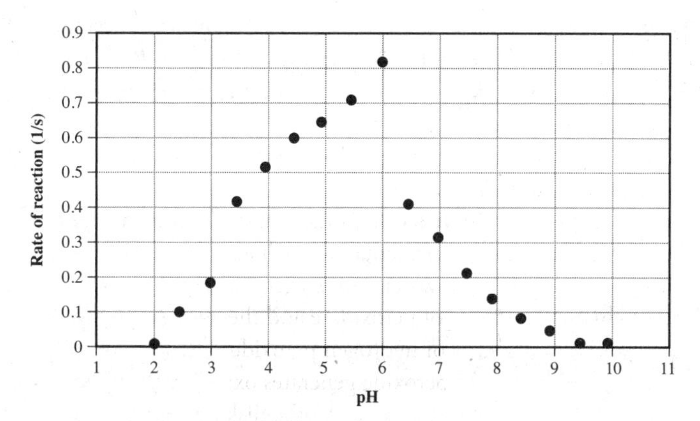

Figure III Altering the pH

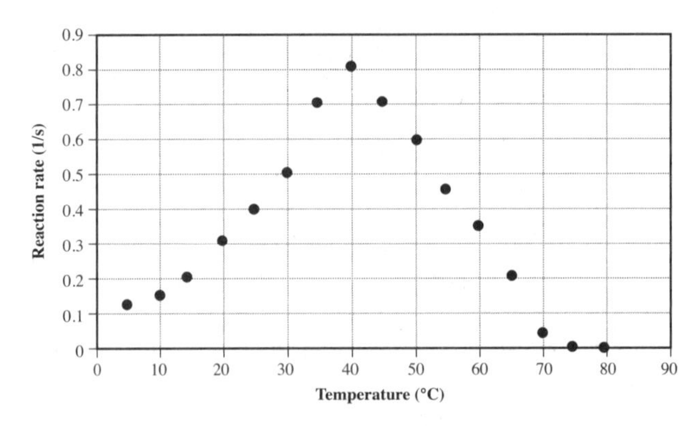

Figure IV Altering the temperature

234. **Which of the following best describes the relationship shown in Figure I?**

(A) The reaction rate increases as the concentration of the enzyme increases.
(B) The reaction rate decreases as the concentration of the enzyme increases.
(C) The reaction rate is independent of the enzyme concentration.
(D) None of these are good descriptions of the relationship shown.

235. **Which of the following best describes the relationship shown in Figure II?**

(A) The reaction rate increases as the concentration of the substrate increases.
(B) The reaction rate decreases as the concentration of the substrate increases.
(C) The reaction rate is independent of the substrate concentration.
(D) None of these are good descriptions of the relationship shown.

236. **Which of the following best explains the relationship shown in Figure I?**

(A) Increasing the concentration of the enzyme makes a larger number of substrate molecules available to react.
(B) Increasing the concentration of the enzyme makes the active sites of each molecule work faster.
(C) Increasing the concentration of the enzyme makes a larger number of active sites available to react.
(D) Increasing the concentration of the enzyme makes each enzyme molecule open up more active sites.

237. **Which of the following best explains the relationship shown in Figure II?**

(A) Increasing the substrate concentration makes more active sites available to react.
(B) Increasing the substrate concentration makes enzymes more active.
(C) Increasing the substrate concentration increases the temperature of the reaction mixture.
(D) Increasing the substrate concentration makes it more likely that a substrate will encounter an enzyme.

238. What is the optimum pH for the peroxidase used in the experiment according to Figure III?

(A) 3
(B) 6
(C) 7
(D) 9.5

239. Which of the following best explains why the rate of reaction falls for pH levels below and above the optimum pH for the enzyme as shown in Figure III?

(A) The phosphate ion concentration affects amino acid side chain interactions, which disrupts the primary structure of the protein.
(B) The hydrogen ion concentration affects the interactions among the amino acid side chains in the protein. This changes the shape of the active site such that the peroxide does not fit as well as it does at the optimum pH.
(C) Both low and high pH levels lead to the digestion of enzymes by macrophages.
(D) Molecules move more slowly at low pH and at high pH. Therefore, more reactions take place at medium pH levels.

240. According to Figure IV, what is the optimum temperature for peroxidase activity in this experiment?

(A) 5°C
(B) 20°C
(C) 40°C
(D) 70°C

241. Which of the following best explains why temperatures higher than the optimum temperature cause enzyme activity to fall?

(A) As the temperature increases, the atoms that make up the protein move faster and with more energy. This movement makes the active site change shape such that it doesn't fit the substrate as well.
(B) As the temperature increases, the atoms move faster, making more collisions occur in the same amount of time.
(C) As the temperature increases, the primary structure of the protein degrades, making the enzyme less able to work efficiently.
(D) As the temperature increases, substrate spontaneously degrades, leaving less hydrogen peroxide available to react.

242. Which of the following best explains why temperatures that are colder than the optimum temperature cause the enzyme activity to fall the colder the temperature gets?

(A) As the temperature decreases, the atoms that make up the protein move more slowly and with less energy. This movement makes the active site change shape such that it doesn't fit the substrate as well.
(B) As the temperature decreases, the atoms move more slowly, making fewer collisions occur in the same amount of time.
(C) As the temperature decreases, the primary structure of the protein degrades, making the enzyme less able to work efficiently.
(D) As the temperature decreases, substrate spontaneously degrades, leaving less hydrogen peroxide available to react.

Questions 243–245

Students conducted an experiment in which they cut leaf disks from plant leaves using a hole punch and placed the disks into a syringe containing a buffer solution. They exposed the disks to a vacuum to evacuate the air from the leaf disks. Since the disks no longer contained gas, they sank. Once the disks had sunk, the students exposed them to light and recorded the amount of time it took for the leaves to float to the top of the 5-cm-long syringe.

In one experiment, students used leaf disks from both sun-grown leaves and shade-grown leaves from the same hybrid poplar tree. They exposed the leaf disks to different amounts of light using window screening to create shade. The data are presented below.

Light (% of Maximum)	Time It Took for Leaf Disks to Float (seconds)		Rate of Rise for Leaf Disks (cm/second)	
	Sun-Grown	**Shade-Grown**	**Sun-Grown**	**Shade-Grown**
10	40	20	0.03	0.05
20	30	10	0.03	0.10
30	20	7	0.05	0.14
40	10	5	0.10	0.20
50	5	4	0.20	0.25
60	2	3	0.50	0.33
70	1	3	1.00	0.33
80	1	3	1.00	0.33
90	1	3	1.00	0.33
100	1	3	1.00	0.33

243. What causes the leaves to float after the gas is evacuated from the tissue?

(A) Photosynthesis generates carbon dioxide gas, which builds up in the tissue.
(B) Photosynthesis generates oxygen gas, which builds up in the tissue.
(C) Nitrogen gas eventually bubbles out of the solution and builds up on the underside of the leaf tissue.
(D) Hydrogen gas generated from the degradation of water in photosynthesis builds up in the xylem.

244. Which type of leaf or leaves performed photosynthesis the fastest?

(A) Shade-grown leaves performed photosynthesis fastest in low light, and sun-grown leaves performed photosynthesis fastest in high light.
(B) Shade-grown leaves performed photosynthesis fastest in high light, and sun-grown leaves performed photosynthesis fastest in low light.
(C) Shade-grown leaves performed photosynthesis fastest in all light conditions.
(D) Sun-grown leaves performed photosynthesis fastest in all light conditions.

245. For each type of leaf, the rate of reaction plateaus. Which of the following is the best explanation for this result?

(A) The reaction rate plateaus when the leaf is exposed to the maximum amount of light.
(B) The reaction rate plateaus because the components of the photosynthetic electron transport chain get damaged from high use.
(C) The reaction rate plateau occurs when all of the photosystems in the leaf are excited at the same time.
(D) At a certain point, the electrons in the photosynthetic electron transport chain cannot move any faster. The plateau is a result of this limit on electron speed.

Questions 246–247

The questions that follow refer to the three transport processes shown below.

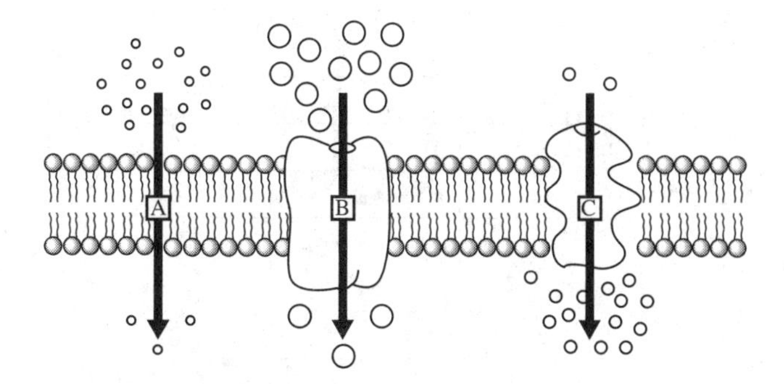

246. Which of these transport processes requires the input of energy?

(A) Process A requires the input of energy because molecules must move rapidly to cross membranes without protein channels.
(B) Process B requires the input of energy because the molecules are flowing up their concentration gradient.
(C) Process C requires the input of energy because the molecules are moving from an area of low concentration to an area of high concentration.
(D) None of these processes require the input of energy because they all occur spontaneously.

247. Which of the following most likely depicts the transport of a nonpolar substance?

(A) The molecules in process A are able to cross through the nonpolar fatty acid tails of the membrane without aid.
(B) The molecules in process B are able to cross through a channel protein, which is necessary because the fatty acid tails are polar.
(C) The molecules in process C are nonpolar because only nonpolar substances can be moved against their concentration gradient.
(D) None of the molecules shown in these processes can be nonpolar.

Questions 248–251

The diagram below depicts the portion of the inner mitochondrial membrane with an ATP synthase embedded in it.

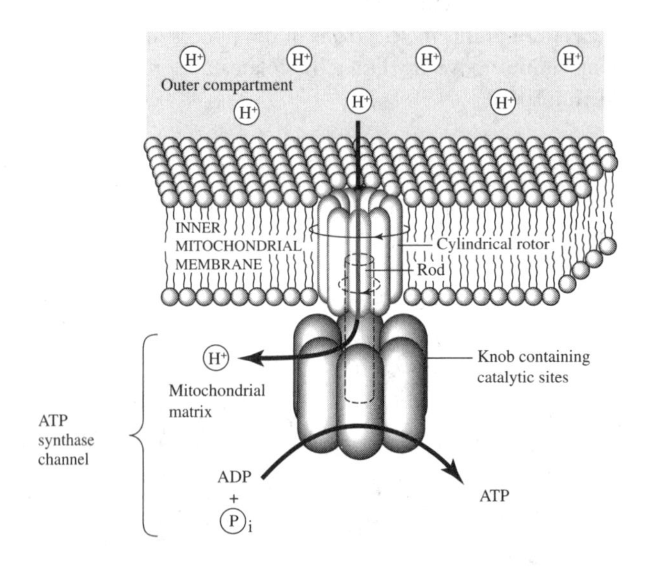

248. Which of the following is the immediate source of energy that drives the production of ATP from ADP and phosphate?

(A) sunlight
(B) the breakdown of glucose into carbon dioxide
(C) hydrogen ions flowing down their concentration gradient
(D) the pull of gravity on hydrogen ions

249. What would happen if ATP synthase were inserted in the opposite orientation?

(A) ATP would be hydrolyzed to ADP, causing the ATP synthase to pump hydrogen ions against their concentration gradient.
(B) ATP would be hydrolyzed to ADP as the hydrogen ions flow down their concentration gradient.
(C) Hydrogen ions would return through the membrane without affecting ATP concentrations.
(D) None of these are correct.

250. Assume that all reactions depicted in the diagram are reversible. What would happen if large amounts of ATP were supplied to the stroma side?

(A) The ATP would be hydrolyzed to ADP, causing the ATP synthase to pump hydrogen ions against their concentration gradient.
(B) ATP would be hydrolyzed to ADP as the hydrogen ions flow down their concentration gradient.
(C) Hydrogen ions would return through the membrane without making ATP.
(D) None of these are correct.

251. What would happen if the knob portion of the ATPase was not attached and only the transmembrane portion of the protein was present in the membrane?

(A) The ATP would be hydrolyzed to ADP, causing the ATP synthase to pump hydrogen ions against their concentration gradient.
(B) ATP would be hydrolyzed to ADP as the hydrogen ions flow down their concentration gradient.
(C) Hydrogen ions would return through the membrane without making ATP.
(D) None of these are correct.

Questions 252–255

The diagrams below show a red blood cell in three different solutions. Each solution has a different concentration of salt. In Diagram I, the arrows are the same size, indicating that the same amount of water is flowing in as is flowing out. In Diagram II, more water is flowing in than out. In Diagram III, more water is flowing out than into the cell.

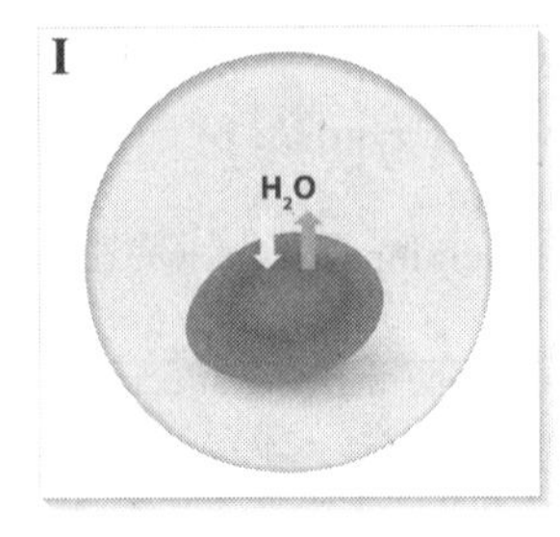

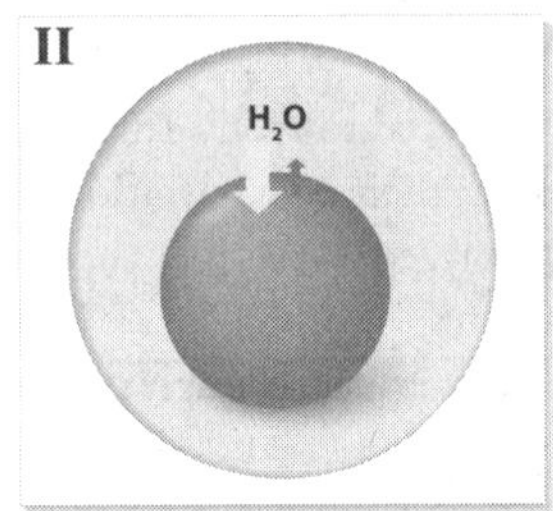

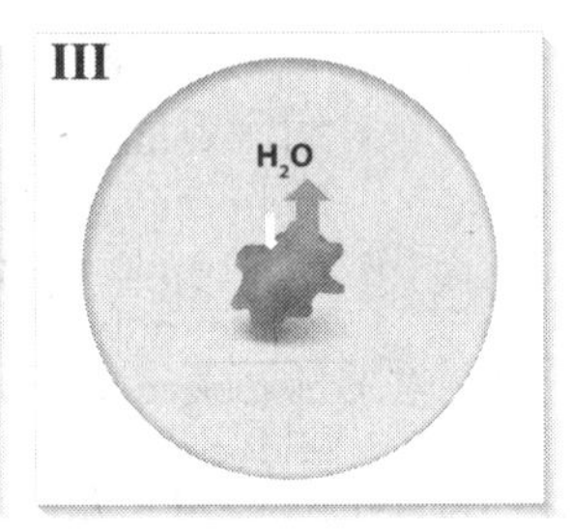

252. Which solution has the same concentration of solutes as the inside of the cell?

(A) Diagram I
(B) Diagram II
(C) Diagram III
(D) It is impossible to tell from these diagrams.

253. Which cell is hypertonic to the solution?

(A) Diagram I
(B) Diagram II
(C) Diagram III
(D) None of the cells are hypertonic.

254. Which diagram shows the normal state of red blood cells in the body?

(A) Diagram I
(B) Diagram II
(C) Diagram III
(D) All of these diagrams show abnormal cells.

255. "Doc," a Green Beret medic, is working in a remote village in Afghanistan. He is often asked to treat farm animals. When diagnosing animals as well as humans, "Doc" does his own laboratory work. In order to save supplies, he dilutes prepared saline for microscopy. While examining the blood cells of a healthy goat, "Doc" sees cells that look like those in Diagram III. Which of the following is the most likely explanation for this?

(A) His saline solution is too concentrated and is hypertonic to the cells.
(B) His saline solution is too dilute and is hypotonic to the cells.
(C) His saline solution is isotonic to the cells.
(D) The goat is suffering from hydrophobia.

Questions 256–260

The diagrams below show a cell of a plant leaf placed into three different solutions of sucrose. The amount of solute inside the cell is the same in each case. The size of each arrow shows the relative amount of water moving in that particular direction.

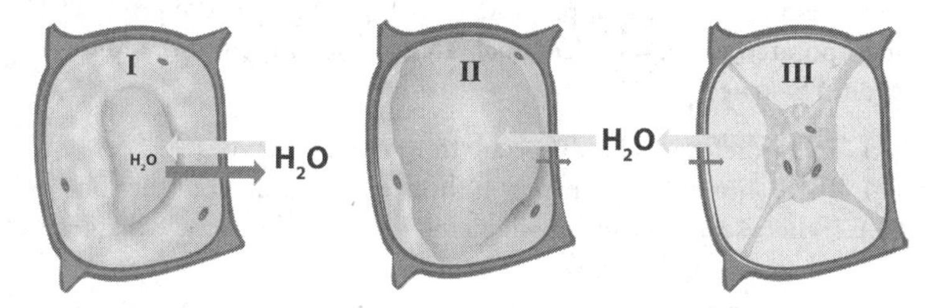

256. Which diagram might depict a cell that was placed into distilled water?

(A) Diagram I
(B) Diagram II
(C) Diagram III
(D) None of these shows a plant cell in distilled water.

257. Which cell has the highest turgor pressure?

(A) Diagram I
(B) Diagram II
(C) Diagram III
(D) The turgor pressure is the same in each case.

258. Which cell represents a typical, healthy plant cell under normal conditions?

(A) Diagram I
(B) Diagram II
(C) Diagram III
(D) All of the images represent cells from healthy plants under normal conditions.

259. Which cell is typical of a plant at the wilting point, when the turgor pressure just reaches zero?

(A) Diagram I
(B) Diagram II
(C) Diagram III
(D) Both Diagram II and Diagram III represent cells from a plant at the wilting point.

260. Which cell has a lower solute water potential than the solution it is in?

(A) Diagram I
(B) Diagram II
(C) Diagram III
(D) All of the cells have the same solute water potential as the solutions they are in.

---

261. Researchers created phospholipid bilayer spheres into which they incorporated purified protein as part of the bilayer. If the protein incorporated was a water channel (aquaporin), which of the following should be true?

(A) The proteins should dissociate from the membrane.
(B) There should be no change in the average diameter of the spheres.
(C) If the researchers added salt to the solution, the average diameter of the spheres should increase.
(D) If the researchers added salt to the solution, the average diameter of the spheres should decrease.

Questions 262–264

Biology students conducted an experiment in which they compared the density of stomata on poplar leaves grown in the sun and those grown in the shade. To do this, they collected 30 healthy leaves exposed to full sunlight and 30 leaves grown in the shade. The students made casts of the stomata using clear nail polish. They then counted the number of stomata on randomly selected microscopic fields of view. The data are presented in the graph that follows. Error bars represent the standard error of the mean.

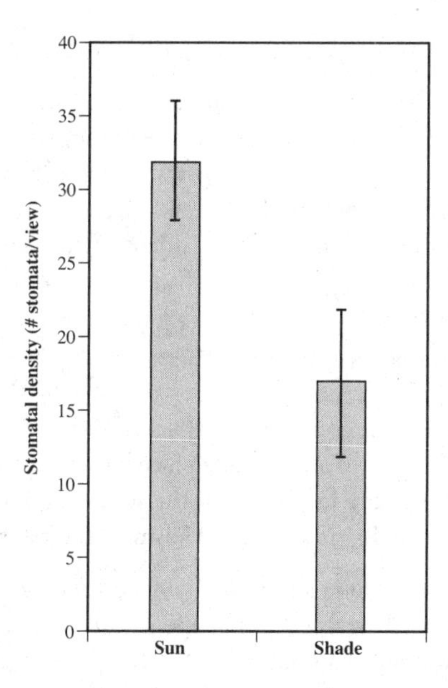

262. Which of the following is consistent with the data presented?

(A) Sun-grown leaves have thicker cuticles than leaves grown in the shade.
(B) Shade-grown leaves have more stomata than leaves grown in the sun.
(C) Sun-grown leaves have more stomata than leaves grown in the shade.
(D) No significant difference exists between the number of stomata on sun-grown leaves and on shade-grown leaves.

263. Which of the following explains the physiological need for a difference in stomatal density in sun-grown and shade-grown leaves?

(A) Shade-grown leaves perform more photosynthesis and therefore need more stomata to perform more gas exchange.
(B) Shade-grown leaves use more water and therefore need fewer stomata to retain the excess water.
(C) Sun-grown leaves perform more photosynthesis and therefore need more stomata to perform more gas exchange.
(D) Sun-grown leaves become hotter during the day and need fewer stomata to prevent excessive water loss.

264. How can a single plant have different stomatal densities on different leaves?

(A) Sun-grown leaves and shade-grown leaves have different DNA sequences.
(B) Sun-grown leaves and shade-grown leaves express different genes.
(C) Plants that touch each other can become grafted so that the mature plant is a chimera consisting of different genotypes.
(D) It is not possible for genetically identical leaves to have different numbers of stomata.

Questions 265–267

Students studying transpiration conducted an experiment in which they filled 50 mL flasks with water, covered them with Parafilm,* and placed sections of *Forsythia* branches (a woody shrub) through the Parafilm into the bottom of the flask. They then subjected the branches to various treatments. The students measured the amount of water lost by measuring the mass of the water in the flasks both before and after treatments.

*Parafilm is a registered trademark of Bemis Company, Inc.

265. One group chose to investigate the effect of wind. To do this, they placed five branches in front of a fan. They placed another five branches in an adjacent area without a fan but with the same temperature and light. Which of the following is a valid criticism of their experimental design?

(A) They did not include a control in which they measured the amount of evaporation that would have happened without the plant material.
(B) A fan is different than natural wind. It would have been better to put some plants outside in the wind and some inside without wind.
(C) Mass is not an appropriate measure of water lost.
(D) The experiment should have included different temperatures to see if that variable had an effect.

266. One group chose to investigate the effect of temperature. To do so, they placed one group of five plants in a refrigerator at 4°C with a fluorescent light and one group on the counter by the window at 24°C. Which of the following is the most valid criticism of their experimental design?

(A) The experiment did not include enough different temperatures.
(B) The sample size was not large enough.
(C) The light conditions were likely to be different.
(D) None of these are valid criticisms.

267. One group chose to manipulate the solute concentration. To do so, they placed different concentrations of salt solutions into different flasks. The salt concentrations were 0 M, 0.125 M, 0.25 M, and 0.5 M. The students then placed pieces of 30-cm-long *Forsythia* into the flasks. Their results are shown below. Data points are the average percent change in water for 5 samples. Error bars represent the standard error of the mean.

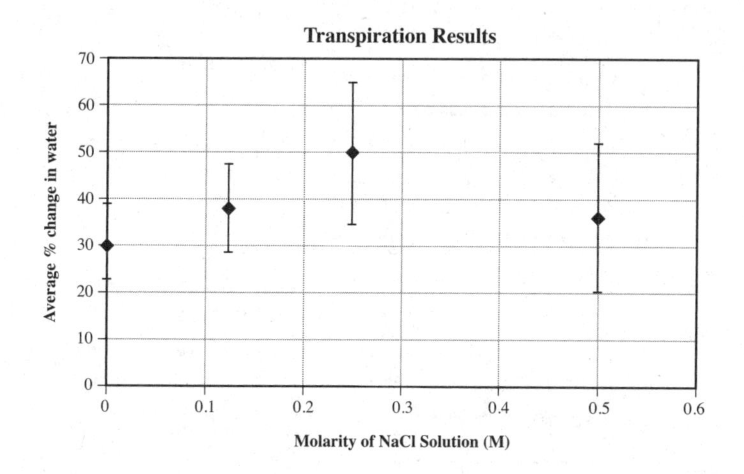

Which of the following is a reasonable assertion based on the data shown?

(A) Plants in the 0.25 M solution had significantly greater change in water than plants in any other concentration solution.
(B) The error in the data is too great to draw any conclusions about the relationship between salt molarity and average percent change in water.
(C) Plants in the 0 M solution had significantly less change in water than plants in any other concentration solution.
(D) None of these are reasonable assertions.

268. In the bacterium *E. coli*, the protein encoded by the *lacY* gene uses the energy from a proton gradient to drive the uptake of lactose against its concentration gradient into the cell. This proton gradient is created when the bacterium uses a proton pump to move hydrogen ions out of the cell. Which of the following diagrams correctly depicts the process performed by the protein encoded by the *lacY* gene?

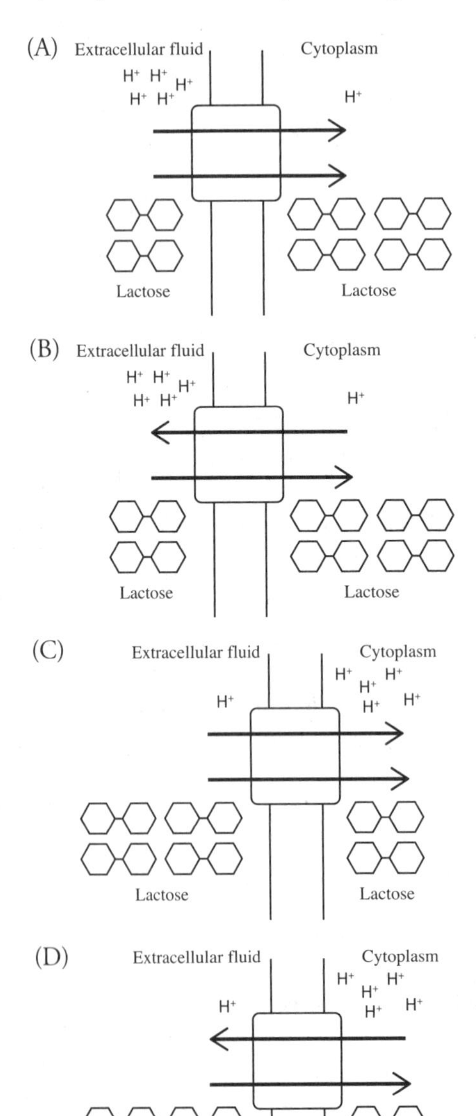

GRID-IN QUESTIONS

Questions 269–270

AP Biology students cut russet potatoes and sweet potatoes into cubes 1 cm on a side. They exposed 34 cubes of each variety to 6 different concentrations of sucrose solution. The solutions varied in concentration from 0 M to 1 M. Each cube was measured before and after 24 hours, and the percent change in mass was calculated. The experiment was conducted at 20°C. The results are shown in the graph below.

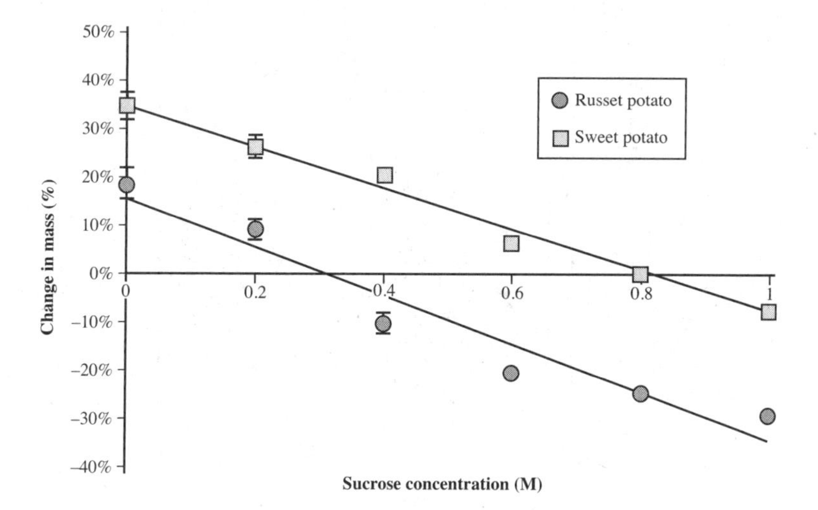

269. **Calculate the solute water potential for the sweet potato to the nearest tenth of a bar.**

270. **Calculate the solute water potential for the russet potato to the nearest tenth of a bar.**

---

271. **A plant cell with a solute water potential of –7.2 bars is placed into a beaker of distilled water that is open to the atmosphere. Calculate the pressure potential at equilibrium. Express your answer to one decimal place.**

272. **A plant cell with a solute water potential of –7.2 bars is placed into a beaker that is open to the atmosphere. The beaker contains a solution with a solute water potential of –3.3 bars. Calculate the pressure potential at equilibrium. Express your answer to one decimal place.**

Questions 273–274

Researchers measured predawn water potential in two different clones of hybrid poplar trees growing in the greenhouse. Trees were grown in 30 boxes, each of which contained 6 individuals of each clone. Half of the boxes were watered 3 times per week (watered group). The others were watered only when the predawn water potential fell below –10 bars (stressed group). The data are presented below.

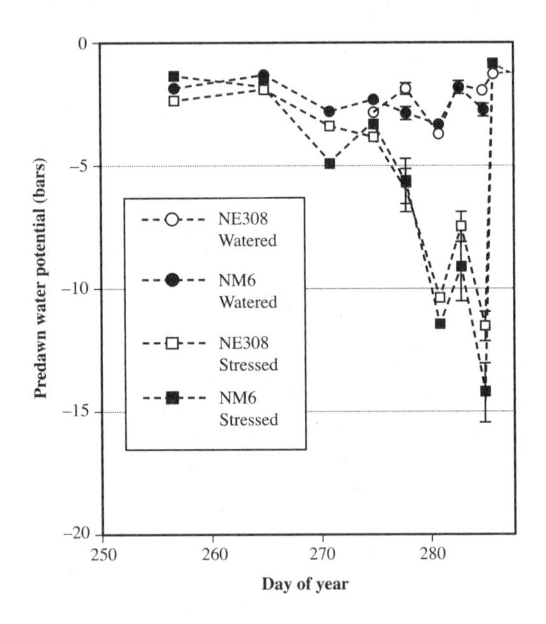

273. Assuming that the predawn water potential values were entirely due to solutes in the leaf, calculate the solute concentration for NM6 when it was under the most serious drought stress. Express your answer to the nearest hundredth of an M. Note that the temperature was 20°C.

274. Assuming that the predawn water potential values were entirely due to solutes in the leaf, calculate the solute concentration for NE308 when it was under the most serious drought stress. Express your answer to the nearest hundredth of an M. Note that the temperature was 20°C.

Questions 275–277

A group of AP Biology students measured dissolved oxygen in samples of pond water incubated at different intensities of light over a period of two days. The data are presented in the graph below.

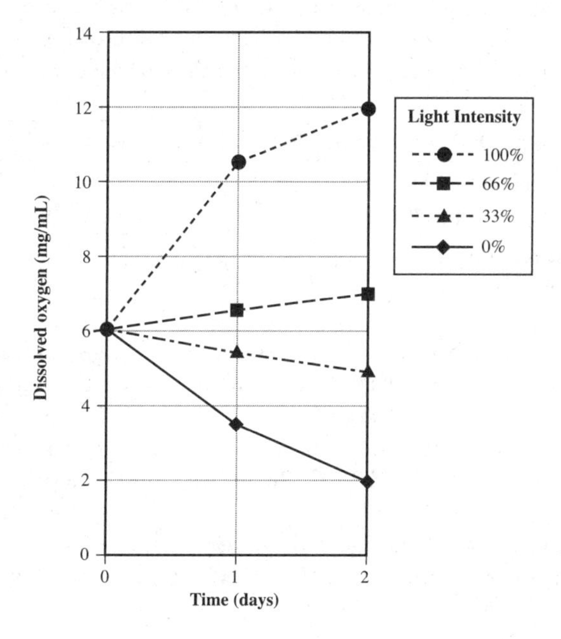

275. Biological oxygen demand (BOD) is a measure of how much oxygen is used in the dark when photosynthesis does not happen. Calculate the BOD for the pond water over the first two days. Express your answer to the nearest tenth of an mg/mL oxygen.

276. Net primary productivity is the amount of oxygen generated by photosynthesis in the light that is in excess of the amount of respiration occurring in the light. Calculate the net primary productivity for the pond water incubated in 100% light over the first two days. Express your answer to the nearest tenth of an mg/mL oxygen.

277. Gross primary productivity is the total amount of oxygen generated by photosynthesis. Some of this oxygen is used by respiration that is occurring at the same time. Calculate the gross primary productivity for the pond water incubated in 100% light over the first two days. Express your answer to the nearest tenth of an mg/mL oxygen.

Questions 278–281

A student constructed respirometers to measure oxygen consumption in small animals at various temperatures. Ten respirometers were used for each temperature listed below. Five chipmunks and five geckos, each weighing approximately 20 g, were used for each temperature. The total oxygen consumed in milliliters was measured for 1 hour for each animal. The averages are reported in the table below.

Temperature (°C)	10	15	20	25	30	35	40	45
Oxygen consumed by the chipmunks (mL/hr)	68	55	42	36	35	37	46	60
Oxygen consumed by the geckos (mL/hr)	5	8	13	16	20	24	28	32

278. Calculate the temperature coefficient (Q_{10}) for the geckos using the data points for 20°C and 30°C. Report your answer to two decimal places.

279. Calculate the temperature coefficient (Q_{10}) for the chipmunks using the data points for 10°C and 20°C. Report your answer to two decimal places.

280. Calculate the rate of change in the oxygen consumed by the chipmunks as the temperature rises from 10°C to 25°C. Report your answer to two decimal places.

281. Calculate the rate of change in the oxygen consumed by the geckos as the temperature rises from 10°C to 25°C. Report your answer to two decimal places.

---

282. Respiration of 1 mole of glucose produces up to 38 moles of ATP, which organisms can use to drive cellular processes. Using the Gibbs free energy values below, calculate how much of the energy released in the breakdown of glucose is NOT converted into ATP. Report your answer to one decimal place.

$$C_6H_{12}O_6 + O_2 \rightarrow H_2O + CO_2 \qquad \Delta G° = -686 \text{ kcal/mol}$$
$$ATP + H_2O \rightarrow ADP + P_i \qquad \Delta G° = -7.3 \text{ kcal/mol}$$

LONG FREE-RESPONSE QUESTIONS

283. A student constructed respirometers to measure oxygen consumption in small animals at various temperatures. Ten respirometers were used for each temperature listed below. Five chipmunks and five geckos, each weighing approximately 20 g, were used for each temperature. The total oxygen consumed in milliliters was measured for 1 hour for each animal. The averages are reported in the table below.

Temperature (°C)	10	15	20	25	30	35	40	45
Oxygen consumed by the chipmunks (mL/hr)	68	55	42	36	35	37	46	60
Oxygen consumed by the geckos (mL/hr)	5	8	13	16	20	24	28	32

(a) Graph the results on the axes provided.

(b) For EACH animal species:

- Describe the relationship between temperature and respiration.
- Explain the physiological basis for the shape of each curve.

(c) For ONE of the two species:

- Identify ONE cost and ONE benefit for the type of metabolism used.
- Describe the conditions under which the benefit outweighs the cost for the organism chosen.

284. A scientist measured the size of the stomatal opening of wild-type and mutant *Arabidopsis* plants in response to light intensity. The following data were collected.

	Width of Stomatal Aperture (μm)	
Light Intensity (lux)	**Wild-Type**	**Mutant**
0	1	1
15	1.3	1.2
30	1.9	1.4
45	2.9	1.5
60	3.2	1.6
75	3.2	1.9
90	3.2	2.1

(a) Use the grid below to construct a graph of the data.

(b) **Describe** and **explain** the physiological reason for the shape of the curve for the wild-type plant between 0 and 60 lux and between 60 and 90 lux.

(c) **Describe** a controlled experiment that would test the idea that the two different plant lines would respond differently to drought stress.

SHORT FREE-RESPONSE QUESTIONS

285. An AP Biology student investigated the ability of the yellow mealworm (*Tenebrio molitor*) to use plastics as an energy source. To do this, he presented mealworms with various plastics, which were ground small enough to be ingested. A small reservoir of water was included in each habitat. The student used bran as a positive control and water only as a negative control. He measured the mass of each mealworm before and after 30 days. There were 7 mealworms in each treatment group. The data presented in the following graph are the average for the 7 mealworms. The mealworms in the negative control group died and are not shown on the graph. The error bars represent the standard error of the mean.

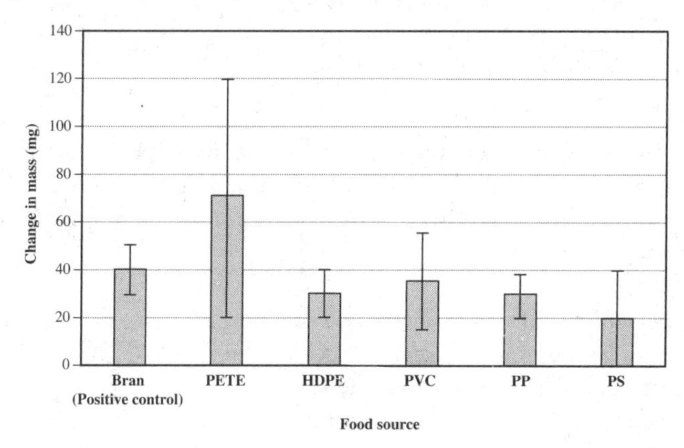

(a) Using the information presented in the graph, **determine** whether there were statistically significant differences among the different treatments shown. **Justify** your answer.

(b) **Determine** whether the mealworms were able to use plastics as an energy source. **Justify** your answer.

286. A *mycorrhiza* is a mutualistic symbiotic relationship between a plant and a fungus in which the fungus provides the plant with mineral nutrients and water in exchange for sugar. A typical plant root hair has a diameter of 15 μm, while the diameter of a fungal cell is 5 μm.

(a) **Calculate** the surface area:volume ratio for a 100 μm long cell of each type and record your answers in the table below. Assume that each cell type is approximated by a cylinder (see the Reference Tables at the end of this book). Use the value 3.14 for π. The surface area is given for both the plant root hair and the fungal cell.

	Diameter (μm)	Surface Area (μm^2)	Volume (μm^3)	Surface Area: Volume (μm^{-1})
Plant root hair	15	4,712		
Fungal cell	5	1,570		

(b) **Compare** the surface area:volume ratio of the two types of cells. **Explain** why this ratio is significant to the relationship between the two organisms.

287. Plant cells are highly organized structures in which the placement of organelles supports their functions. The diagram to the right shows a typical arrangement of the chloroplast, mitochondrion, and peroxisome.

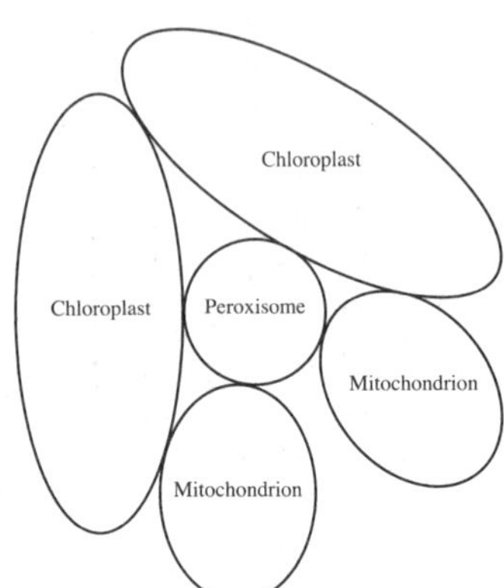

(a) **Identify** the primary functions of the chloroplast and the mitochondrion. **Explain** TWO reasons why these two organelles might be situated close together.

(b) The peroxisome contains the enzyme peroxidase that breaks down toxic hydrogen peroxide (H_2O_2) into water and oxygen. **Propose** a plausible explanation for locating the peroxisome near the chloroplast and the mitochondrion.

288. The diagram to the right shows two proteins that can be found in the inner mitochondrial membrane. ATP synthase uses the H^+ gradient produced by the electron transport chain to make ATP. Uncoupling protein (UCP) makes the inner mitochondrial membrane freely permeable to hydrogen ions, but it does not make ATP.

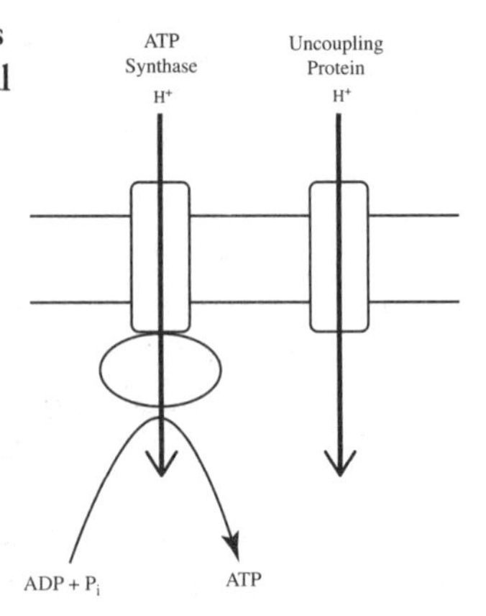

(a) **Describe** the cost of expressing UCP.

(b) UCP is present in the mitochondria of brown fat in human infants and in chipmunks during the winter. **Identify** one possible adaptive function of expressing this protein, and **explain** why that function would be beneficial.

Answers and explanations for all of the questions in this chapter can be found on pages 309–346.

Big Idea 3—Information

CHAPTER 3

Answers and explanations can be found on page 347.

BIG IDEA 3

Living systems store, retrieve, transmit, and respond to **information** essential to life processes.

For the complete list of big ideas, learning objectives, enduring understandings, essential knowledge, and science practices, refer to the "AP Biology Course and Exam Description" from the College Board:

https://secure-media.collegeboard.org/digitalServices/pdf/ap/ap-biology-course-and-exam-description.pdf

MULTIPLE-CHOICE QUESTIONS

Questions 289–292

Students at two schools collaborated on a research project. One school was located in Maine at 44.8° N latitude, and the other was located in Florida at 27° N latitude. Students at both schools collected wild samples of the fruit fly *Drosophila melanogaster*. The students tested the flies for the ability to enter diapause by exposing them to short day length conditions (10 hours of light and 14 hours of darkness). Diapause is a slowing of development that is associated with unfavorable growth conditions. The Florida strain (FL) was unable to enter diapause. All the flies from Maine (ME) were able to enter diapause under the short day length conditions. The strains were crossed to produce F1 offspring. The F1 hybrids were backcrossed to the original parental (P) strains to produce F2 progeny. Fruit fly progeny were tested for their ability to enter diapause under short day length conditions. The results are shown in the graph on page 136.

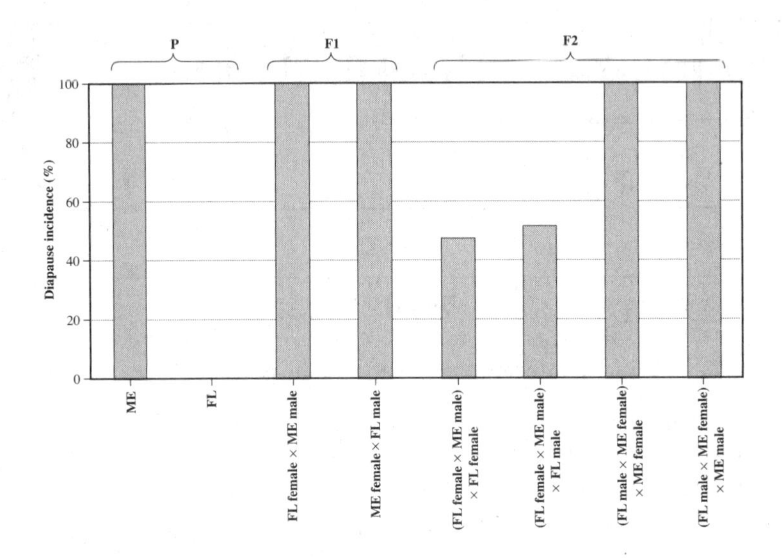

289. Which of the backcrosses produced a ratio of approximately 1:1 of fruit flies that could enter diapause to those that could not enter diapause?

(A) crosses in which the F1 were mated with individuals from Maine
(B) crosses in which the F1 were mated with individuals from Florida
(C) crosses in which the F1 were mated with females
(D) crosses in which the F1 were mated with males

290. Based on the data presented for the F1 offspring, which of the following is the most likely mechanism of inheritance for diapause?

(A) The ability to enter diapause is sex-linked recessive.
(B) The inability to enter diapause is sex-linked recessive.
(C) The ability to enter diapause is autosomal recessive.
(D) The inability to enter diapause is autosomal recessive.

291. All the F1 flies were capable of entering diapause even though one of the parents could not. Which of the following is the best explanation for this phenomenon?

(A) The diapause plus allele encodes a protein important in the signal transduction pathway that controls diapause genes. One functional copy of this gene is sufficient to allow the pathway to function.
(B) The diapause plus allele segregates independently from the diapause minus allele.
(C) The diapause minus allele encodes a protein that inhibits diapause genes. One copy of the diapause minus allele is sufficient to turn off the pathway.
(D) Mate choice prevented flies that couldn't enter diapause from mating.

292. Which of the following is the best explanation for the results of the backcross to the ME female and male shown in the F2 progeny?

(A) Backcrossing hybrids to a homozygous dominant P produces a 1:1 ratio of homozygous dominant to heterozygous individuals, all of whom show the dominant phenotype.
(B) Backcrossing hybrids to a homozygous dominant P produces all homozygous dominant individuals.
(C) This cross produces a 9:3:3:1 ratio, which is what is seen in the F2 progeny.
(D) This cross produces a 3:1 ratio, which is what is seen in the F2 progeny.

Questions 293–303

The table below shows the amino acid specified by triplets of mRNA bases.

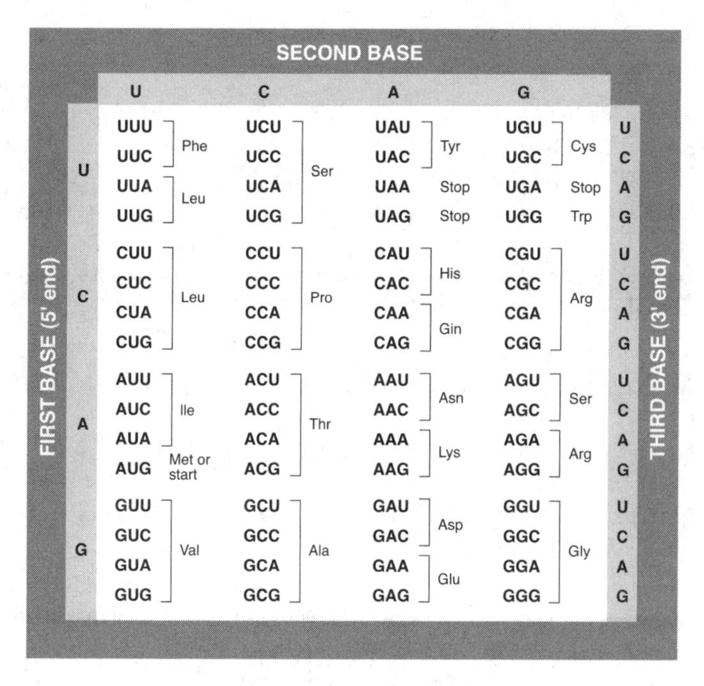

FIRST BASE (5' end)	SECOND BASE: U		SECOND BASE: C		SECOND BASE: A		SECOND BASE: G		THIRD BASE (3' end)
U	UUU	Phe	UCU	Ser	UAU	Tyr	UGU	Cys	U
U	UUC	Phe	UCC	Ser	UAC	Tyr	UGC	Cys	C
U	UUA	Leu	UCA	Ser	UAA	Stop	UGA	Stop	A
U	UUG	Leu	UCG	Ser	UAG	Stop	UGG	Trp	G
C	CUU	Leu	CCU	Pro	CAU	His	CGU	Arg	U
C	CUC	Leu	CCC	Pro	CAC	His	CGC	Arg	C
C	CUA	Leu	CCA	Pro	CAA	Gin	CGA	Arg	A
C	CUG	Leu	CCG	Pro	CAG	Gin	CGG	Arg	G
A	AUU	Ile	ACU	Thr	AAU	Asn	AGU	Ser	U
A	AUC	Ile	ACC	Thr	AAC	Asn	AGC	Ser	C
A	AUA	Ile	ACA	Thr	AAA	Lys	AGA	Arg	A
A	AUG	Met or start	ACG	Thr	AAG	Lys	AGG	Arg	G
G	GUU	Val	GCU	Ala	GAU	Asp	GGU	Gly	U
G	GUC	Val	GCC	Ala	GAC	Asp	GGC	Gly	C
G	GUA	Val	GCA	Ala	GAA	Glu	GGA	Gly	A
G	GUG	Val	GCG	Ala	GAG	Glu	GGG	Gly	G

Figure I The genetic code

293. Based on Figure I, which of the following is an mRNA sequence that codes for Met-Pro-Leu-Stop?

(A) 5′-AUGCCCAAUGUUGA-3′
(B) 5′-ACAUGCCCCUCUAAUUCAC-3′
(C) 5′-ACATGCCCCTCTAATTCAC-3′
(D) 5′-GUUCAUGAAAGCUACAUAG-3′

294. Based on Figure I, which of the following is an mRNA sequence that codes for Met-Pro-Asn-Val?

(A) 5′-AUGCCCAAUGUUGA-3′
(B) 5′-ACAUGCCCCUCUAAUUCAC-3′
(C) 5′-ACATGCCCCTCTAATTCAC-3′
(D) 5′-GUUCAUGAAAGCUACAUAG-3′

295. Based on Figure I, which of the following is an mRNA sequence that codes for Met-Lys-Ala-Thr-Stop?

(A) 5′-AUGCCCAAUGUUGA-3′
(B) 5′-ACAUGCCCCUCUAAUUCAC-3′
(C) 5′-ACATGCCCCTCTAATTCAC-3′
(D) 5′-GUUCAUGAAAGCUACAUAG-3′

296. Based on Figure I, which of the following is a DNA sequence that codes for Met-His-Phe-Leu-Stop?

(A) 3′-CCGATACCTTACAAGTGGCACT-5′
(B) 3′-TTCCUACUAACGGAUAAUUGGAUA-5′
(C) 3′-TTACGTAAAAGACATCGGTGT-5′
(D) 3′-CCCTACCGCCTCCCCGACACTGTC-5′

297. Based on Figure I, which of the following is a DNA sequence that codes for Met-Glu-Cys-Ser-Pro-Stop?

(A) 3′-CCGATACCTTACAAGTGGCACT-5′
(B) 3′-TTCCUACUAACGGAUAAUUGGAUA-5′
(C) 3′-TTACGTAAAAGACATCGGTGT-5′
(D) 3′-CCCTACCGCCTCCCCGACACTGTC-5′

298. Based on Figure I, which of the following is a DNA sequence that codes for Met-Ile-Ala-Tyr-Stop?

(A) 3′-CCGATACCTTACAAGTGGCACT-5′
(B) 3′-TTCCUACUAACGGAUAAUUGGAUA-5′
(C) 3′-TTACGTAAAAGACATCGGTGT-5′
(D) 3′-CCCTACCGCCTCCCCGACACTGTC-5′

299. **Based on Figure I, which of the following double-stranded DNA molecules could code for both Met-Met-Met and for STOP-STOP-STOP?**

(A) 5′-TAATAATAATAATAATAA-3′
3′-ATTATTATTATTATTATT-5′
(B) 5′-ATTATTATTATTATTATT-3′
3′-TAATAATAATAATAATAA-5′
(C) 5′-ATCATCATCATCATCATC-3′
3′-TAGTAGTAGTAGTAGTAG-5′
(D) 5′-TAGTAGTAGTAGTAGTAG-3′
3′-ATCATCATCATCATCATC-5′

300. **The anticodon on a tRNA is 3′-GAA-5′. Based on Figure I, which of the following is the matching codon on the mRNA?**

(A) 5′-AAG-3′
(B) 5′-CUU-3′
(C) 5′-ACC-3′
(D) 5′-UGG-3′

301. **Based on Figure I (a table of mRNA codons), which amino acid does a tRNA with 3′-GAA-5′ specify?**

(A) Gln
(B) Leu
(C) Thr
(D) Trp

302. **Several experimenters have successfully modified a certain tRNA. This modified triplet now specifies an amino acid that is not naturally used by organisms. In other words, in addition to the 20 amino acids listed in Figure I, researchers were able to add a 21st amino acid. Which of the following triplets could NOT be reassigned and still result in proteins that incorporate 21 different amino acids?**

(A) AUG
(B) UGA
(C) CUA
(D) UAU

303. The genetic triplet code presented in the genetic code is identical in almost all organisms. There are a few exceptions. For example, *Mycoplasma* use GUG as a start codon in addition to AUG. These exceptions cluster within groups of organisms that researchers have decided were closely related based on other characteristics. For example, *Mycoplasma* are a group of bacteria that lack cell walls. Which of the following is a reasonable conclusion based on this information?

(A) The exceptions in the genetic code suggest that life evolved independently in those lineages and that any similarity is due to convergent evolution.
(B) The overwhelming similarity of the genetic code across all organisms supports the idea of common ancestry. The fact that the exceptions cluster within closely related organisms supports the idea that the code has undergone evolution by natural selection.
(C) These exceptions are a significant problem for evolutionary theory because the theory cannot explain how these variations occur.
(D) These variations are possible because the genetic code is degenerate.

---

304. When heat is applied to a double-stranded DNA (dsDNA) molecule, the two strands separate. This "melting" temperature depends on the sequence of bases in the strand. Recall that AT pairs share 2 hydrogen bonds, while GC pairs share 3 hydrogen bonds. Two double-stranded DNA molecules are shown below.

I. 5′-ATTATTATTATTATTATT-3′
3′-TAATAATAATAATAATAA-5′
II. 5′-ATCATCATCATCATCATC-3′
3′-TAGTAGTAGTAGTAGTAG-5′

Which of the following is true concerning the melting point of these two dsDNA strands?

(A) Sequence I has a higher melting point because it has more AT pairs than sequence II.
(B) Sequence II has a higher melting point because it has more GC pairs than sequence I.
(C) The melting points are identical because the melting point is an intrinsic property of DNA and is not affected by the sequence of DNA.
(D) Unlike RNA, DNA does not melt because it is double-stranded.

Questions 305–308

Frederick Griffith performed one of the classic experiments that led to the conclusion that DNA stores the information responsible for heredity. He used the bacterium *Streptococcus pneumoniae*, which causes pneumonia and death in mice. Griffith had two strains of the bacterium. One produced smooth colonies in culture and was pathogenic. The other strain produced "rough" colonies and did not cause disease. He injected various treatments into mice, observed whether the mice lived or died, and then spread a sample of each mouse's blood onto agar media to see if he could grow the bacteria (reisolation). Table I summarizes his experimental results.

Table I The Effects of Injections of Different Strains of Bacteria into Mice and the Results from the Reisolation of Bacteria from Blood

Treatment Injected (in saline solution)	Health of Mouse	Reisolation of Bacteria from Mouse Blood
Live smooth bacteria	Dead	Smooth bacteria
Live rough bacteria	Live	No bacteria
No bacteria (saline only)	Live	No bacteria
Heat-killed smooth bacteria	Live	No bacteria
Heat-killed rough bacteria	Live	No bacteria
Heat-treated saline	Live	No bacteria
Heat-killed rough bacteria + Live smooth bacteria	Dead	Smooth bacteria
Heat-killed smooth bacteria + Live rough bacteria	Dead	Smooth bacteria

Oswald Avery, Colin MacLeod, and Maclyn McCarty followed up on Griffith's work to try to determine what molecule was responsible for transforming live rough bacteria into smooth bacteria. To do this, they mixed heat-killed smooth bacteria with live rough bacteria in a way that led to transformation in the culture. They then applied different treatments to the mixture, as summarized in Table II. Note that DNase, RNase, and proteinase break down DNA, RNA, and protein, respectively.

Table II The Effect of Treatment with Enzymes on Transformation

Treatment	Result in Culture
No addition (transformation solution alone)	Smooth
DNase	No bacteria
RNase	Smooth
Proteinase	Smooth

305. What was the purpose of the saline-only injection into mice in Griffith's experiment?

(A) to ensure that some of the mice lived for further experimentation
(B) to confirm that the saline solution alone had no effect on the health of the mouse
(C) to practice the injection procedure
(D) to study the osmotic conditions in the mouse

306. Why was Griffith unable to reisolate rough bacteria from the mice he injected with rough bacteria?

(A) Other bacteria in the bloodstream outcompeted *Streptococcus pneumoniae.*
(B) He isolated from the wrong part of the mouse. *Streptococcus pneumoniae* grows on the teeth, not in the blood.
(C) The blood was the wrong pH for reisolating bacteria.
(D) The mouse's immune system recognized the bacteria and killed them.

307. Which of the following best explains the result in which a combined treatment of dead smooth bacteria and live rough bacteria yielded dead mice and the reisolation of smooth bacteria in Griffith's experiment?

(A) The life force from the live rough bacteria reanimated the dead smooth bacteria.
(B) The heat treatment did not kill 100% of the smooth bacteria.
(C) The rough bacteria began to express genes they already contained in their genome when presented with smooth bacteria.
(D) Some molecule in the dead smooth bacteria contained the information that was able to turn the rough bacteria into smooth bacteria.

308. In the Avery, MacLeod, and McCarty experiment, which of the following best supports the idea that protein is not the genetic material?

(A) Transformation did not occur in the presence of DNase.
(B) Transformation occurred in the presence of proteinase, which breaks down protein.
(C) Transformation occurred in the presence of RNase.
(D) Transformation occurred when no enzyme was added.

Questions 309–313

In Wisconsin Fast Plants (*Brassica rapa*), anthocyanin is a pigment that produces purple stems. Nonpurple stems (a) are recessive to the purple allele (A). Green leaves (G) are dominant to yellow leaves (g). Students crossed pure-breeding purple stem, green leaf plants with pure-breeding nonpurple stem, yellow leaf plants to produce an F1 hybrid generation. F1 were self-crossed (crossed with themselves) in one experiment. F1 were backcrossed with pure-breeding nonpurple stem, yellow leaf plants in a second experiment.

309. Which of the following represents the cross between the pure-breeding parents that produced the F1 offspring?

(A) AAGG × aagg
(B) Aagg × aaGg
(C) AaGg × AaGg
(D) AaGg × aagg

310. Which of the following represents the cross between the F1 offspring and the pure-breeding nonpurple stem, yellow leaf plants?

(A) AAGG × aagg
(B) Aagg × aagg
(C) AaGg × AaGg
(D) AaGg × aagg

311. Which of the following are the expected offspring in the experiment in which the F1 were self-crossed (crossed with themselves)?

(A) 9 purple stem–green leaf : 3 purple stem–yellow leaf : 3 nonpurple stem–green leaf : 1 nonpurple stem–yellow leaf
(B) 1 purple stem–green leaf : 3 purple stem–yellow leaf : 3 nonpurple stem–green leaf : 9 nonpurple stem–yellow leaf
(C) 6 purple stem–green leaf : 1 purple stem–yellow leaf : 1 nonpurple stem–green leaf : 6 nonpurple stem–yellow leaf
(D) 1 purple stem–green leaf : 1 purple stem–yellow leaf : 1 nonpurple stem–green leaf : 1 nonpurple stem–yellow leaf

312. Which of the following are the expected offspring in the experiment in which the F1 were backcrossed to a double homozygous recessive individual?

(A) 9 purple stem–green leaf : 3 purple stem–yellow leaf : 3 nonpurple stem–green leaf : 1 nonpurple stem–yellow leaf
(B) 1 purple stem–green leaf : 3 purple stem–yellow leaf : 3 nonpurple stem–green leaf : 9 nonpurple stem–yellow leaf
(C) 6 purple stem–green leaf : 1 purple stem–yellow leaf : 1 nonpurple stem–green leaf : 6 nonpurple stem–yellow leaf
(D) 1 purple stem–green leaf : 1 purple stem–yellow leaf : 1 nonpurple stem–green leaf : 1 nonpurple stem–yellow leaf

313. If instead of the expected results for the cross in which the F1 were backcrossed to a double homozygous recessive individual, you actually obtained the following results in the table below, which of the following conclusions would most likely explain these results?

Stem Color	**Purple**	**Purple**	**Nonpurple**	**Nonpurple**
Leaf Color	**Green**	**Yellow**	**Green**	**Yellow**
	843	142	147	851

(A) The results do not differ significantly from what would be expected by random chance in a dominant/recessive system.
(B) The genes for stem color and for leaf color assort independently.
(C) Gene regulation turns on purple stems in the presence of green leaves.
(D) The genes for stem color and for leaf color are located close to each other on the same chromosome.

Questions 314–318

Ferns are plant species that spend a portion of their life cycle as multicellular diploids and another portion as multicellular haploids. *Ceratopteris richardii* is a fern species that is easily grown in the laboratory. Spores were collected from an individual plant and then sown on agar plates on days 1, 3, and 5 as shown in the diagram below.

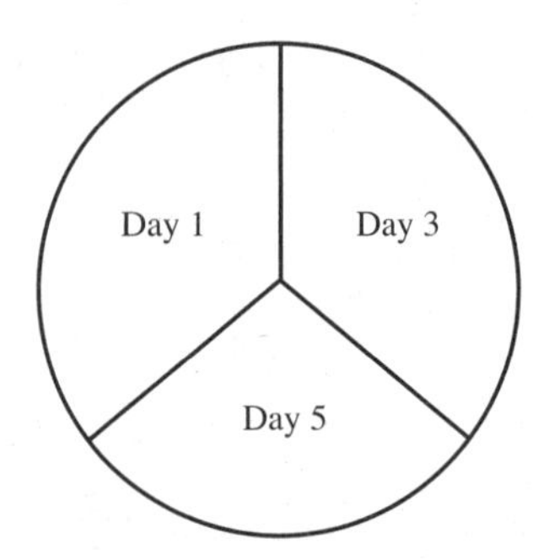

The spores grew into multicellular haploid plants. After 21 days, the students collected data on the phenotype of plants sown on each of the three sections of the plate. In each of the three sections of the plate, they counted the number of polka dotted males, solid males, polka dotted hermaphrodites, and solid hermaphrodites. These data are presented in Table I. The haploid plants then produced haploid sperm and eggs. Fertilization produced diploid plants. The students counted the diploid plants as either polka dotted or solid. These data are presented in Table II.

Table I Haploid Plants

	Polka Dotted	**Solid**	**Total**
Day 1 Male	57	61	118
Day 1 Hermaphrodite	145	149	294
Day 3 Male	98	106	204
Day 3 Hermaphrodite	103	96	199
Day 5 Male	187	181	368
Day 5 Hermaphrodite	19	23	42
Total	609	616	

Table II Diploid Plants

Polka Dotted	**Solid**
67	218

314. **Which of the following best explains the ratio of polka dotted haploid plants to solid haploid plants seen in Table I?**

(A) The diploid plant that produced the haploid spores was heterozygous for the polka dot trait.
(B) The diploid plant that produced the haploid spores mated with a plant that was heterozygous for the polka dot trait.
(C) A diploid polka dotted plant mated with a haploid solid plant.
(D) Haploid plants, by definition, have a 1:1 ratio for any trait because half of them have to express either the dominant or the recessive allele.

315. Which of the following explains the ratio of hermaphrodites to males in the haploid plants on days 1 and 5? (Refer to Table I.)

(A) Hermaphroditism is dominant. Therefore, it appears more frequently in the population.
(B) Since the sex ratio changes with increasing time, the ratio is probably determined by signals from the environment.
(C) Since the ratio of hermaphrodites to males is approximately 3:1, only homozygous recessive individuals are male.
(D) Gender is epigenetically determined by the diploid plant that produced the spores.

316. Early-colonizing *C. richardii* spores develop into hermaphrodites more frequently than do late-colonizing spores. Which of the following best describes the benefit of early colonizers developing into hermaphrodites?

(A) When late colonizers arrive, egg abundance is high. Therefore, there is an advantage in making small gametes to fertilize these eggs.
(B) When late colonizers arrive, egg abundance is low. Therefore, there is an advantage in making large gametes to fertilize the sperm.
(C) Eggs are scarce early in colonization. Therefore, there is a strong advantage to producing eggs as well as sperm to fertilize them.
(D) Eggs are abundant early in colonization. Therefore, there is a strong advantage in avoiding producing sperm.

317. Late-colonizing *C. richardii* spores develop into males more frequently than do early-colonizing spores, which favor hermaphroditism. Which of the following best describes the benefit of late colonizers developing into males?

(A) When late colonizers arrive, egg abundance is high. Therefore, there is an advantage in making small gametes to fertilize these eggs.
(B) When late colonizers arrive, egg abundance is low. Therefore, there is an advantage in making large gametes to fertilize the sperm.
(C) Eggs are scarce early in colonization. Therefore, there is a strong advantage to producing eggs as well as sperm to fertilize them.
(D) Eggs are abundant early in colonization. Therefore, there is a strong advantage in avoiding producing sperm.

318. Which of the following best explains the ratio of polka dotted to solid in the diploid plants as shown in Table II?

(A) Since the haploid plants are in a 3:1 ratio of polka dotted to solid, the diploids show the same ratio.
(B) Since the haploid plants are in a 1:1 ratio of polka dotted to solid, the sperm and eggs are also in a 1:1 ratio of polka dotted to solid. This is analogous to two heterozygous diploids mating and producing a 3:1 ratio.
(C) Diversifying selection leads to genetic variability like the 3:1 ratio seen in the diploids.
(D) Polka dots are sex-linked. Therefore, they show up less frequently in males.

Questions 319–321

Ceratopteris richardii is a species of fern that is easily grown in the laboratory. Students flooded plates of adult *C. richardii* gametophytes with sperm release buffer and collected sperm with a pipette. They placed a drop containing sperm onto a microscope slide. The students tested the effects of various chemicals by dipping a toothpick into the test chemical and then touching the toothpick to the side of the drop. For some compounds, the sperm showed no reaction. For others, the sperm moved toward the toothpick. The students rated each of the five compounds based on the strength of the response on a scale of 0–5, where 0 represents no response and 5 represents an immediate swarming of the toothpick. The results are shown in the table below.

Chemical	**Strength of Response (0–5)**
Water	1
W	1
X	4
Y	5
Z	2

319. Which of the following is a reasonable interpretation for the response shown by chemical *W*?

(A) Since the response was different from the water control, we can conclude that chemical *W* has strong attractant activity.
(B) Since there was a relatively strong reaction, chemical *W* probably binds strongly to a receptor on the surface of the sperm cell.
(C) Since there was no reaction, chemical *W* probably inhibits the ability of surface receptors on the sperm from detecting the attractant.
(D) Since the response was not different from the water control, we can conclude that chemical *W* has no attractant activity.

320. Which of the following is most likely to be true about chemicals *X* and *Y*?

(A) Since *X* and *Y* both produce relatively strong reactions, they probably have different shapes and charges and they probably bind to different receptors.
(B) Since *X* and *Y* both produce relatively strong reactions, they probably have similar shapes and charges, which allow them to bind to the same receptor.
(C) Since *X* and *Y* both produce reactions similar to water, they probably use active transport to cross the membrane and act as transcription factors in the nucleus.
(D) Since *X* and *Y* both produce reactions similar to chemical *W*, they probably have the same shape and charge as *W*.

321. Which of the following describes a plausible model that would cause *C. richardii* sperm to swim toward chemical *Y*?

(A) Chemical *Y* diffuses passively across the plasma membrane of the sperm and into the nucleus. Once there, it acts as a transcription factor to produce flagella proteins that are inserted in the portion of the membrane where the chemical entered the cell.
(B) Receptors in the mitochondria of the sperm detect chemical *Y* by binding to it and are stimulated to create more ATP in areas where concentrations of chemical *Y* are high.
(C) Chemical *Y* binds to surface receptors on the *C. richardii* sperm. Receptors near the flagella inhibit motion while receptors located on the opposite side of the sperm promote flagellar motion.
(D) Chemical *Y* binds to surface receptors on the *C. richardii* sperm. Receptors near the flagella promote flagellar motion.

Questions 322–324

The *TAS* gene encodes a protein that is a receptor for a particular chemical called DA, which many people perceive as a butterscotch taste. A diagram of this protein is shown below. DA binding causes a conformational (shape) change in the protein that exposes phosphorylation sites on the cytosolic domain. This, in turn, leads to a signal transduction pathway that leads to a nerve impulse being sent to the brain.

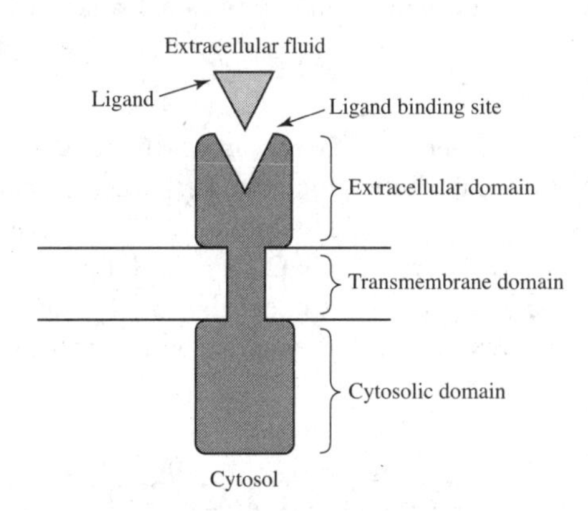

322. Which of the following could lead to the inability to taste DA?

(A) inheriting a single copy of the *TAS* gene that has mutations in the ligand-binding domain that prevent ligand binding and in the cytosolic domain that prevent kinase activity
(B) inheriting one copy of the *TAS* gene that has a mutation in the ligand-binding domain that prevents ligand binding and inheriting one copy of the *TAS* gene that has a mutation in the cytosolic domain that prevents kinase activity
(C) inheriting one copy of the *TAS* gene that has a mutation in the extracellular domain that does not prevent ligand binding and inheriting one copy of the *TAS* gene that has a mutation in the cytosolic domain that prevents kinase activity
(D) inheriting two copies of the mutation-free *TAS* gene

323. What is the likely effect of a mutation in the ligand-binding domain that alters the shape of the binding site so that it is able to bind to a different chemical?

(A) The new ligand will be perceived as a taste corresponding to the molecule bound by the receptor.
(B) The ability to detect the new compound will have to be learned by association with good or bad experiences.
(C) The new ligand will be perceived as butterscotch.
(D) The new ligand will be perceived to be tasteless.

324. What is the likely effect of a frameshift mutation in the transmembrane region that changes a series of hydrophobic amino acids to hydrophilic and charged amino acids?

(A) There will be no effect.
(B) The protein will no longer be able to bind to its ligand.
(C) The protein will no longer embed into the membrane properly.
(D) The kinase portion of the protein will be stuck in the "on" position so that a signal transduction will be constantly produced.

Questions 325–329

Students conducted a laboratory exercise in which they detected a common nucleotide substitution mutation in the *TAS* gene using the polymerase chain reaction (PCR) and restriction digestion. In this method, PCR is used to amplify DNA isolated from subjects. The resulting 700 base pair (bp) piece of DNA is incubated with the restriction enzyme *Eco*R1, which recognizes and cuts at the following sequence.

5′...G▼A A T T C...3′
3′...C T T A A▲G...5′

Portions of the DNA of the *TAS* gene are shown below.

Wild-type DNA	5′ . . . GCATTC . . . 3′
Mutant DNA	5′ . . . GAATTC . . . 3′

The diagram on page 151 shows an electrophoresis gel of the results for four different people. For each person, both uncut (U) and cut (C) samples were loaded and run on the gel. The leftmost lane includes DNA standards (ladder), which are labeled in 100s of base pairs. The sizes of the pieces in the ladder range from 100 to 1,000 base pairs long.

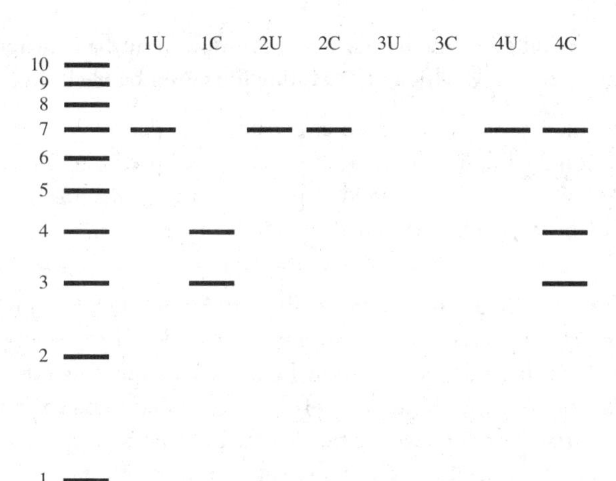

325. Person 1 has a single band in the uncut lane but two in the cut lane. Which of the following explains these data?

(A) Electrophoresis amplified more fragments when ligase was included in the PCR mixture.
(B) Two samples were loaded in the same lane.
(C) Ligase fused the 400 bp piece to the 300 bp piece.
(D) *Eco*R1 digested the 700 bp piece into 400 bp and 300 bp pieces.

326. According to the information in the gel, should person 2 be able to taste butterscotch?

(A) yes, unless the person's PCR reaction did not work correctly
(B) no, unless the person's PCR reaction did not work correctly
(C) yes, unless the person's restriction enzyme did not work correctly
(D) no, unless the person's restriction enzyme did not work correctly

327. Which of the following is NOT a reasonable interpretation of the results for person 3?

(A) This person's DNA was not loaded into the gel.
(B) This person lacks either copy of the *TAS* gene.
(C) This person's PCR reaction did not amplify.
(D) This person's DNA isolation did not work properly.

328. Which of the following is a reasonable interpretation of person 4's genotype?

(A) homozygous for the wild-type allele
(B) homozygous for the mutant allele
(C) heterozygous at this locus
(D) It is not possible to determine a reasonable interpretation from the results shown.

329. Assuming that all of the laboratory techniques were done properly for persons 1, 2, and 4, which of the following could be true?

(A) Person 4 could be the child of person 1 and person 2.
(B) Person 2 could be the child of person 1 and person 4.
(C) Person 1 could be the child of person 2 and person 4.
(D) Any of these occurrences are possible.

---

330. Huntington's disease is an inherited disorder caused by a dominant allele that encodes a protein called huntingtin. A student proposed a research project in which she planned to make transgenic zebra fish that contained the *Huntingtin* gene. A member of the school board was concerned about the potential for the spread of the disease into the student population. How should the student respond to these concerns?

(A) The student should instead work on a recessive disorder such as sickle cell anemia because it is not as potent.
(B) The student should abandon the project due to the risk of the spread of genetic disorders.
(C) The student should explain how zebra fish *Huntingtin* is different from human *Huntingtin*.
(D) The student should explain why genetic diseases are not contagious.

331. Humans have 23 pairs of chromosomes, while chimpanzees have 24 pairs of chromosomes. Human chromosome 2 is homologous to two chromosomes in chimpanzees as shown in the diagram below.

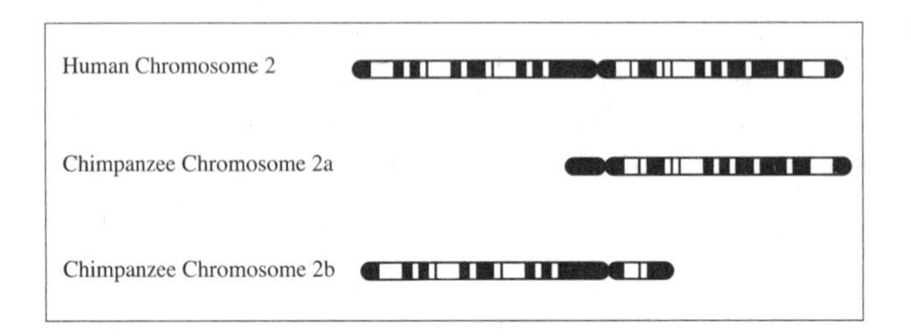

Which of the following would best support the hypothesis that human chromosome 2 arose by the fusion of two chromosomes in an ancestor of chimpanzees and humans?

(A) the ability to fuse the two chimpanzee chromosomes experimentally in the laboratory
(B) the fact that bonobos, gorillas, and orangutans all have 24 pairs of chromosomes
(C) the discovery of homologous genes on both chimpanzee and human chromosomes
(D) the presence of DNA sequences similar to telomeres in the middle of human chromosome 2 as well as a second centromere-like region

Questions 332–333

Seedless watermelon plants are produced by crossing a tetraploid watermelon with a diploid watermelon. The seeds that result are planted and produce fruit without seeds.

332. Which of the following would best describe the results of the cross between a tetraploid and a diploid?

(A) The offspring are tetraploid.
(B) The offspring are diploid.
(C) The offspring are triploid.
(D) The offspring are pentaploid.

333. Which of the following best explains why crossing these two types of watermelon plants produces offspring that grow fruit without seeds?

(A) Triploid sperm are not capable of fertilizing triploid eggs.
(B) Triploid pollen grains are incapable of forming pollen tubes and therefore cannot deliver sperm to the egg.
(C) The 3 copies of each chromosome do not segregate correctly in mitosis.
(D) The 3 copies of each chromosome do not segregate correctly in meiosis.

Questions 334–344

Operons are groups of genes that are regulated by an operator. This is a typical form of gene regulation found in bacteria. Two examples of operons are shown in both their off and on states in Figures I–IV. The *trp* operon controls the production of the enzymes that synthesize the amino acid tryptophan, while the *lac* operon controls the production of the enzymes involved in using lactose as an energy source.

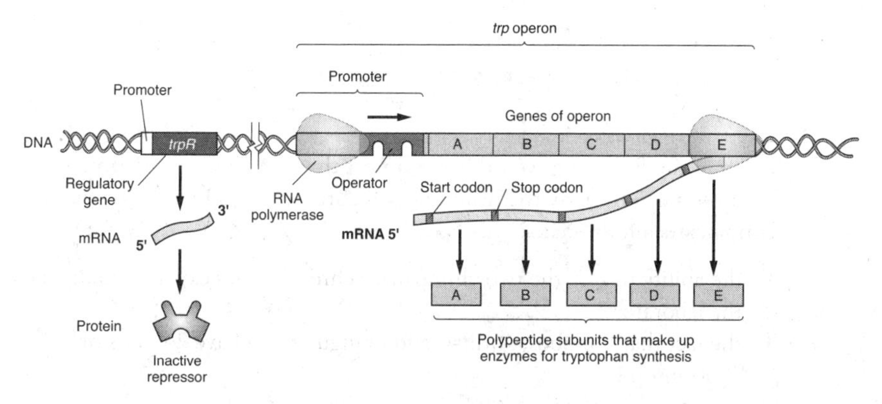

Figure I *trp* operon off

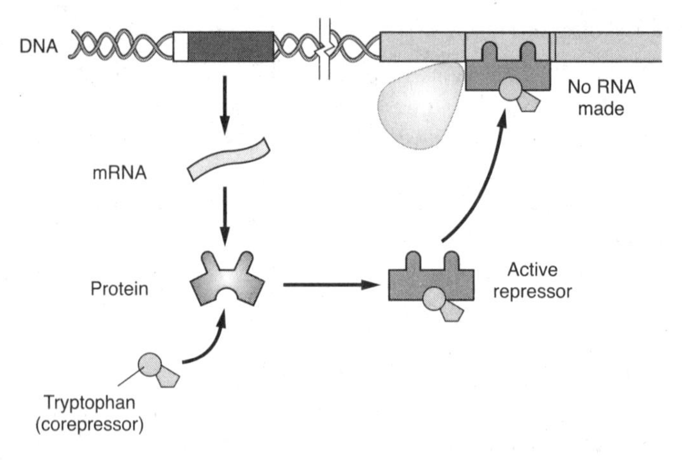

Figure II *trp* operon on

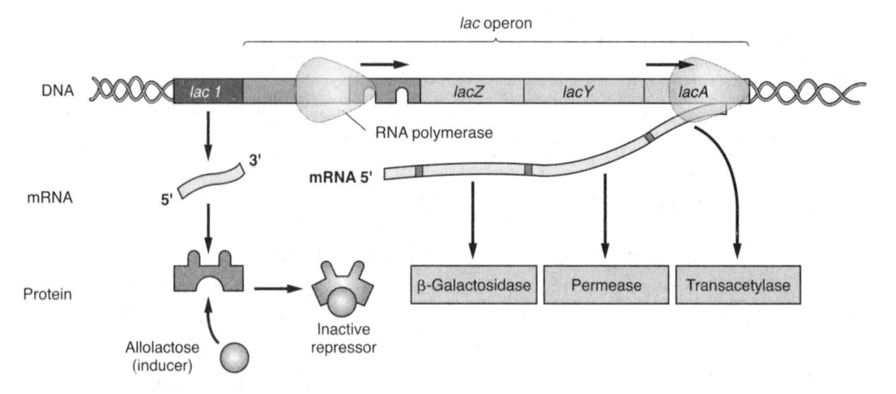

Figure III *lac* operon on

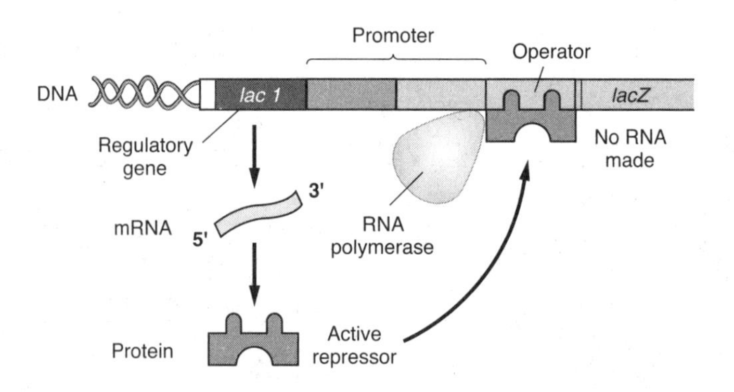

Figure IV *lac* operon off

The two diagrams below show chimeric operons in which the control elements from one operon were fused with the structural genes from the other.

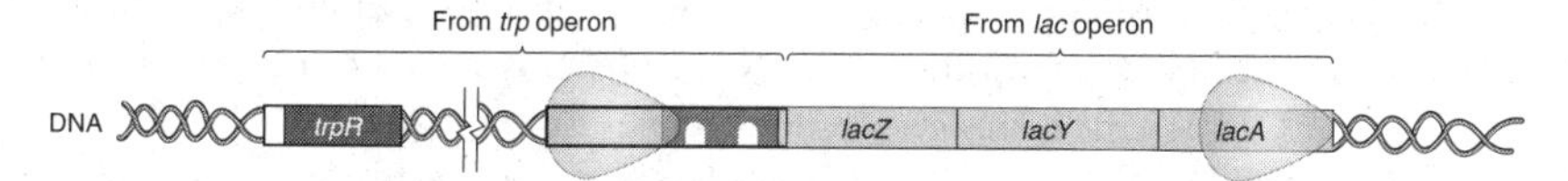

Figure V *trp/lac* chimera

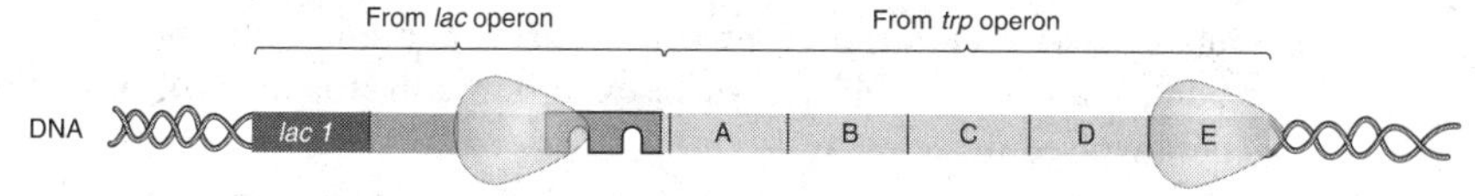

Figure VI *lac/trp* chimera

334. Based on Figures I and II, which of the following accurately describes the regulation of the *trp* operon?

(A) When tryptophan is present, the tryptophan binds to the repressor and inactivates it. Since the repressor can no longer bind to the operator region of the DNA, RNA polymerase cannot transcribe the mRNA for the genes involved in tryptophan production.

(B) When tryptophan is present, the tryptophan binds to the repressor and activates it. Since the repressor can now bind to the operator region of the DNA, RNA polymerase can transcribe the mRNA for the genes involved in the breakdown of tryptophan.

(C) When tryptophan is absent, the repressor is unable to bind to the operator region of the DNA. RNA polymerase can transcribe the mRNA for the genes involved in tryptophan production.

(D) When tryptophan is absent, the repressor binds to the operator region of the DNA, which makes it possible for RNA polymerase to transcribe the mRNA for the genes involved in the breakdown of tryptophan.

335. **Based on Figures III and IV, which of the following accurately describes the regulation of the *lac* operon?**

(A) When lactose is present, an isomer of lactose called allolactose binds to the repressor and inactivates it. Since the repressor can no longer bind to the operator region of the DNA, RNA polymerase can transcribe the mRNA for the genes involved in lactose use.
(B) When lactose is present, an isomer of lactose called allolactose binds to the repressor and activates it. Since the repressor can now bind to the operator region of the DNA, RNA polymerase can transcribe the mRNA for the genes involved in lactose use.
(C) When lactose is absent, the repressor is unable to bind to the operator region of the DNA. RNA polymerase can transcribe the mRNA for the genes involved in lactose use.
(D) When lactose is absent, the repressor binds to the operator region of the DNA, which makes it possible for RNA polymerase to transcribe the mRNA for the genes involved in lactose use.

336. **Which of the following best explains the advantage of organizing *lacZ*, *lacY*, and *lacA* as an operon controlled by a single mechanism? (Refer to Figures III and IV.)**

(A) Eukaryotes don't use this system, so there is no real benefit to organizing genes as an operon.
(B) Since these genes are needed under the same conditions, organizing them as an operon allows them to be expressed at the same time.
(C) Organizing genes as an operon allows the organism to adjust the transcription of mRNA for each protein independently.
(D) Organizing genes as an operon allows the organism to use the same DNA sequence to produce more than one protein.

337. **Which of the following would be the result for the *trp/lac* chimera shown in Figure V if lactose was introduced into the culture?**

(A) The concentration of lactose would decrease.
(B) There would be no effect because lactose would not interact with the repressor.
(C) The concentration of tryptophan would increase.
(D) The cell would convert lactose into tryptophan.

338. Which of the following would be the result for the *trp/lac* chimera shown in Figure V if tryptophan was introduced into the culture?

(A) The *lac* genes would be transcribed but would have no substrate on which to act, and the concentration of tryptophan would stay the same.
(B) The concentration of tryptophan would increase, and the transcription of the *lac* genes would be turned off.
(C) There would be no transcription of the *lac* genes, and the tryptophan concentration would stay the same.
(D) The concentration of tryptophan would increase, and the transcription of the *lac* genes would be turned on.

339. Which of the following would be the result for the *trp/lac* chimera shown in Figure V if both lactose and tryptophan were introduced into the culture?

(A) The concentration of lactose would decrease, and the concentration of tryptophan would stay the same.
(B) The concentration of tryptophan would increase, and the concentration of lactose would stay the same.
(C) The concentration of both lactose and tryptophan would stay the same.
(D) The concentration of tryptophan would increase, and the concentration of lactose would decrease.

340. Which of the following would be the result for the *lac/trp* chimera shown in Figure VI if lactose was introduced into the culture?

(A) The concentration of lactose would decrease, and no tryptophan would be produced.
(B) The concentration of tryptophan would increase, and the concentration of lactose would stay the same.
(C) The concentration of lactose would stay the same, and no tryptophan would be produced.
(D) The concentration of tryptophan would increase, and the concentration of lactose would decrease.

341. Which of the following would be the result for the *lac/trp* chimera shown in Figure VI if tryptophan was introduced into the culture?

(A) The *lac* genes would be transcribed, and the concentration of tryptophan would stay the same.
(B) The concentration of tryptophan would increase, and no lactose would be produced.
(C) The concentration of tryptophan would stay the same.
(D) The concentration of tryptophan would increase, and the concentration of lactose would decrease.

342. Which of the following would be the result for the *lac/trp* chimera shown in Figure VI if both lactose and tryptophan were introduced into the culture?

(A) The concentration of lactose would decrease, and the concentration of tryptophan would stay the same.
(B) The concentration of tryptophan would increase, and the concentration of lactose would stay the same.
(C) The concentration of both lactose and tryptophan would stay the same.
(D) The concentration of tryptophan would increase, and the concentration of lactose would decrease.

343. The lacI protein is a repressor that binds to DNA as long as the lacI protein is not bound to allolactose (which is present if lactose is present). Imagine that you could mutate the *lacI* gene so that instead of binding to DNA when not bound to allolactose, the lacI protein did the opposite. The mutated version would bind to DNA only when bound to allolactose, and it would not be able to bind if allolactose was not present. Which of the following would be the result?

(A) All of the lactose present would be broken down because the enzymes would already be present in the cell.
(B) Some lactose would be broken down but not all because enzymes are used up in the reactions.
(C) No lactose would be broken down because the presence of lactose would turn off the transcription of lactose breakdown genes.
(D) Depending on how much lactose was added and how long the cell took to recycle proteins, some or all of the lactose would be broken down.

344. The trpR protein is a repressor molecule that binds to DNA when tryptophan is present. Imagine that you could mutate the *trpR* gene so that the protein binds to DNA in the absence of tryptophan but not in the presence of tryptophan. Which of the following would be the result?

(A) The bacteria would not be able to grow because they could not make tryptophan if they didn't have tryptophan to turn on the operon.
(B) The bacteria would substitute some other amino acid for tryptophan because there would be no way to synthesize tryptophan.
(C) The bacteria would produce large amounts of tryptophan because the breakdown of existing proteins would release tryptophan, which would trigger a positive feedback cycle to make more tryptophan.
(D) The bacteria would produce lactose instead.

345. The pedigree diagram below shows the inheritance of a rare disorder over three generations. Circles represent females, and squares represent males. Shaded symbols indicate individuals with the disorder.

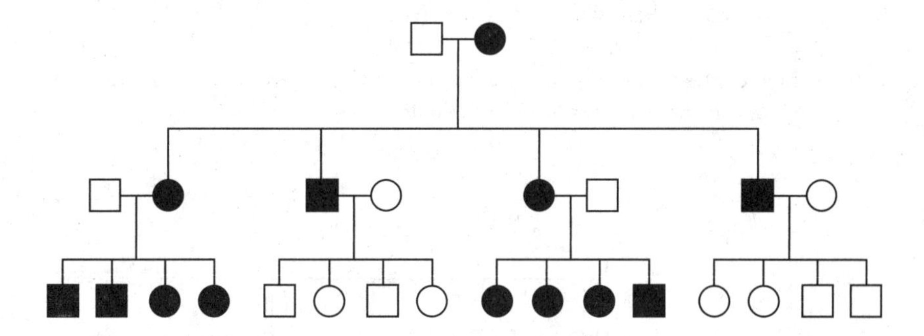

An individual with a rare disorder investigated the incidence of the disorder in his family and drew the pedigree diagram above. He contends that the gene for the disorder is on the mitochondrial chromosome, which is passed from mother to child in the cytoplasm of the egg. The sperm does not donate mitochondria. Is he correct? Why or why not?

(A) He is correct. The pedigree supports mitochondrial inheritance. All children of affected women have the disorder. No children of nonaffected women have the disorder.
(B) He is incorrect. The pedigree supports sex-linked recessive inheritance because the diagram shows that males are more likely to have the disorder than females.
(C) He is incorrect. The pedigree supports autosomal dominant inheritance because the disorder shows up in every generation.
(D) He is correct. The pedigree supports mitochondrial inheritance. Some males are carriers, which explains why some of the males do not display the trait.

Questions 346–350

In the pedigree diagrams given in each of the questions below, squares represent males and circles represent females. Shaded symbols represent individuals who display the trait.

346. Which of the following must be true about the inheritance of the trait depicted in the pedigree diagram below?

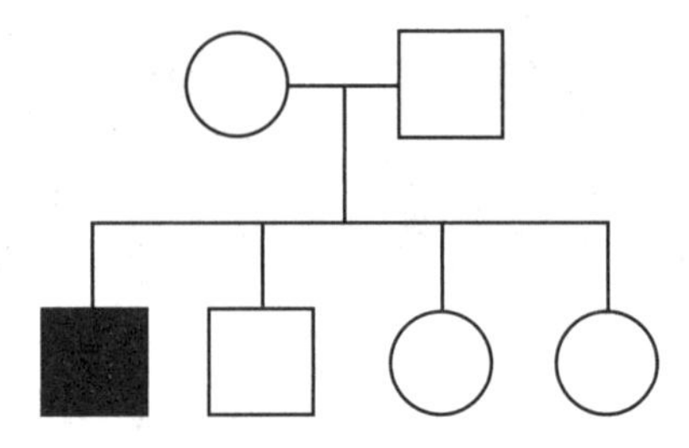

(A) It is recessive.
(B) It is dominant.
(C) It is recessive, and it is on the X chromosome.
(D) There is not enough information to determine the mechanism of inheritance.

347. Which of the following cannot be true about the inheritance of the trait depicted in the pedigree diagram below?

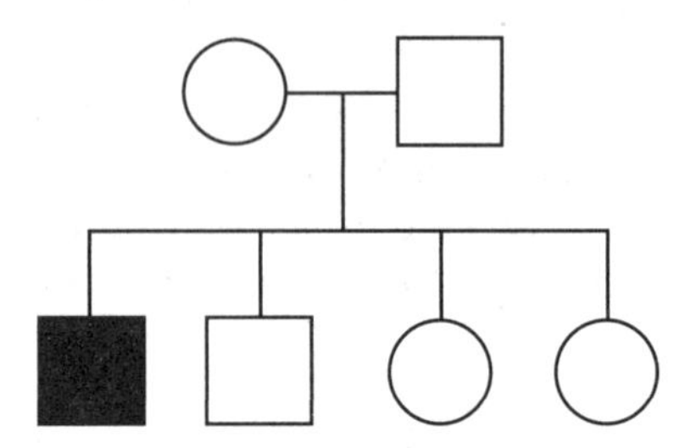

(A) It is recessive.
(B) It is dominant.
(C) It is on the X chromosome.
(D) It is recessive, and it is on the X chromosome.

348. Which of the following must be true about the inheritance of the trait depicted in the pedigree diagram below?

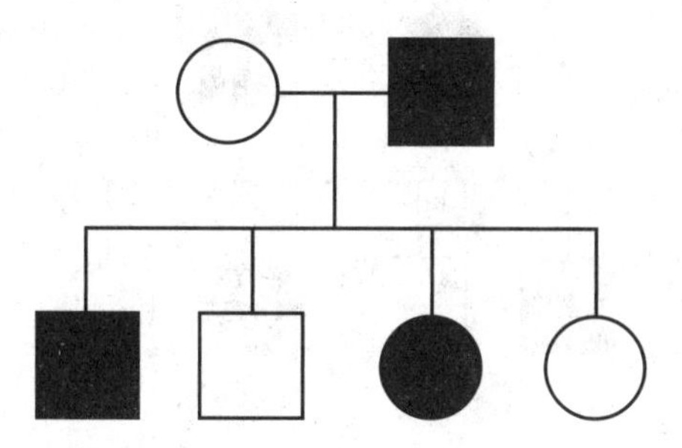

(A) It is recessive.
(B) It is dominant.
(C) It is on the X chromosome.
(D) There is not enough information to determine the mechanism of inheritance.

349. Which of the following cannot be true about the inheritance of the trait depicted in the pedigree diagram below?

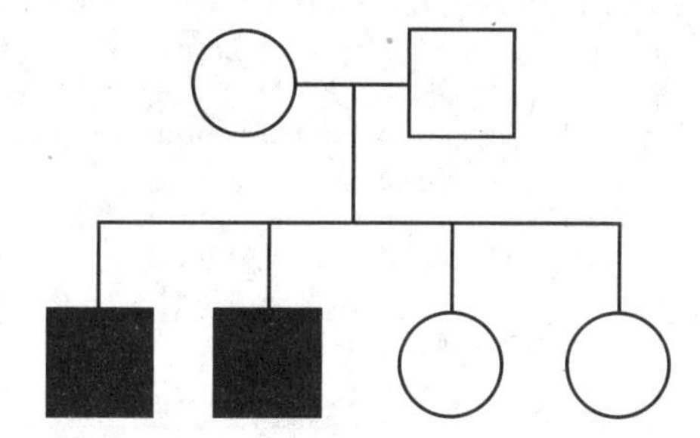

(A) It is recessive.
(B) It is dominant.
(C) It is on the X chromosome.
(D) There is not enough information to determine the mechanism of inheritance.

350. Which of the following cannot be true about the inheritance of the trait depicted in the pedigree diagram below?

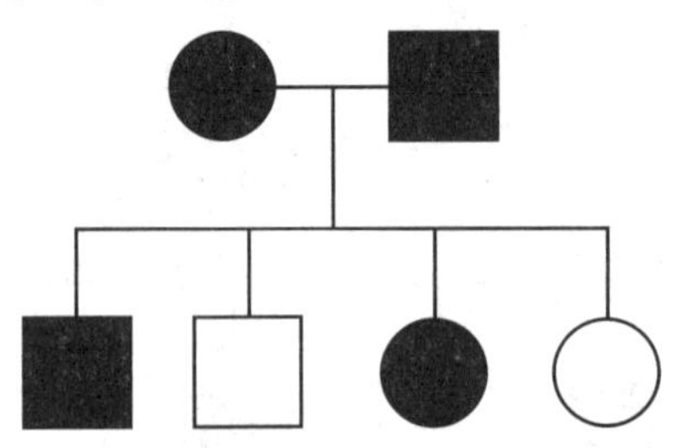

(A) It is recessive.
(B) It is dominant.
(C) It is on the X chromosome.
(D) There is not enough information to determine the mechanism of inheritance.

Questions 351–353

Sertraline is a drug that is commonly used to treat depression. It acts on nerves that use the neurotransmitter serotonin. The figure below shows how serotonin behaves at a synapse in the absence of sertraline (top) and in the presence of sertraline (bottom).

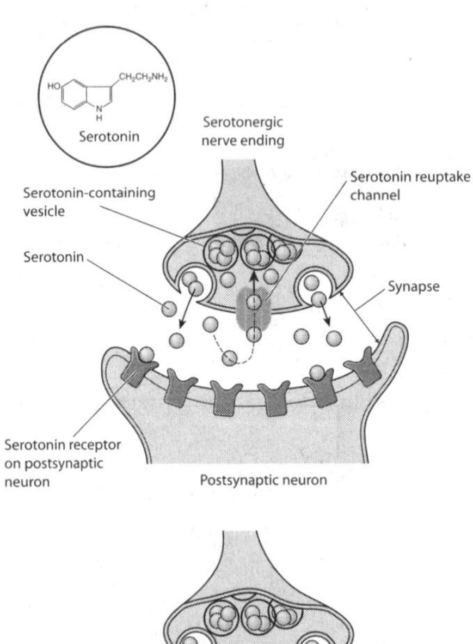

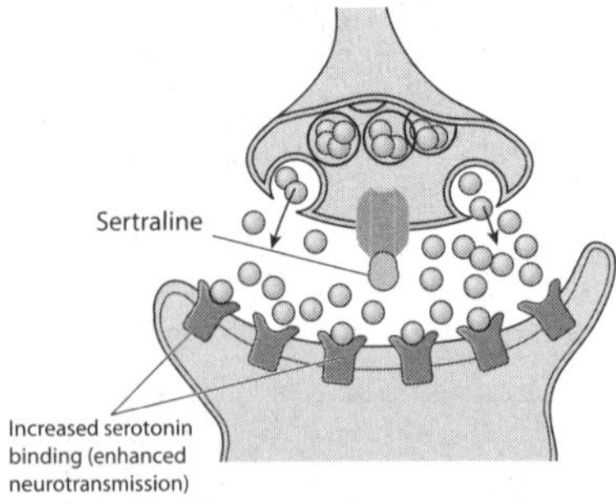

351. Why does serotonin interact with the receptor on the postsynaptic neuron?

(A) Serotonin has the correct shape and charge to fit in the ligand-binding site.
(B) The presynaptic neuron fires the serotonin with enough force that it physically triggers the serotonin receptor.
(C) Serotonin reacts with the cytoplasmic domain of the protein.
(D) Since serotonin is hydrophobic, it passes through the channel protein and activates transcription in the postsynaptic neuron.

352. According to the diagram, how is serotonin secreted from the cell?

(A) exocytosis of vesicles that contain serotonin
(B) endocytosis of vesicles that contain serotonin
(C) direct translation of serotonin protein into the extracellular fluid
(D) facilitated diffusion through membrane channels

353. Sertraline binds to the serotonin reuptake channel, as shown in the diagram. What effect does this have on the level of serotonin in the synapse?

(A) The serotonin levels become higher.
(B) The serotonin levels become lower.
(C) There is no change in the serotonin levels because a positive feedback loop causes less serotonin to be made.
(D) There is no change in the synaptic serotonin levels because additional serotonin is absorbed by the postsynaptic neuron.

Questions 354–355

Cancer is characterized by uncontrolled cell division. Many different mutations can contribute to cancer. Several of them affect the MAP kinase pathway, as shown in the figure below.

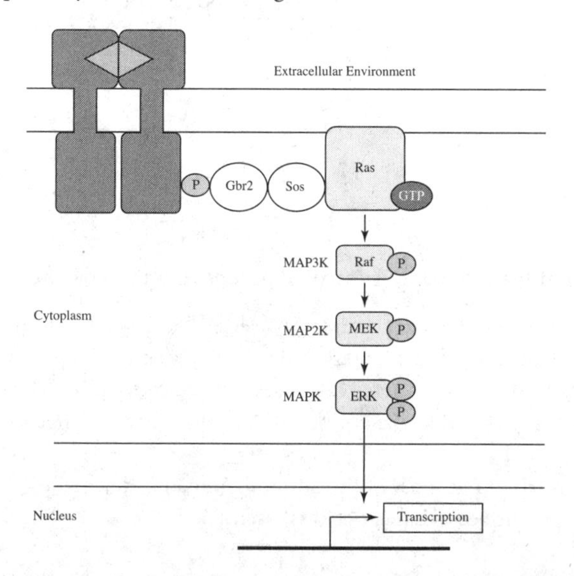

354. Which of the following would produce the greatest effect on transcription?

(A) doubling the number of receptors
(B) doubling the amount of Raf
(C) doubling the rate at which Raf phosphorylates MEK
(D) doubling the rate at which MEK phosphorylates ERK

355. The *Ras* gene is mutated in many forms of cancer. Given that this pathway leads to the transcription of genes involved in mitosis, which of the following would most likely lead to cancerous cell proliferation?

(A) mutation of *Ras* so that the protein it encodes constantly phosphorylates Raf
(B) mutation of *Ras* so that the protein it encodes no longer phosphorylates Raf
(C) mutation of *Ras* so that the protein it encodes phosphorylates MAPK instead of MAP3K
(D) mutation of *Ras* so that the protein it encodes cannot interact with Sos

Questions 356–358

One promising new group of cancer drugs focuses on the use of monoclonal antibodies (which bind to a single antigen) attached to cytotoxic chemicals to target cancer cells. The diagram below shows how these drugs are intended to work.

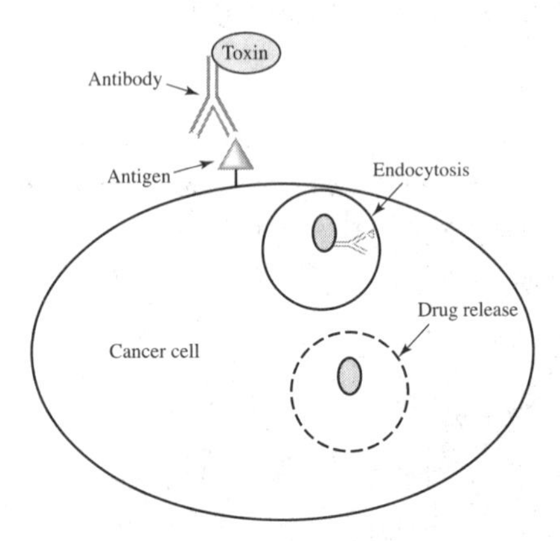

356. Which of the following is essential in reducing the side effects of this drug?

(A) The antibody must be directly inserted into cancer cells using a syringe.
(B) The antibody must be targeted to an antigen that is present in high amounts on the surface of cancer cells but not on healthy cells.
(C) Making sure that the antibody reacts with a wide range of antigens is essential.
(D) Ensuring that most of the antibody remains attached to antigens on the outer surface of cancer cells is essential.

357. Which of the following is NOT a good choice for the type of toxin, assuming that the drug delivery mechanism works as shown in the diagram?

(A) an agent that binds RNA polymerase and prevents transcription
(B) a toxin that cross-links two DNA strands and prevents strand separation
(C) a chemical that binds to mitogen receptors on the cell membrane
(D) a substance that breaks down microtubules and prevents attachment to centromeres

358. Many of the compounds being investigated as payloads attached to the antibodies are extremely toxic chemicals. For example, one toxin produced by some fungi in the genus *Amanita* (the death angel fungus) causes coma and death. Which of the following explains why these extremely toxic compounds are being considered?

(A) As a general rule, the more toxic the compound used, the fewer side effects it is likely to cause.
(B) Extremely small doses are used because they can be delivered directly to cancer cells.
(C) Since the chemicals are widely distributed in the body, the effect on any given tissue is low.
(D) Noncancerous cells are able to detect the toxin and avoid absorbing it.

---

359. A flower grower attempting to develop new varieties of a particular species took pure-breeding white flowers and crossed them with pure-breeding red flowers. The F1 offspring were all pink. The flower grower hypothesized that this was the result of incomplete dominance. In incomplete dominance, two alleles are expressed. The phenotype that results is intermediate between the two. In this case, she hypothesized that R is a red allele and that W is a white allele, so a heterozygote would be pink. Which of the following would refute the hypothesis that the inheritance of flower color in this case was incomplete dominance?

(A) self-crossing the F1 flowers and obtaining offspring that differed significantly from a 1:2:1 ratio of red:pink:white
(B) backcrossing the F1 flowers to the white parent and obtaining offspring that differed significantly from a 1:2:1 ratio of red:pink:white
(C) self-crossing the F1 flowers and obtaining offspring that differed significantly from a 1:1 ratio of red:pink
(D) self-crossing the F1 flowers and obtaining any red or pink flowers

Questions 360–364

Coat color in Labrador Retrievers is primarily controlled by two genes. The gene for pigment production has a dominant allele that produces a black pigment (B) and a recessive allele that produces a brown pigment (b). A second gene locus controls the expression of pigment in the fur. The dominant allele (E) for coat color deposition encodes a transmembrane receptor that triggers the expression of eumelanin in response to the melanocyte-stimulating hormone (MSH). The dominant allele causes pigment expression in fur, while the recessive allele (e) does not. Fur is yellow unless either a black or a brown pigment is present.

360. Which of the following shows all of the possible genotypes for a black dog?

(A) BBEE, BBEe, BbEE, BbEe
(B) BBEE, BBEe, BbEE, BbEe, BBee
(C) BBee, Bbee
(D) Bbee

361. Two black dogs mate. They have a litter of puppies that includes black, brown, and yellow dogs. Which of the following is most likely to be the genotypes of the parent dogs?

(A) BbEe × BbEe
(B) BBee × bbEE
(C) bbee × BBEE
(D) Bbee × bbee

362. A brown dog and a yellow dog mate. They have a litter of puppies that includes black, brown, and yellow dogs. Which of the following is most likely to be the genotypes of the parent dogs?

(A) BbEe × BbEe
(B) BBee × bbEE
(C) bbee × BBEE
(D) Bbee × bbEe

363. Which of the following mutations would most likely produce an MSH receptor that is unable to trigger pigment deposition in the fur?

(A) an inversion mutation of 18 nucleotides in an intron of the MSH receptor gene
(B) a single nucleotide substitution in the 5′ untranslated region of the MSH receptor gene
(C) a single nucleotide insertion upstream of the promoter region of the eumelanin gene
(D) a single nucleotide deletion in the first exon of the MSH receptor gene

364. Which of the following best explains why a single copy of the E allele, which codes for a functional MSH receptor, would be sufficient for triggering color expression in hair?

(A) The E allele is dominant, and dominant alleles mask the presence of recessive alleles.
(B) The E allele encodes a functional MSH receptor that can detect MSH and activate a signal transduction pathway, leading to the expression of eumelanin.
(C) The cell uses reverse transcriptase to make a second DNA copy of the functional receptor protein.
(D) A single functional copy of the receptor gene directly encodes eumelanin. Therefore, if the gene binds to MSH, eumelanin will be produced.

Questions 365–369

The Cavendish variety of bananas represents 99% of bananas sold in the United States. As a triploid, it does not produce seeds and is propagated vegetatively by separating roots. It is currently attacked by several pathogens including *Pseudocercospora fijiensis*, which causes black lesions on leaves. The fungus is controlled by regular fungicide applications (up to 50 applications per year) and by the removal of dead leaves.

365. Which of the following describes a physiological consequence of a *P. fijiensis* infection?

(A) Leaf removal is labor intensive and time consuming.
(B) Fungicide application is expensive.
(C) Consumers will not buy discolored bananas.
(D) Black lesions reduce the amount of surface area for photosynthesis and therefore reduce the growth rate.

366. Which of the following is a likely result of the regular fungicide applications?

(A) Discoloration of the fruit will occur.
(B) The evolution of fungicide-resistant *P. fijiensis* strains will occur.
(C) Eventually, the fungus will be eradicated.
(D) The evolution of fungus-resistant banana strains will occur.

367. Which of the following is a consequence of vegetatively propagating Cavendish bananas?

(A) The seeds become vitally important.
(B) Banana cultivation becomes more expensive.
(C) Increased genetic diversity occurs as a result of natural selection.
(D) A lack of genetic diversity occurs as a result of clonal reproduction.

368. Which of the following would be a good source of genes for resistance to the fungus?

(A) wild relatives of bananas that produce seeds
(B) Cavendish bananas grown in a greenhouse
(C) the fungal genome
(D) animals

369. Which of the following strategies would most likely produce a fungus-resistant banana that tastes the same as a Cavendish banana?

(A) combining the nuclei of a Cavendish banana and of a *P. fijiensis*–resistant variety in tissue culture and then screening the resulting plants for resistance and taste
(B) screening a large number of wild banana plants for resistance to *P. fijiensis* and then performing double-blind taste tests
(C) identifying a gene that confers resistance to *P. fijiensis* in wild varieties of bananas and using genetic engineering techniques to insert it into Cavendish banana plants
(D) fertilizing Cavendish bananas with pollen from wild varieties of bananas and then screening the resulting plants for resistance and taste

Questions 370–372

The sodium-potassium pump, shown below, pumps 3 sodium ions out of a nerve cell for every 2 potassium ions it pumps in. The pump is powered by the hydrolysis of ATP to ADP plus inorganic phosphate.

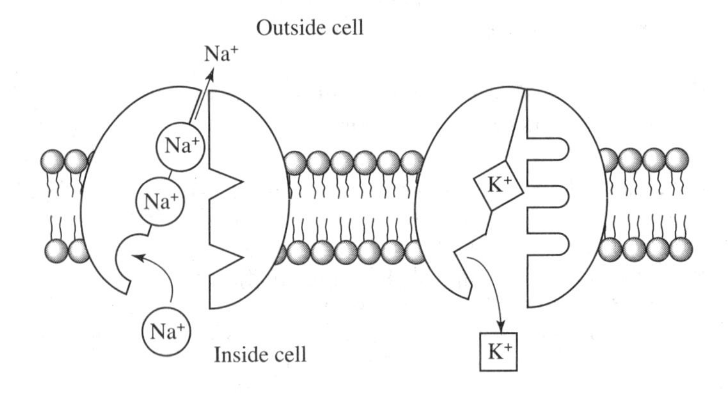

370. Based on the information presented, which of the following is true about the sodium-potassium pump?

(A) The sodium-potassium pump performs passive transport because 3 sodium ions are transported out of the cell for every 2 potassium ions that are transported in.
(B) The sodium-potassium pump performs active transport because it uses ATP to move sodium and potassium ions against their concentration gradients.
(C) The sodium-potassium pump creates energy by moving ions up their concentration gradient.
(D) The sodium-potassium pump likely uses the same binding site for sodium and potassium because they both have a positive charge.

371. Which of the following best explains how the sodium-potassium pump creates an electrical gradient across the membrane of the nerve cell?

(A) Since the pump moves single electrons as well as ions, an electrical charge builds up due to friction.
(B) The breakdown of ATP into ADP and inorganic phosphate releases electrons captured in the electron transport chain.
(C) Since the pump moves 3 positive ions out of the cell for every 2 it moves in, the inside of the membrane becomes negatively charged.
(D) Hydrolysis of ATP generates hydrogen ions on one side of the membrane and oxygen ions on the other side of the membrane.

372. Which of the following is true about the movement of sodium and potassium by the sodium-potassium pump with respect to each ion's electrical gradient?

(A) Potassium is moved against its electrical gradient (toward +), while sodium is moved down its electrical gradient (toward –).
(B) Sodium and potassium are both moved against their electrical gradients (toward +).
(C) Sodium is moved against its electrical gradient (toward +), while potassium is moved down its electrical gradient (toward –).
(D) Sodium and potassium are both moved down their electrical gradients (toward –).

Questions 373–377

The graph below depicts the membrane potential at a single location inside the axon of a neuron over a period of time (from earliest, I, to most recent, VI). At the beginning of the time shown, sodium levels outside the membrane are high while potassium levels are low due to the action of the sodium-potassium pump. The charge inside the cell is negative relative to the outside.

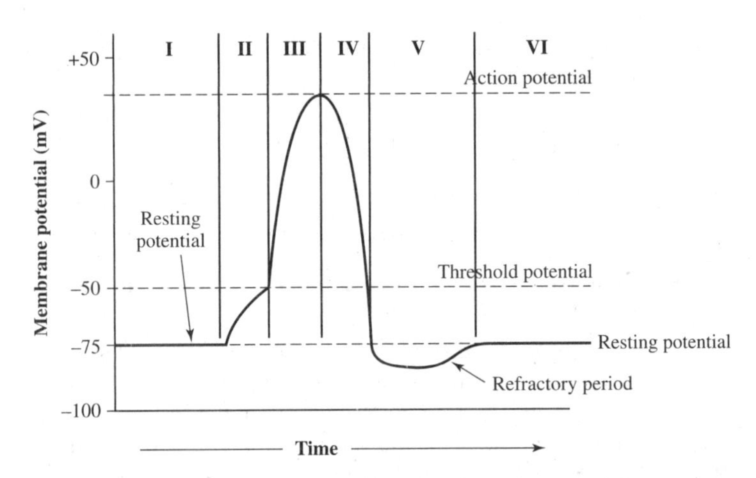

The table below shows the concentrations of sodium and potassium ions at resting potential.

	Concentration (nM)	
Ion	**Outside Cell**	**Inside Cell**
Na^+	145	12
K^+	4	155

373. Which of the following is a reasonable explanation for the change in voltage during the time period labeled III on the graph?

(A) Potassium channels open, allowing all of the potassium to move inside the cell.
(B) Sodium channels open, allowing sodium levels to become more equal.
(C) Potassium channels open, allowing potassium levels to become more equal.
(D) Sodium channels open, allowing all of the sodium ions to move inside the cell.

374. Which of the following is a reasonable explanation for the change in voltage during the time period labeled IV on the graph?

(A) Potassium channels open, allowing all of the potassium to move inside the cell.
(B) Sodium channels open, allowing sodium levels to become more equal.
(C) Potassium channels open, allowing potassium levels to become more equal.
(D) Sodium channels open, allowing all of the sodium ions to move inside the cell.

375. No action potential can be generated during the refractory period, which is labeled V on the graph. Which of the following is a model that would explain this?

(A) The sodium-potassium pump is deactivated during this period, and it is not possible to generate an ion charge separation.
(B) Since both the sodium and potassium channels are stuck in the open conformation, a stimulus cannot cause them to open further.
(C) During the refractory period, potassium channels are in a conformation that does not permit them to open.
(D) During the refractory period, sodium channels are in a conformation that does not permit them to open.

376. Which of the following is consistent with time period II on the graph?

(A) Stimulatory signals are causing some sodium channels to open. This period ends when the nerve becomes depolarized enough that the voltage-gated ion channels open.
(B) Inhibitory signals are causing some sodium channels to open. This ends during the refractory period when the same signals cause the potassium channels to close.
(C) Presynaptic neurons are firing, leading to a buildup of electrons in the synapse. This polarizes the membrane and prevents action potential.
(D) Electrical signals jump between the nodes of Ranvier, slowing the action potential.

377. Which of the following is true about time period V on the graph?

(A) Rapid opening and closing of sodium and potassium channels cause water to diffuse across the membrane, dehydrating the cell and leading to a voltage change.
(B) Secretion of neurotransmitters into the intercellular space causes the cell to express genes associated with action potentials.
(C) The sodium-potassium pump uses ATP to reestablish a resting voltage across the membrane.
(D) Translation of mRNA generates proteins that bind sodium and potassium ions. This decreases the electrical potential across the inner mitochondrial membrane.

Questions 378–379

Muscle contraction is stimulated by the release of acetylcholine at the neuromuscular junction. When an action potential reaches the end of a motor neuron, vesicles containing acetylcholine fuse with the plasma membrane and release the neurotransmitter. Acetylcholine binds to receptors on the muscle cell and causes depolarization of the membrane. When depolarization reaches a threshold value, an action potential is generated and the muscle contracts. This mechanism is shown in the diagram below.

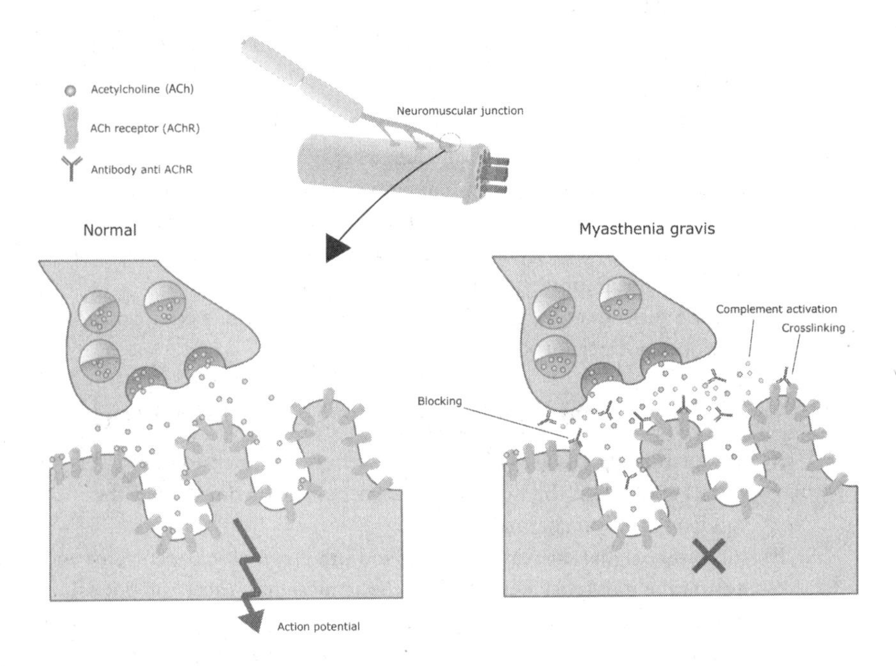

378. In myasthenia gravis, the immune system produces antibodies that bind to acetylcholine receptors. Which of the following is the most likely result of this binding?

(A) Since the receptors are bound by antibodies, acetylcholine cannot access the ligand-binding site on the receptor and no action potential can be generated. This prevents muscle contraction.
(B) Since the antibodies bind to the receptors, the membrane becomes depolarized and action potentials cause constant muscle contraction.
(C) Antibody binding of receptors prevents acetylcholine reuptake by the motor neuron. As a result, action potentials are continuously generated and the muscle remains permanently contracted.
(D) Antibody binding signals leukocytes to remove bacteria from the neuromuscular junction.

379. Antibodies that are bound to antigens activate the complement system. This system is a set of proteins that make holes in the membranes of cells displaying the antigen. Which of the following is most likely to be the result of this process in myasthenia gravis?

(A) the death of antibody-producing cells
(B) the release of additional acetylcholine into the neuromuscular junction
(C) the death of motor neurons
(D) the death of muscle cells

Questions 380–381

The diagram below shows the thyroid hormone T_4 signaling.

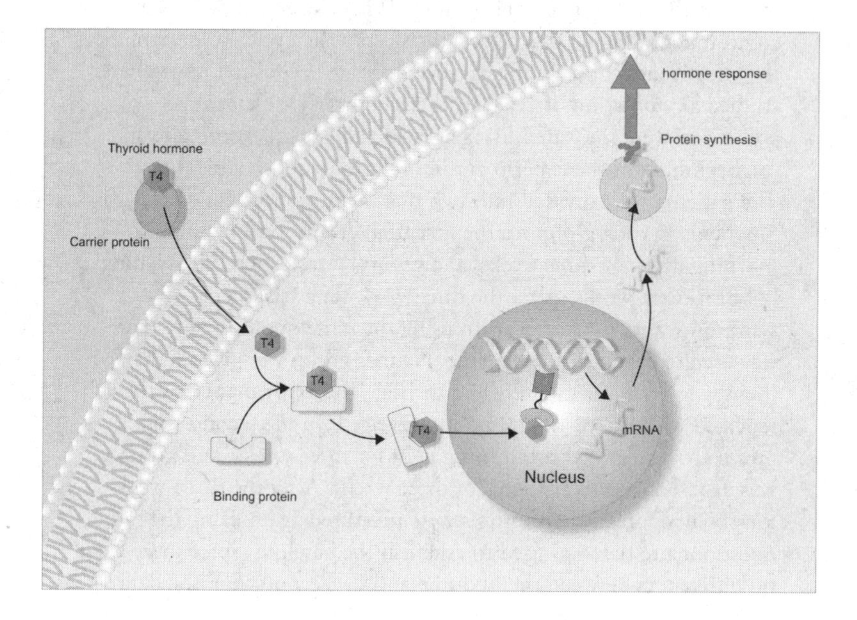

380. Thyroid hormone T_4 is transported in the blood by a carrier protein. Once it passes through the membrane, T_4 binds to another protein in the cytoplasm. Which of the following explains why T_4 is transported in this way?

(A) Carrier proteins use ATP to move molecules rapidly through the bloodstream.
(B) The thyroid hormone T_4 is charged. As a result, it stably interacts with fatty acid tails of the phospholipids in the membrane, but it needs carrier proteins to transport it in aqueous solutions.
(C) Carrier proteins are needed for T_4 because it is always traveling up its concentration gradient.
(D) Thyroid hormone T_4 is nonpolar. As a result, it does not dissolve in aqueous solutions such as the blood or the cytoplasm.

381. According to the diagram, which of the following is the result of thyroid hormone signaling?

(A) transcription of mRNA and protein synthesis
(B) replication of DNA
(C) transformation of T_4 molecules into mRNA molecules
(D) regulation of the metabolic rate

Questions 382–384

Light therapy is a treatment for seasonal affective disorder (SAD), a condition in which patients feel depressed during the winter months. Patients undergoing light therapy are exposed to bright light in either the morning or the evening. This light exposure is a key environmental signal that affects the sleep/wake cycle. To determine the effectiveness of morning and evening light exposure, researchers studied a population of 10 individuals with SAD. Depressive symptoms were measured using a self-diagnostic questionnaire at the beginning of the experiment in order to establish a baseline. The patients were divided into two treatment groups, I and II, with 5 patients in each group. In the first trial, group I was exposed to morning light for three weeks, and group II was exposed to evening light for three weeks. After the three-week-long trial, depressive symptoms were measured again using the same questionnaire that was used to establish the baseline. Neither group was given light therapy for two weeks after the first trial. The experiment was then repeated with the treatment groups reversed. In the second trial, group II was exposed to morning light for three weeks, and group I was exposed to evening light for three weeks. After the three-week-long trial, depressive symptoms were measured again using the same questionnaire that was used to establish the baseline. Since there were no differences between the first trial and the second trial, the data for each treatment were combined. Each bar in the diagram on page 175 represents the average of 10 individuals. Error bars represent the standard error of the mean.

The researcher then developed a method by which regions of the fruit fly's brain that are responsible for taste could be selectively inhibited (off) or left uninhibited (on). She performed the single exposure protocol. The researcher measured the response of the flies after 5 minutes with different regions of the taste lobe of the brain turned on or turned off. The data are presented in Figure II.

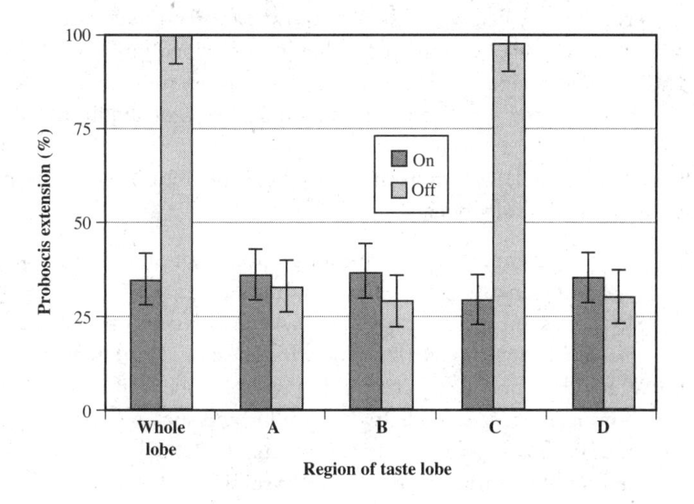

Figure II Aversive learning in regions of the fruit fly's brain turned on or off

388. Which of the following is true of the untrained flies according to Figure I?

(A) Untrained flies never fully extended their feeding proboscis.
(B) Untrained flies responded the same as the trained flies in the first 15 minutes of the trial.
(C) Untrained flies fully extended their feeding proboscis nearly 100% of the time.
(D) Untrained flies extended their proboscis less frequently than trained flies after 20 minutes.

389. Which of the following is true of the flies that were trained with a single exposure to quinine according to Figure I?

(A) The trained flies retained memory of the averse-tasting chemical 100 minutes after exposure.
(B) The trained flies retained memory of the averse-tasting chemical for at least 25 minutes.
(C) The trained flies retained memory of the averse-tasting chemical for approximately 30 seconds.
(D) The trained flies did not differ from the untrained flies in the first 15 minutes of the experiment.

390. Which of the following is true regarding the comparison of flies trained with a single exposure and those exposed for 25 minutes according to Figure I?

(A) Fruit flies trained for 25 minutes remained averse to sucrose longer than those exposed once.
(B) Fruit flies trained for 25 minutes did not differ from flies trained with a single exposure.
(C) Fruit flies trained with a single exposure retained their memory of the averse-tasting chemical longer than those trained for 25 minutes.
(D) Fruit flies trained with a single exposure avoided the quinine taste, while those exposed for 25 minutes sought out additional quinine exposure.

391. Based on Figure II, which of the following is the best interpretation of the results for the whole lobe?

(A) When the whole taste lobe is uninhibited (on), the fruit fly is incapable of forming an association between quinine and sucrose.
(B) Since the same response is elicited when the lobe is off as when it is on, this lobe is not involved in associating the averse taste with sucrose.
(C) When the whole taste lobe is inhibited (off), the fruit fly is incapable of forming an association between quinine and sucrose.
(D) Since proboscis extension is 100% when the taste lobe is turned on, this lobe is not essential for associative learning.

392. Based on Figure II, which of the following is a reasonable conclusion about region A?

(A) Since proboscis extension is 100% when this portion of the taste lobe is turned off, this region is essential for associative learning.
(B) Since approximately the same response is elicited when this region is off as when it is on, this region is not involved in associating the averse taste with sucrose.
(C) Region A controls averse taste learning because proboscis extension is approximately the same when it is on or off.
(D) Since proboscis extension is 0% when this portion of the taste lobe is turned off, this region is not necessary for associative learning.

393. Based on Figure II, which of the following is a reasonable conclusion about region C?

(A) Region C controls averse taste learning because proboscis extension is the same when the region is on or off.
(B) Since the results of region C do not differ significantly from that of the whole lobe, this region is not involved in taste memory.
(C) Since the same response is elicited when this region is off as when it is on, this region is not involved in associating the averse taste with sucrose.
(D) Since proboscis extension is approximately 100% when this portion of the taste lobe is turned off, this region is essential for associative learning.

Questions 394–395

In order to study phototropism, students placed plant seeds onto moistened soil in darkened chambers. Each chamber had a green, red, and blue light-emitting diode (LED) of the same intensity located equally distant from each other. Germinated seedlings invariably grew toward the blue LED.

394. Which of the following is a likely explanation of how plants sense differences in the color of light?

(A) Different wavelengths of light affect rubisco differently. Rubisco is the enzyme that fixes carbon dioxide.
(B) A light receptor protein changes shape when exposed to blue light. This conformation change exposes kinase sites which activate a signal transduction cascade that changes gene expression.
(C) Different colors of light have different amounts of energy. This leads to a change in soil temperature. Plant roots measure the kinetic energy in the soil and respond by bending toward the cooler blue light.
(D) Light absorbed through the stomata affects the rate at which transpiration occurs. Hydrostatic receptors in the guard cells trigger the plant to bend in the direction of the most intense light.

395. Which of the following models would explain the mechanism by which the plants bended toward the blue light?

(A) Cells detecting the blue light transmit cell-lengthening signals to the cells directly below them. These cells directly below expand, and the stem bends toward the light.
(B) Cells detecting the blue light transmit cell-lengthening signals laterally while inhibiting cell lengthening locally. This causes cells stimulated by the least light to elongate, bending the plant toward the blue light.
(C) Cells detecting the blue light transmit cell-lengthening signals locally while inhibiting cell lengthening laterally. This causes cells stimulated by the most light to elongate, bending the plant toward the blue light.
(D) Blue light causes more photosynthesis in the cells on the side of the plant closest to the light. The increased sugar causes these cells to elongate, bending the plant toward the light.

Questions 396–398

A group of AP Biology students performed an experiment in which they exposed growing onion roots to distilled water, lectin, and various concentrations of caffeine. After a week of exposure, the students stained the cells and examined them under a microscope. They counted the number of cells in interphase and the number of cells undergoing mitosis. Their data are presented in the table below.

	Number of Cells in Interphase	Number of Cells Undergoing Mitosis	Total Number of Cells Examined	% Mitotic
Distilled Water	424	40	464	8.6
0.5 mM Lectin	417	209	626	33.4
1 mM Caffeine	549	26	575	4.5
5 mM Caffeine	449	20	469	4.3
10 mM Caffeine	330	12	342	3.5

396. The students want to perform a statistical test to determine whether the treatments had any effect on the number of cells undergoing mitosis. Which of the following would provide the best expected values to use in a chi-square analysis?

(A) Use the ratio of interphase to mitosis for the distilled water to generate expected values.
(B) Use an average of all of the numbers shown. All expected numbers in a chi-square should be equal.
(C) Find the proportion of cells undergoing mitosis in each treatment, and take the average of those values.
(D) Count the number of columns, and subtract 1.

397. The students performed a chi-square analysis on the 10 mM caffeine data using the values shown below.

	Interphase	**Mitosis**
Observed	330.0	12.0
Expected	312.5	29.5

$$\chi^2 = 11.3$$

Which of the following is a correct interpretation of this result?

(A) On 4 degrees of freedom at the 0.01 cutoff, 11.3 > 6.64. Therefore, accept the null hypothesis and conclude that these observations are not different from what is expected from random chance.
(B) On 2 degrees of freedom at the 0.01 cutoff, 11.3 > 6.64. Therefore, the null hypothesis has been proven correct.
(C) On 1 degree of freedom at the 0.01 cutoff, 11.3 > 6.64. Therefore, accept the null hypothesis and conclude that these observations are not different from what is expected from random chance.
(D) On 1 degree of freedom at the 0.01 cutoff, 11.3 > 6.64. Therefore, reject the null hypothesis and conclude that these observations are different from what is expected from random chance.

398. Using chi-square analysis, the students determined that the values for 1 mM caffeine were significantly different from those of distilled water. Which of the following must also be true?

(A) The lectin must not be significantly different from distilled water because fewer cells exposed to lectin were counted and because the difference in % mitotic cells is also lower for cells exposed to lectin.
(B) The 5 mM caffeine must not be significantly different from distilled water because this concentration is higher than the 1 mM caffeine concentration and would have led to a smaller chance of a difference.
(C) The lectin must also be significantly different from distilled water because more cells exposed to lectin were counted and because the difference in % mitotic cells is also higher for cells exposed to lectin.
(D) It is not possible to draw any additional conclusions from this information.

Questions 399–401

AP Biology students performed an experiment in which they crossed a strain of the fungus *Sordaria fimicola* that produces black spores with a strain of *S. fimicola* that produces tan spores. *S. fimicola* is haploid for most of its life cycle. When mating occurs between two haploids, their nuclei fuse, creating a diploid that undergoes meiosis shortly thereafter. Due to the way their centromeres assort in meiosis, the spores in a cross form patterns of 8 spores in a sac (called an ascus). These patterns indicate whether or not crossing-over occurred between the centromere and the spore color gene. The patterns are shown in the figure below. Each circle represents a spore. Shaded circles represent black spores. Unshaded circles represent tan spores. Each row represents the arrangement of spores in a particular sac. The patterns in the first two rows are for sacs in which no crossing-over occurred. The remaining four patterns indicate that crossing-over did occur during meiosis.

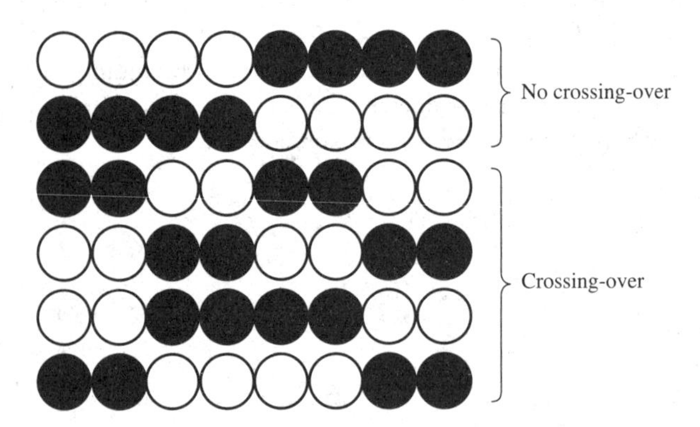

The distance from the centromere to the spore color gene (in map units) can be calculated by determining the percentage of asci showing crossing-over and then dividing by 2. We divide by 2 because only four of the eight spores in each sac come from the two chromatids that swapped alleles. The other four of the eight spores came from the two sister chromatids that did not swap alleles.

399. Meiosis produces four genetically different nuclei, yet the spores in every pattern shown in the figure are in pairs. What is the most reasonable explanation for this observation?

(A) A round of mitosis occurs after meiosis to produce four pairs of spores.
(B) Meiosis I occurs twice, resulting in diploid spores.
(C) During meiosis, crossing-over causes chromosomes to break into smaller pieces. Each of these pieces is incorporated into a separate nucleus.
(D) Nondisjunction of chromosomes creates polyploidy and a doubling of nuclei.

400. Students in the AP Biology class evaluated the frequency of asci showing crossing-over in a cross between tan and wild-type *S. fimicola*. Assuming that the published value of 26 map units is correct and that the students scored 200 asci, which of the following is closest to the number of asci that should show crossing-over?

(A) 26
(B) 52
(C) 104
(D) 208

401. Students in the AP Biology class obtained a third strain of *S. fimicola* that produced red spores. They evaluated the frequency of asci showing crossing-over in a cross between red and wild-type *S. fimicola*. Of the 340 asci the students scored, 95 showed a pattern consistent with crossing-over. According to these data, how many map units is the spore color gene from the centromere?

(A) 7
(B) 14
(C) 28
(D) 56

Questions 402–408

Students performed an experiment in which they transformed *E. coli* cells using the pGlo plasmid shown below. In this plasmid, the green fluorescent protein (GFP) is under the control of the arabinose promoter. The *ara*C protein interacts with arabinose, turning GFP expression on when arabinose is present and turning GFP expression off when arabinose is absent.

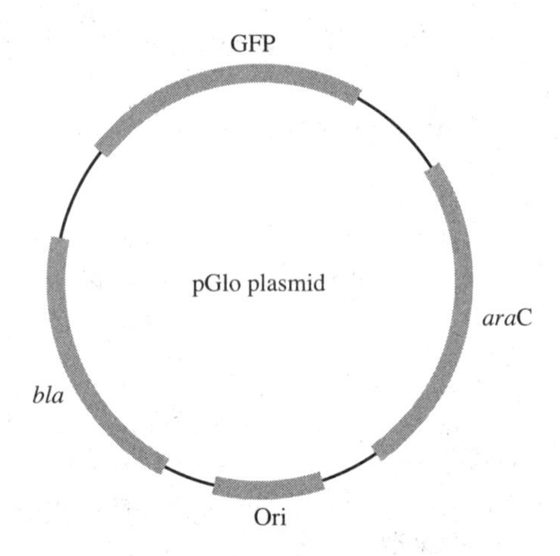

When reviewing this diagram, keep in mind the following:

- GFP = green fluorescent protein
- Ori = origin of replication
- *bla* = the gene for beta lactamase, which confers resistance to ampicillin
- *ara*C = the gene that encodes the protein that controls GFP expression

To accomplish the transformation performed in this experiment, the students mixed 12-hour-old *E. coli* cells with pGlo plasmids, heat shocked the cells, and then plated them on different agar media. A second tube of 12-hour-old *E. coli* cells was subjected to the same treatment but was not mixed with pGlo plasmids. The results are shown in the table below.

	Type of Media		
	LB Agar No Ampicillin No Arabinose	**LB Agar Ampicillin No Arabinose**	**LB Agar Ampicillin Arabinose**
No pGlo	Plate 1	Plate 2	Plate 3
pGlo	Plate 4	Plate 5	Plate 6

402. Which of the following best describes the results expected for the LB agar plates without ampicillin or arabinose (plates 1 and 4)?

(A) There will be lots of growth on both plates 1 and 4 because the presence of the plasmid will not have any effect.
(B) There will be no growth on these plates because the plasmid is not present.
(C) There will be no growth on plate 1 but lots of growth on plate 4 because the plasmid is present on plate 1 but not on plate 4.
(D) There will be no growth on plate 4 but lots of growth on plate 1 because the bacteria are omitted from plate 4 but are present on plate 1.

403. Which of the following best describes the results expected for the LB agar plates with ampicillin but no arabinose (plates 2 and 5)?

(A) There will be no growth on the no pGlo plate (plate 2), but nonglowing colonies will grow on the pGlo plate (plate 5).
(B) There will be no growth on the no pGlo plate (plate 2), but glowing colonies will grow on the pGlo plate (plate 5).
(C) There will be lots of growth on all 6 plates, but none will glow.
(D) The bacteria will grow and glow on all 6 plates.

404. Which of the following best describes the results expected for the LB agar plates with ampicillin and arabinose (plates 3 and 6)?

(A) There will be no growth on the no pGlo plate (plate 3), but nonglowing colonies will grow on the pGlo plate (plate 6).
(B) There will be no growth on the no pGlo plate (plate 3), but glowing colonies will grow on the pGlo plate (plate 6).
(C) There will be lots of growth on both plates, but none will glow.
(D) The bacteria will grow and glow on both plates.

405. Imagine that the person who prepared the agar plates forgot to add ampicillin to the media. Which of the following best describes the expected results?

(A) There will be no growth on the no pGlo plates (plates 1, 2, and 3), but nonglowing colonies will grow on the pGlo plates (plates 4, 5, and 6).
(B) There will be no growth on the no pGlo plates (plates 1, 2, and 3), but glowing colonies will grow on the pGlo plates (plates 4, 5, and 6).
(C) There will be lots of growth on all 6 plates, but none will glow.
(D) The bacteria will grow and glow.

406. Students transferred bacteria from a colony on plate 5 to a Petri plate containing LB agar, ampicillin, and arabinose and then incubated it overnight. Which of the following is expected?

(A) There will be no growth.
(B) The bacteria will grow but not glow.
(C) The bacteria will grow and glow.
(D) It is impossible to predict what is expected without more information.

407. The pGlo plasmid is 5,371 base pairs long. DNA nucleotides have an average molecular mass of 327 g/mol. What is the approximate molar mass of the pGlo plasmid?

(A) 3.51×10^5 g/mol
(B) 1.76×10^6 g/mol
(C) 3.51×10^6 g/mol
(D) 1.76×10^7 g/mol

408. Students calculated their transformation efficiency to be approximately 1.8×10^3 transformants/μg plasmid DNA. If they counted 90 colonies on a plate, how many μg of plasmid did the students plate?

(A) 0.025 μg
(B) 0.05 μg
(C) 0.1 μg
(D) 1 μg

Questions 409–417

The plasmid pTrx, depicted in Figure I, is 3.6 kb (3,600 bp) long and contains an origin of replication (Ori), an ampicillin resistance gene (*bla*), and the *lacZ* gene. Located in the *lacZ* gene is a site called the Multiple Cloning Site (MCS). Pieces of DNA that a researcher is interested in, such as unknown pieces from another organism, can be inserted in the MCS. The important details of the MCS are shown below the plasmid in Figure I. These details include two sites where the restriction enzyme *Eco*R1 cuts as well as binding sites for polymerase chain reaction (PCR) primers. The forward PCR primer (PF) and the reverse PCR primer (PR) make it possible to use PCR to amplify any DNA that lies between the two primers. The *lacZ* gene encodes a protein that metabolizes X-gal into a blue product. When growing on LB agar, *E. coli* colonies appear white. If X-gal is in the medium and the bacteria are expressing *lacZ,* the colonies appear blue.

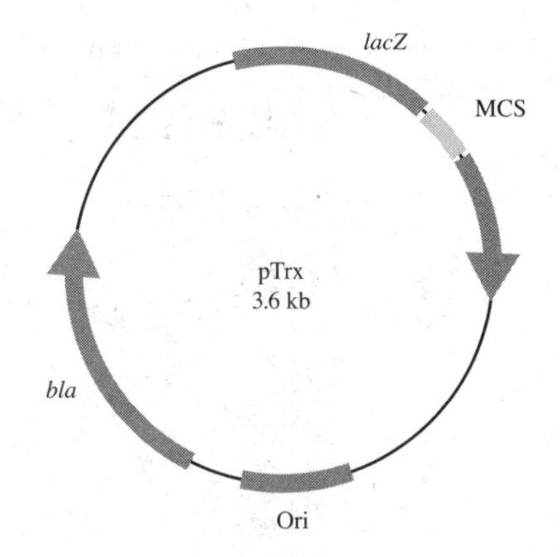

Detail of the Multiple Cloning Site (MCS)

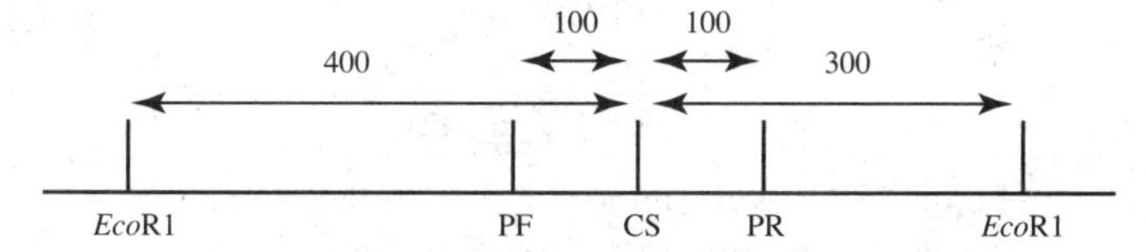

Figure I Map of the pTrx plasmid and a close up of the Multiple Cloning Site (MCS)

When reviewing Figure I, keep in mind the following:

- *Eco*R1 = a restriction enzyme that cuts at the indicated locations
- PF = the binding site for the forward PCR primer
- PR = the binding site for the reverse PCR primer
- CS = the cloning site where DNA will insert

In one experiment, students cut the CS and cDNA from a plant by using a restriction enzyme that specifically cuts in the middle of the CS (the specific enzyme used is not shown in Figure I). They treated the plasmids with ligase, which seals the nicks in the DNA strands. The students then transformed bacteria with the plasmids and plated the transformed bacteria on LB agar plates containing ampicillin. Some of the colonies appeared blue, while others appeared white. The students selected several of the white colonies and performed two different experiments to determine the size of the inserted DNA for each colony. In one experiment, the students ran PCR followed by electrophoresis, and the results are produced in Figure II. Lane M contains marker (also called ladder) DNA fragments that are of known sizes. The size of some of these pieces is labeled in the number of base pairs. Lane 1 shows the results from a bacterial colony identified as clone 1. Lane 2 shows the results from a second bacterial colony identified as clone 2.

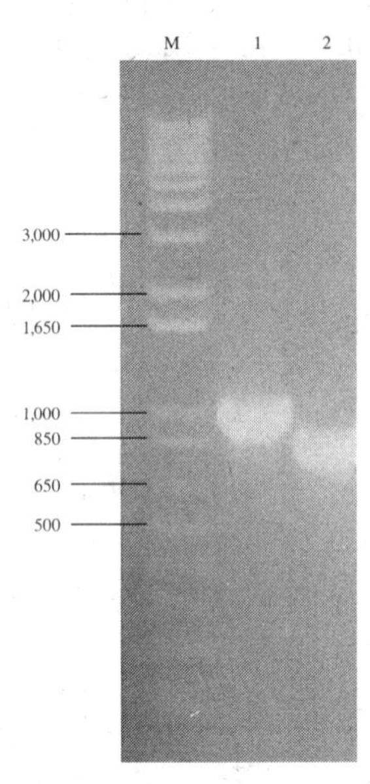

Figure II An electrophoresis gel of PCR results from the amplification of DNA from transformed bacteria

Note that, in Figure II, lane M contains marker DNA, which is used as a standard against which to compare DNA sizes in other lanes. The lengths of some of the pieces are labeled to the left of the diagram. Lane 1 contains DNA from bacterial clone 1. Lane 2 contains DNA from bacterial clone 2.

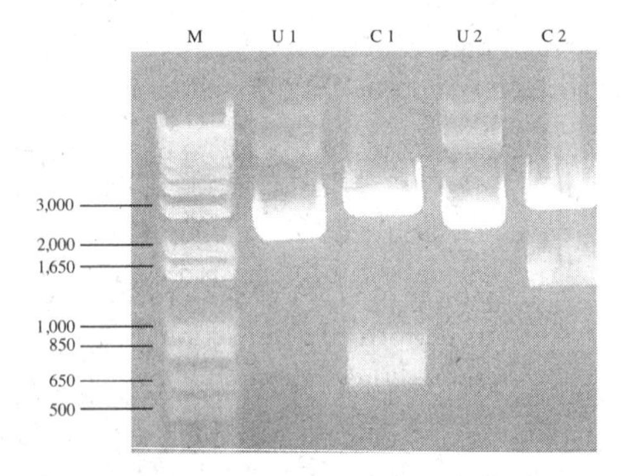

Figure III An electrophoresis gel of restriction digest results for bacterial clones 1 and 2

Note that, in Figure III, the first lane, labeled M, contains marker DNA that serves as size standards. The lane marked U1 is uncut DNA from clone 1. The lane marked C1 is cut DNA from clone 1. Lane U2 contains uncut DNA from clone 2. Lane C2 contains cut DNA from clone 2.

409. Based on Figure I, which of the following identifies the colonies most likely to contain DNA inserted in the CS and provides a correct explanation?

(A) The DNA inserted in the CS coded for a transcription factor that turned on the *lacZ* gene, so these bacteria were not able to produce the blue color from X-gal.
(B) The DNA inserted in the CS disrupted the *lacZ* gene, so these bacteria were not able to produce the white color from X-gal.
(C) The DNA inserted in the CS disrupted the *lacZ* gene, so these bacteria were not able to produce the blue color from X-gal.
(D) The DNA inserted in the CS repaired the *lacZ* gene, so these bacteria were able to produce the blue color from X-gal.

410. Imagine that the students had forgotten to add ligase in the experiment described. Based on Figure I, which of the following would most likely be the result?

(A) There would be lawn growth on the media containing ampicillin because ligase degrades the antibiotic.
(B) There would be lots of white colonies on the media containing ampicillin.
(C) There would be a mixture of blue and white colonies on the media containing ampicillin.
(D) Very few, if any, colonies would grow on the media containing ampicillin, and they would all be blue.

411. Imagine that a 900 bp piece of DNA was inserted in the CS. Based on Figure I, if the students performed PCR, how many and what size pieces should be produced?

(A) two 900 bp pieces
(B) one 1,100 bp piece
(C) one 900 bp piece
(D) three 550 bp pieces

412. Imagine that a 900 bp piece of DNA was inserted in the CS. Based on Figure I, if the students performed a restriction digest using *Eco*R1, how many pieces of DNA should they expect and how long should those pieces be?

(A) one 1,600 bp piece
(B) one 2,900 bp piece and one 900 bp piece
(C) one 900 bp piece
(D) one 1,600 bp piece and one 2,900 bp piece

413. Refer to Figure II. Compare the band of DNA labeled 1,000 bp in the lane labeled M to the band of DNA in the lane labeled 1, which is next to the band labeled 1,000 bp. Which of the following makes a correct statement about the amount of DNA present?

(A) Since the band in lane 1 is brighter and thicker than the 1,000 bp DNA band in lane M, there is less DNA in lane 1 than in the 1,000 bp DNA band in lane M.
(B) Since the 1,000 bp band in lane M is brighter and thicker than the DNA in the lane labeled 1, there is more DNA in the band in lane M.
(C) Since the band in lane 1 is the same brightness and thickness as the 1,000 bp band in lane M, there is the same amount of DNA in both lanes.
(D) Since the band in lane 1 is brighter and thicker than the 1,000 bp DNA band in lane M, there is more DNA in lane 1 than in the 1,000 bp band in lane M.

414. Compared to the DNA in lane 1 in Figure II, which of the following is true about the size of the DNA in lane 2?

(A) The DNA in lane 1 is longer than the DNA in lane 2.
(B) The DNA in lane 1 is shorter than the DNA in lane 2.
(C) The DNA in lane 1 is the same size as the DNA in lane 2.
(D) It is not possible to determine the size of the DNA using an electrophoresis gel.

415. Lane U1 in Figure III contains uncut DNA from the pTrx plasmid. Which of the following pairs an accurate observation from Figures I and III with a correct explanation? (Ignore the faint bands at the top of the lane in Figure III.)

(A) The plasmid pTrx is 3.6 kb long. It is expected to be above the 3,000 bp standard in lane M, but instead it is below that standard. This is probably because circular DNA is less compact than linear DNA and therefore moves through agarose more slowly.
(B) The plasmid pTrx is 3.6 kb long. It is expected to be below the 3,000 bp standard in lane M, but instead it is above that standard. This is probably because uncut plasmid DNA is longer than circular plasmid DNA that has been cut in one location.
(C) The plasmid pTrx is 3.6 kb long. It is expected to be above the 3,000 bp standard in lane M, but instead it is below that standard. This is probably because circular DNA is more compact than linear DNA and therefore moves through agarose more rapidly.
(D) The plasmid pTrx is 2.2 kb long and is at the expected location on the gel.

416. A student comparing the results of clone 2 from the PCR (see Figure II) and restriction digest data (see Figure III) decides that his PCR band is about 900 bp long, and his restriction digest band is 1,400 bp long. He concludes that he must have mixed up his clones when doing the experiments. Which of the following explains the discrepancy? (Refer to Figure I for a map of the plasmid.)

(A) The student is correct. He must have mixed up the samples. There is no other possible explanation.
(B) The DNA on the bacterial plasmid had introns that were excised before the mature mRNA was produced.
(C) The student forgot to subtract 200 bp from the PCR results and 700 bp from the restriction digest results. Doing so gives a consistent prediction of a fragment that is 700 bp long.
(D) The student forgot to subtract 200 bp from the restriction digest results and 700 bp from the PCR results. Doing so yields a consistent prediction of a fragment that is 500 bp long.

417. A student compares his results from the PCR (see Figure II) and the restriction digest data (see Figure III) for clone 1. From the PCR data, he concludes that his insert is 800 bp long. In the restriction digest, he expected a fragment that was 1,500 bp long, but instead he got a thick band of DNA that was about 650–800 bp long. Which of the following is consistent with the data? (Refer to Figure I for a map of the plasmid.)

(A) There is a PCR primer site in the 1,500 bp long fragment. This generated two fragments of approximately the same length that migrated to nearly the same place on the gel.
(B) There is an *Eco*R1 cut site in the 1,500 bp long fragment. This generated two fragments of approximately the same length that migrated to nearly the same place on the gel.
(C) RNases that digest RNA preferentially targeted this sequence in the gel.
(D) Electrophoresis used RNA, which is always smaller than the DNA sequence from which it was translated.

Questions 418–420

The comet assay is a method for analyzing DNA damage in single cells. Cells are embedded on slides covered with agarose and then lysed to release the DNA from the nucleus. Electrophoresis is then performed. DNA with double-stranded breaks migrates faster than the intact DNA. In the comet assay below, the intact DNA forms the head and the damaged DNA forms the tail.

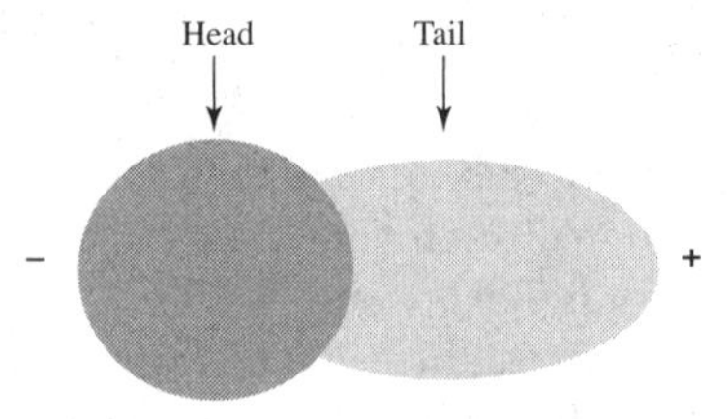

Figure I The comet assay—damaged DNA migrates toward the positive electrode and forms the tail, while intact DNA remains in the head

418. Which of the following represents a cell with little to no DNA damage?

(D) A cell with little to no DNA damage would not show up in a comet assay.

419. Which of the following represents a cell with some intact DNA, but with more DNA damage than that of Figure I?

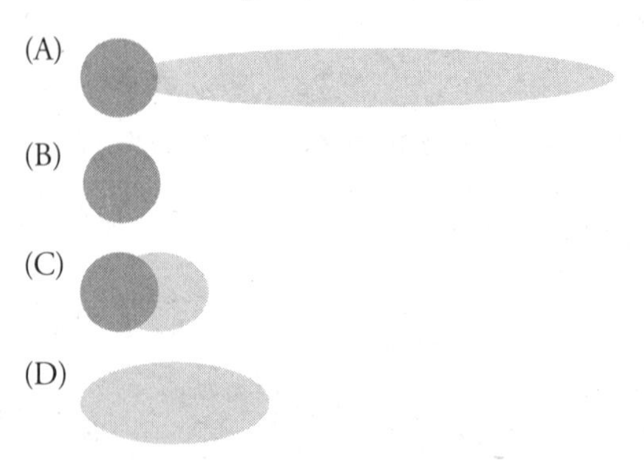

420. Which of the following is NOT a useful application of the comet assay?

(A) detecting cancerous cells
(B) detecting cells with extra chromosomes
(C) detecting sperm with DNA damage
(D) detecting cells treated with mutagenic chemicals

---

421. B-cells are immune cells that initially make a membrane-bound form of an antibody. Once this antibody binds to an antigen, the B-cell produces a secreted form of the antibody. The only difference between these two forms of the antibody is in the C-terminus of the heavy chain. Which of the following would explain this difference?

(A) The C-terminus of the membrane-bound form is hydrophilic, but the C-terminus of the secreted form is hydrophobic.
(B) The C-terminus of the membrane-bound form is hydrophobic, but the C-terminus of the secreted form is hydrophilic.
(C) The C-terminus of the membrane-bound form uses DNA, but the C-terminus of the secreted form uses RNA.
(D) The C-terminus of the membrane-bound form is composed of antigens, but the C-terminus of the secreted form uses T-cells.

Questions 422–429

The process that generates the vast variety of antibodies in the immune system involves producing a protein that uses one of 40 V segments and one of 5 J segments. Two genes, *rag-1* and *rag-2*, that are expressed only in lymphocytes such as B-cells are responsible for creating double-stranded DNA breaks between two randomly selected V and J segments. A portion of the process is shown in the diagram below. In this case, V2 is joined to J4 in the mature mRNA. In other B-cells, different V and J segments are randomly selected.

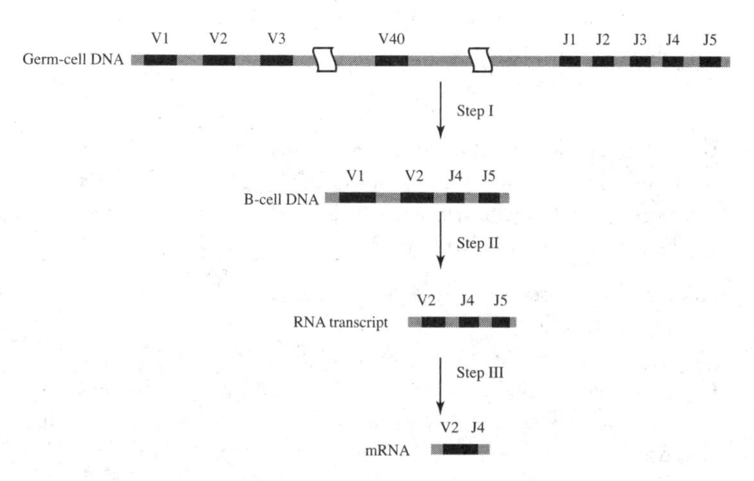

422. Which of the steps permanently alters the genome of the B-cell?

(A) Step I
(B) Step II
(C) Step III
(D) All cells in the body contain the same genome.

423. Which of the steps uses RNA polymerase?

(A) Step I
(B) Step II
(C) Step III
(D) Both Steps II and III use RNA polymerase.

424. Which of the steps involves RNA splicing?

(A) Step I
(B) Step II
(C) Step III
(D) Steps I and II both involve RNA splicing.

425. Step I in the process shown often leads to the insertion of one or more nucleotides into the DNA sequence between the joined segments. Assume that the extra DNA inserted is between 0 and 8 nucleotides long, and each length is equally likely. Also assume that these errors are still present in the mRNA. Under these conditions, which of the following will likely be true?

(A) Approximately two-thirds of the mRNAs produced will contain frameshift mutations that will be likely to make the antibody nonfunctional.
(B) Approximately one-half of the mRNAs produced will contain nonsense mutations that will be likely to make the antibody nonfunctional.
(C) Approximately one-third of the mRNAs produced will contain inversions that will be likely to have no effect on the protein produced.
(D) All of the mRNAs produced will contain an early stop codon so that no protein will be produced.

426. An experiment was conducted in which researchers artificially expressed *rag-1* and *rag-2* in nonimmune cells. The altered cells were able to rearrange antibody gene segments. Which of the following is a reasonable conclusion given this information?

(A) The double-stranded DNA breaks are rejoined by a process that occurs only in B-cells.
(B) The single-stranded DNA breaks are fixed by ligase.
(C) The proteins rag-1 and rag-2 break and repair DNA strands.
(D) The double-stranded DNA breaks were rejoined by processes that occur in nonimmune cells.

427. Individuals with mutated *rag-1* or *rag-2* are unable to complete Step I in the process shown. Which of the following would be true of such individuals?

(A) They would be unable to produce functional antibodies and would be very susceptible to infections.
(B) Other proteins would mutate to assume the function of *rag-1* and *rag-2*.
(C) Their B-cells would skip Step I but would still perform Step II. They would have longer mRNAs and normal antibodies.
(D) As long as one of the two proteins was functional, the individual would produce functional antibodies.

428. B-cell maturation is a process in which newly produced B-cells express antibodies as membrane-bound receptors and are presented with self-antigens in the bone marrow before being released into the rest of the body. This process is essential in the immune system's ability to distinguish between self and nonself molecules. Assuming that the antibody of a particular B-cell was able to bind to one of these self-antigens, which of the following does NOT pair a hypothetical scenario to a correct outcome?

(A) If the B-cell was released from the bone marrow, it would cause an autoimmune response.
(B) If the binding led to reediting of the DNA, a different antibody would be produced.
(C) If the binding led to apoptosis of the B-cell, there would be no autoimmune response.
(D) If the binding led to B-cell proliferation, the antigen would be removed from the body.

429. When B-cells are activated to proliferate by binding to an antigen, the mutation rate of the V segment of the antibody genes increases to about 100,000-fold higher than the regular mutation rate. This produces antibodies with binding sites that are similar to that of the original but with minor variations. At the same time, the binding affinity of antibody to antigen increases. Which of the following best explains the significance of this information?

(A) This is a result of the pathogen that produces the antigen attempting to evade the immune system. The result is increased disease.
(B) An increased mutation rate leads to more B-cells being unable to reproduce. This leads to weaker immune responses to other antigens.
(C) The increased mutation rate produces diversity in the antibodies. Those cells with better-fitting antibodies bind more readily and therefore proliferate more rapidly.
(D) The increased binding affinity causes more friction between antigens and antibodies. This increases the mutation rate as a result of increased temperature.

Questions 430–431

Allergies are the effect of a nonthreatening antigen triggering mast cells to release histamines. Histamines bind to H1 receptors, which are expressed in smooth muscle, in neurons, and in several other tissues. Activation of H1 leads to smooth muscle contraction and the increased firing of neurons in the brain that are associated with alertness. One group of antihistamines works by blocking the H1 receptor.

430. **Antihistamines are often given to patients with allergies to counter the effect of histamine release from mast cells. Which of the following best explains why drowsiness is a side effect of these drugs?**

(A) By contracting smooth muscles in the respiratory system, less carbon dioxide reaches the lungs, making the patient drowsy.
(B) By blocking H1 in nerve cells, antihistamines decrease alertness.
(C) By preventing the release of histamine from mast cells, antihistamines trigger a cascade that upregulates melatonin, the sleep hormone.
(D) By binding to H1 in nerve cells, antihistamines increase the activity of alertness-related neurons, leading to the release of the sleep hormone melatonin.

431. **Some antihistamines do not cause drowsiness. Which of the following is a reasonable hypothesis that explains this result?**

(A) These antihistamines do not pass the blood-brain barrier and therefore do not reach the neurons involved with alertness.
(B) These antihistamines bind to histamine molecules, not to the H1 receptor.
(C) These antihistamines prevent histamine release from mast cells.
(D) All of these are reasonable hypotheses that would explain the antiallergy effects as well as the lack of drowsiness.

Questions 432–434

Researchers interested in RNA performed an experiment in which they injected different pieces of mRNA into cells of the nematode *Caenorhabditis elegans*. They selected a gene important to muscle function for their experiments. Mutants with a nonfunctional copy of the gene exhibited a characteristic twitching behavior. In the experiment, the researchers injected three types of RNA into the gonads of adult nematodes. Message sense RNA is the same as the mRNA that produces a functional protein. Antisense RNA is complementary to message sense RNA. The researchers also used double-stranded RNA (dsRNA). The results of their experiment are summarized in the table on page 199.

Injected into parent	Message sense RNA	Antisense RNA	dsRNA
Effect in offspring	No effect	No effect	Twitching

432. **Which of the following is a reasonable interpretation of the results?**

(A) Injecting double-stranded RNA into the parent turned on expression of the muscle function protein in the offspring.
(B) Injecting double-stranded RNA into the parent turned off expression of the muscle function protein in the offspring.
(C) Injecting double-stranded RNA into the offspring turned off expression of the muscle function protein in the offspring.
(D) Injecting message sense or antisense RNA into the parent turned off expression of the muscle function protein in the offspring.

433. **The researchers who performed this experiment won a Nobel Prize for this and for other supporting work. From a research point of view, which of the following best explains why this work was important enough to be awarded such a prize?**

(A) It allowed researchers to see the sequence of the DNA that is involved in muscle twitching directly.
(B) It allowed researchers to remove specific segments of DNA and replace them with other artificial sequences.
(C) It provided a way to delete specific genes from the genome permanently.
(D) It provided a way to turn off particular genes experimentally without having to delete them from the genome entirely.

434. **DNA molecules will bind to one another if they are complementary. Which of the following would bind to a DNA strand with the sequence 5′-GAATTC-3′?**

(A) 5′-CTTAAG-3′
(B) 5′-GAATTC-3′
(C) 3′-CTTAAG-5′
(D) Both choices (B) and (C) are correct.

Questions 435–436

One promising new strategy for treating certain genetic disorders relies on drugs that cause translational readthroughs. These drugs cause ribosomes to slip a small percentage of the time when they encounter a stop codon. This causes the ribosomes to continue to translate past the stop codon.

435. **Which of the following types of mutations could be treated with this type of drug?**

(A) silent mutations in which the amino acid sequence is not altered
(B) missense mutations in which a different amino acid sequence is produced
(C) nonsense mutations that produce truncated proteins
(D) Any of these types of mutations could be treated with this type of drug.

436. **Assuming that the stop codon occurs in the middle of an exon, approximately what percentage of proteins would be expected to have the correct sequence following the readthrough?**

(A) 25%
(B) 33%
(C) 50%
(D) 100%

Questions 437–444

Arabidopsis thaliana is a vascular plant that displays circadian rhythms. The genes *CCA1* and *TOC1* produce the proteins CCA1 and TOC1. These proteins are involved in regulating the daily cycles of activity. CCA1 is a transcription factor that turns on genes expressed in the morning. It also inhibits both *TOC1* transcription and genes expressed in the evening. *CCA1* is upregulated (turned on) by TOC1 and by sunlight. Both proteins are degraded at a constant rate so a lack of expression leads to a decrease in protein levels. Figure I summarizes this information.

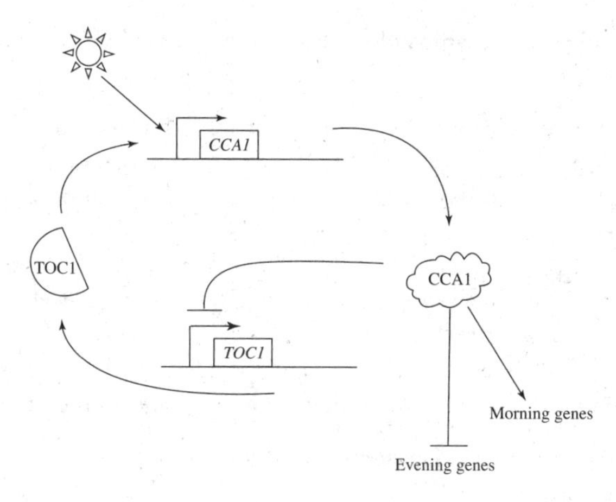

Figure I Regulation of circadian rhythms in *A. thaliana*

Figure II depicts the levels of the proteins CCA1 (open circles) and TOC1 (filled circles) in a photoperiod of 12 hours of light followed by 12 hours of darkness. The shaded bars represent periods of light.

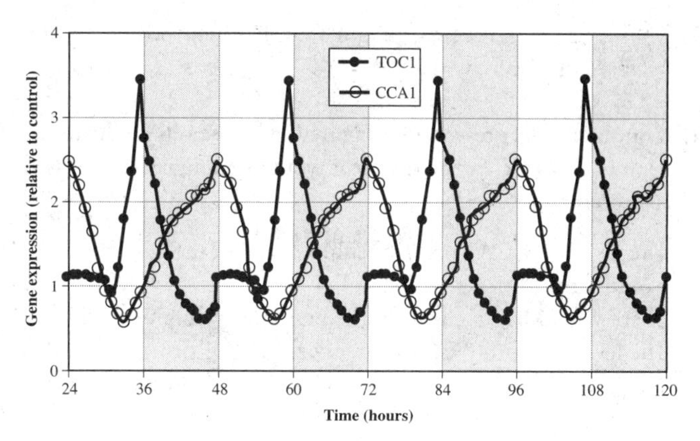

Figure II CCA1 and TOC1 protein levels over 4 days in *A. thaliana*

437. **Based on Figure II, which of the following is true about *CCA1* gene expression?**

(A) *CCA1* gene expression begins approximately 6–7 hours after the light is turned on.
(B) *CCA1* gene expression begins approximately 6–7 hours before the light is turned on.
(C) *CCA1* gene expression begins approximately 2–3 hours after the light is turned on.
(D) *CCA1* gene expression begins approximately 2–3 hours before the light is turned on.

438. Based on Figure II, which of the following is true about the levels of the TOC1 protein?

(A) TOC1 protein levels begin to rise within 6 hours before CCA1 protein levels rise.
(B) TOC1 protein levels begin to rise within 6 hours after CCA1 protein levels rise.
(C) TOC1 protein levels rise only while CCA1 protein levels are rising.
(D) TOC1 protein levels rise only when CCA1 protein levels are falling.

439. Based on the model presented in Figure I, what would be the result of a mutation in the *CCA1* gene so that it no longer responded to light?

(A) The levels of CCA1 protein and TOC1 protein would both rise indefinitely.
(B) The levels of CCA1 protein would fall, and the levels of TOC1 protein would rise indefinitely.
(C) The levels of CCA1 protein and TOC1 protein would continue to oscillate, but their cycles would drift so that they would no longer be on a 24-hour cycle.
(D) The levels of CCA1 protein and TOC1 protein would continue to oscillate indefinitely.

440. Based on the model presented in Figure I, what would be the result of a mutation in the *TOC1* gene so that it would no longer be repressed by CCA1 protein?

(A) The levels of CCA1 protein would rise, and the levels of TOC1 protein would fall.
(B) There would be no effect on the periodic cycling of the two proteins.
(C) The levels of both CCA1 protein and TOC1 protein would fall to zero.
(D) The levels of both CCA1 protein and TOC1 protein would rise and stay high.

441. What is the benefit of turning on morning genes before dawn?

(A) There is no benefit to expressing these genes until the light turns on.
(B) Moonlight is strongest just before dawn, so turning on these genes allows the plant to exploit the moon as a light source.
(C) In nature, the plant would be prepared to begin absorbing light the instant that light became available.
(D) This is an adaptation to a modern urban environment. It allows the plant to take advantage of light supplied by cars during the morning commute.

442. Which of the following would result if the light was left on continuously?

(A) *CCA1* gene expression would be repressed continuously.
(B) *TOC1* gene expression would be repressed continuously.
(C) CCA1 protein levels would decrease, but the periodic cycles of TOC1 protein levels would continue indefinitely.
(D) TOC1 protein levels would increase, leading to a loss of periodic cycles of *CCA1* gene expression.

443. What would be the effect of turning the light off and leaving it off?

(A) The levels of both CCA1 protein and TOC1 protein would rise.
(B) The levels of CCA1 protein would fall, and the levels of TOC1 protein would rise permanently.
(C) The levels of both CCA1 protein and TOC1 protein would continue to oscillate, but their cycles would drift so that they would no longer be on a 24-hour cycle.
(D) The levels of both CCA1 protein and TOC1 protein would continue to oscillate indefinitely.

444. The appropriate photoperiodic timing of metabolic events is critical for unicellular photosynthetic organisms as well as for vascular plants. Which of the following is likely true of the genes involved in these daily cycles?

(A) The sequences for the genes controlling the molecular clock are probably very similar, indicating that the genes derived from the same ancestor.
(B) The genes controlling the molecular clock probably evolved at different times to perform the same function.
(C) The genes controlling the molecular clock are likely to show little similarity.
(D) The genes controlling the molecular clock probably evolved at the same time in response to the same conditions but from different ancestors.

Questions 445–449

The ABC model of organ identity is one model that explains how the genes for making different flower organs are regulated. Flower organs are produced in the following order: sepals, petals, stamens, and then carpels. (Only half of a flower is shown in the figure below.)

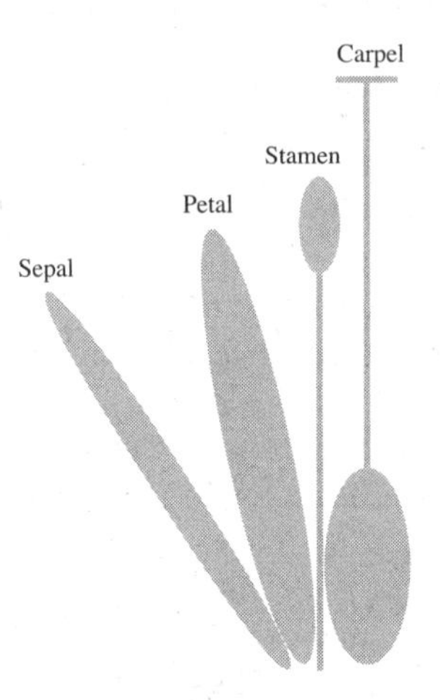

Various genes code for transcription factors that turn on the expression of other genes responsible for making a particular organ. These transcription factors and organs are shown in the table below.

Organ	Sepal	Petal	Stamen	Carpel
Transcription Factor(s)	A	A and B	B and C	C

Another transcription factor, E, is turned on when environmental conditions are right for flowering. Transcription factor E is necessary for all of the others to work.

445. **According to the ABC model, which gene would be turned on last?**

(A) A
(B) B
(C) C
(D) They would all be turned on at the same time.

446. Which organ(s) would mutants that do not express transcription factor B be able to make?

(A) sepals and petals only
(B) stamens and carpels only
(C) sepals and carpels only
(D) They would be able to make all of the organs.

447. Which organ(s) would mutants that do not express transcription factor A be able to make?

(A) sepals and petals only
(B) stamens and carpels only
(C) sepals and carpels only
(D) They would be able to make all of the organs.

448. Which organ(s) would mutants that do not express transcription factor C be able to make?

(A) sepals and petals only
(B) stamens and carpels only
(C) sepals and carpels only
(D) They would be able to make all of the organs.

449. A flower breeder wants to produce showier flowers with a larger number of attractive organs. Which of the following would be a mechanism by which she could achieve this goal?

I. Select mutants that do not express transcription factor C.
II. Place a colorful pigment gene under the control of a promoter that is turned on in the presence of transcription factor A alone.
III. Turn factor E on in all plant tissues.

(A) I only
(B) III only
(C) I and II only
(D) I, II, and III

Questions 450–457

Circadian rhythms are the cycles that an organism goes through each day, including waking and sleeping. Figure I shows the mechanism by which internal clocks of fruit flies (*Drosophila melanogaster*) work. CLK and CYC are transcription factors that are necessary to induce the production of the proteins TIM and PER. CRY is activated by light and degrades TIM when active. PER is degraded unless it is bound to TIM. When PER is bound to TIM, the complex is transported to the nucleus. The PER/TIM complex binds to CLK and CYC, removing them from the promoter. Over time, the PER/TIM complex in the nucleus breaks down, allowing CLK and CYC to bind to the promoter again. This process leads to a daily cycling of TIM and PER expression, which is entrained (set) by light.

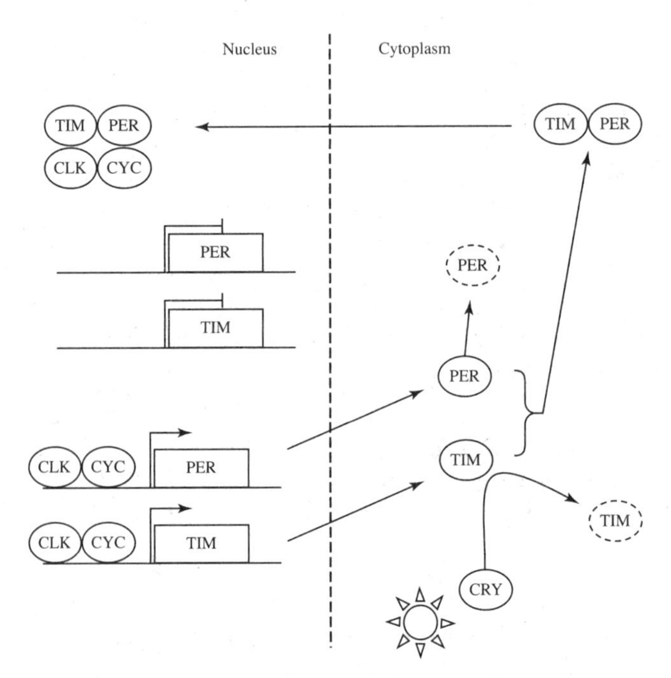

Figure I Regulation of circadian rhythms in *D. melanogaster*

Figure II shows the rise and fall of protein and mRNA for PER and TIM over 24 hours in the fruit fly *Drosophila melanogaster*. Light was supplied for the first 12 hours (unshaded) and the 12 hours after that were in darkness (shaded).

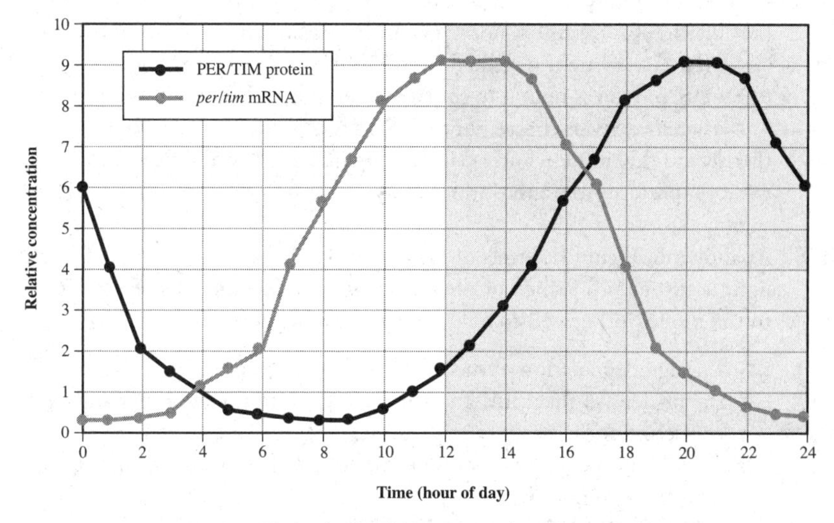

Figure II Cycling of protein and mRNA levels over a period of 24 hours in *D. melanogaster*

450. According to Figure II, approximately how long after mRNA levels peak do protein levels peak?

(A) 1–2 hours
(B) 2–4 hours
(C) 6–8 hours
(D) 8–12 hours

451. According to Figure II, at approximately what time does the rate of translation of mRNA into PER and TIM protein equal the rate of degradation of PER and TIM protein?

(A) at hour 1 and again at hour 13
(B) at hour 4 and again at hour 17
(C) at hour 8 and again at hour 20
(D) The rate of mRNA into protein never equals the rate of protein degradation.

452. **Use the information about mRNA and protein concentrations at hour 8 in Figure II to interpret the model presented in Figure I. Which of the following statements about CLK and CYC is true at hour 8?**

(A) Both CLK and CYC proteins are bound to the promoter, so mRNA is being transcribed from DNA.
(B) Neither CLK protein nor CYC protein is bound to the promoter, so no mRNA is being transcribed from DNA.
(C) CLK protein is bound to the promoter, but CYC protein is not. Therefore, twice the amount of mRNA is produced.
(D) Both CLK protein and CYC protein are bound to the PER/TIM complex, so no transcription occurs.

453. **As shown in Figure II, levels of *per* and *tim* mRNA begin rising before the light is turned off. Which of the following best explains this result? (Refer to the model in Figure I to help you answer this question.)**

(A) Enough degradation of the PER/TIM protein complex has happened in the nucleus so that CLK and CYC can bind to the promoter and begin transcription.
(B) CLK and CYC are still bound to the PER/TIM protein complex, so mRNA is produced.
(C) Enough degradation of the PER/TIM protein complex has happened in the cytoplasm so that CLK and CYC are not bound to the promoter, shutting off transcription.
(D) Sufficient PER and TIM protein has been produced to lead to the buildup of CLK and CYC. This increases the amount of transcription of *per* and *tim* mRNA.

454. **Which of the following would be most likely to happen if the fruit flies were kept in total darkness?**

(A) Since TIM would not be degraded in the dark, the levels of TIM protein would increase earlier in the day. Cycling would continue, but the oscillations would be shorter.
(B) Since TIM would be degraded in the dark, the levels of TIM protein would decrease earlier in the day. Cycling would continue, but the oscillations would be longer.
(C) PER would be degraded faster in the dark instead if TIM were not degrading in the light. This would keep the flies cycling normally even in the absence of light.
(D) The flies would immediately begin sleeping and would not wake up because there would be no cycling of TIM and PER.

455. Which of the following would be most likely to happen if the fruit flies were kept in continuous light?

(A) Since TIM would not be degraded due to the light, protein levels would increase earlier in the day. Cycling would continue, but the oscillations would be shorter.
(B) TIM would be degraded due to the light, so protein levels would remain low for a longer period of time. Cycling would continue, but the oscillations would be longer.
(C) PER would be degraded due to the presence of light in addition to TIM degrading in the light. This would keep the flies cycling normally even in the presence of constant light.
(D) The flies would not sleep because there would be no cycling of TIM and PER.

456. Researchers mutated the *tim* gene so that the mRNA sequence had an early stop codon. Which of the following best explains the result of this mutation?

(A) The fruit fly would still produce a functional TIM protein. There would be no effect on the phenotype.
(B) The fruit fly would no longer produce a functional TIM protein. PER alone would bind to CLK and CYC. The fruit fly would have a functional internal clock that would be a little less reliable.
(C) The fruit fly would produce TIM protein with decreased functionality. The result would be a longer cycle.
(D) The fruit fly would no longer produce a functional TIM protein. The PER/TIM complex would no longer form and bind to CLK and CYC. The fruit fly would have a nonfunctional internal clock.

457. Researchers placed the promoter that controls the genes *tim* and *per* in front of the gene for green fluorescent protein (GFP) and transformed fruit flies with this chimeric DNA. Which of the following best describes the likely result of this experiment?

(A) The flies would lose the ability to sense time due to the effect of GFP on PER degradation.
(B) The flies would either glow continuously if GFP was not degraded rapidly, or they would show a daily cycle if GFP was degraded rapidly.
(C) The flies would blink rapidly because of the negative feedback cycle that was produced.
(D) The flies would glow increasingly brighter in response to a positive feedback cycle in which GFP stimulates TIM and PER, which stimulates the production of more GFP.

Questions 458–463

Researchers used ultraviolet light to destroy the nucleus of an embryonic stem cell (germ cell) of a frog. They then transferred a nucleus from a fully differentiated somatic cell into the nucleus-free germ cell, as shown in Figure I. The cell with the transplanted nucleus was capable of growing into a tadpole and eventually into an adult frog.

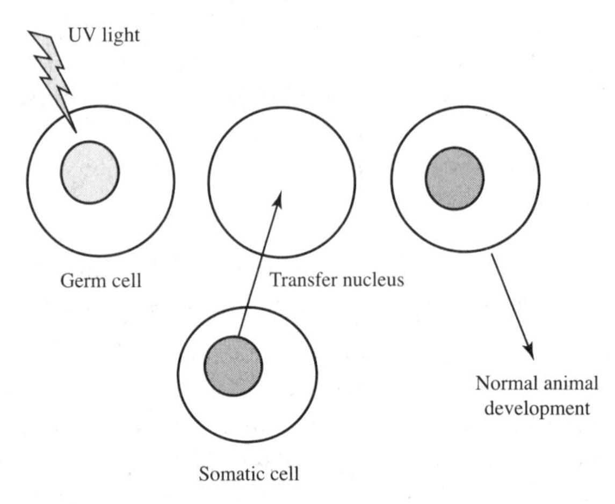

Figure I The transfer of a nucleus from a somatic cell into a germ cell in which the nucleus was destroyed

Researchers identified 24 candidate genes that were strongly expressed in germ cell lines but not in somatic cells. When they expressed these 24 genes in somatic cells, the researchers were able to cause the cells to dedifferentiate to an earlier state of development. Figure II shows data from a test in which the researchers screened 8 of the genes for their ability to induce dedifferentiation. Cells that grew in culture had dedifferentiated and had an induced pluripotent state. Each bar in Figure II represents a mix of 8 factors minus 1 factor. The bars are labeled by the factor that was omitted from each mixture. The researchers also included a treatment in which all 8 factors were present and one treatment in which none of the factors were present (mock).

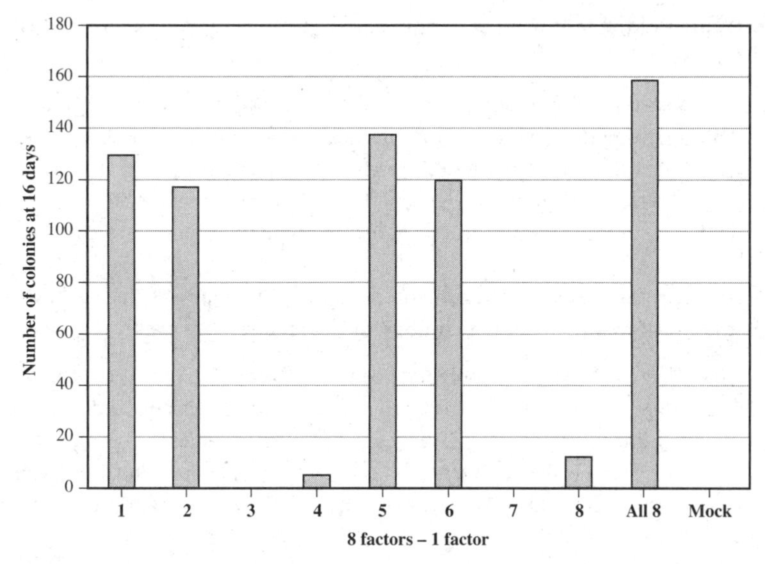

Figure II The effect of different combinations of factors on a cell's ability to dedifferentiate and grow

458. Which of the following is a reasonable conclusion from the experiment illustrated in Figure I?

(A) The destruction of the original nucleus released genetic material that remained in the cell and changed the transplanted nucleus into a pluripotent germ cell nucleus.
(B) The cytoplasm of the somatic cell contained mRNAs that kept the nucleus from dedifferentiating. Transplanting the nucleus removed all mRNAs and therefore reset the nucleus.
(C) The cytoplasm of the germ cell contained the appropriate signals to change the transplanted nucleus into a pluripotent germ cell nucleus.
(D) Transplanting the nucleus caused the expression of all of the genes present in the new nucleus.

459. A scientist used a germ cell from a frog with white eyes, and a somatic cell from a frog with red eyes to perform the process shown in Figure I. After growing the cell that resulted into a frog, which of the following would be expected?

(A) The frog should have red eyes, the same as that of the donor of the nucleus.
(B) The frog should have white eyes, the same as that of the frog where the germ cell came from.
(C) There should be a 3:1 ratio of red-eyed to white-eyed frogs.
(D) The frog should have one red eye and one white eye.

460. Which of the following is an appropriate control for the experiment shown in Figure I?

(A) a treatment in which somatic cells are exposed to UV light
(B) a treatment in which germ cell nuclei are transferred into somatic cells
(C) a treatment in which the researchers do everything to the cell as described in the passage except actually transfer a nucleus
(D) No control group is necessary for this type of experiment.

461. In the second experiment presented (Figure II), what was the purpose of the mock treatment in which none of the factors were used?

(A) This was a negative control. If cells had developed in this treatment, we could conclude that none of the factors were important and that the treatment process alone was responsible for producing colonies.
(B) This was a positive control. If cells had not developed in this treatment, something would have been wrong with the cells or with the treatment process itself.
(C) This was a negative control. If cells had not developed in this treatment, the experiment would need to be repeated.
(D) This was a positive control. If cells had developed in this treatment, the variability would be suspect and there would be reason to discard outliers.

462. In the second experiment presented (Figure II), what was the purpose of including a treatment in which all 8 factors were used?

(A) This was a negative control. If cells had developed in this treatment, we could conclude that none of the factors were important and that the treatment process alone was responsible for producing colonies.
(B) This was a positive control. If cells had not developed in this treatment, something was wrong with the cells or with the treatment process itself.
(C) This was a negative control. If cells had not developed in this treatment, the experiment would need to be repeated.
(D) This was a positive control. If cells had developed in this treatment, the variability would be suspect and there would be reason to discard outliers.

463. Which of the following pairs a reasonable hypothesis based on the experiment shown in Figure II with an experiment that would test that hypothesis?

(A) **Hypothesis:** Factors 1, 2, 5, and 6 are necessary to induce pluripotent cells. **Experiment:** Test factors 1, 2, 5, and 6 in combination and with one factor removed to see if they can cause dedifferentiation and growth.
(B) **Hypothesis:** Factors 1, 2, 5, and 6 promote cell differentiation. **Experiment:** Inject factors 1, 2, 5, and 6 into a somatic cell from a frog and see if the frog grows.
(C) **Hypothesis:** Factors 3, 4, 7, and 8 are necessary to induce pluripotent cells. **Experiment:** Test factors 3, 4, 7, and 8 in combination and with one factor removed to see if they can cause dedifferentiation and growth.
(D) **Hypothesis:** Factors 3, 4, 7, and 8 inhibit cell differentiation. **Experiment:** Inject factors 3, 4, 7, and 8 into germ cells from a frog and see if the frog grows.

---

464. All of the cells shown in the diagram below were produced by mitosis and cell division from the same individual. Which of the following best explains how cells with these diverse shapes and functions resulted?

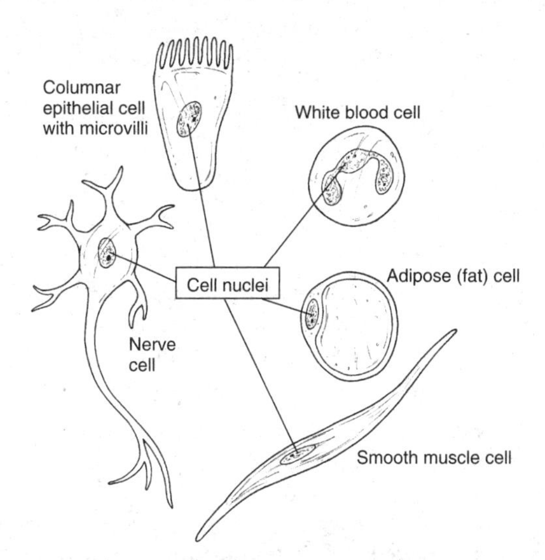

(A) Each of the different types of cells has different DNA that directs the shape the cell takes.
(B) Each of the different types of cells expresses a different set of transcription factors. These turn on the specific genes necessary in the cell that has them.
(C) These shapes are random. Cells take the shape of the space they have.
(D) Cells evolve as the organism grows so that the correct types are left behind.

Questions 465–471

The diagram below depicts processes involved in gene expression in a eukaryotic cell.

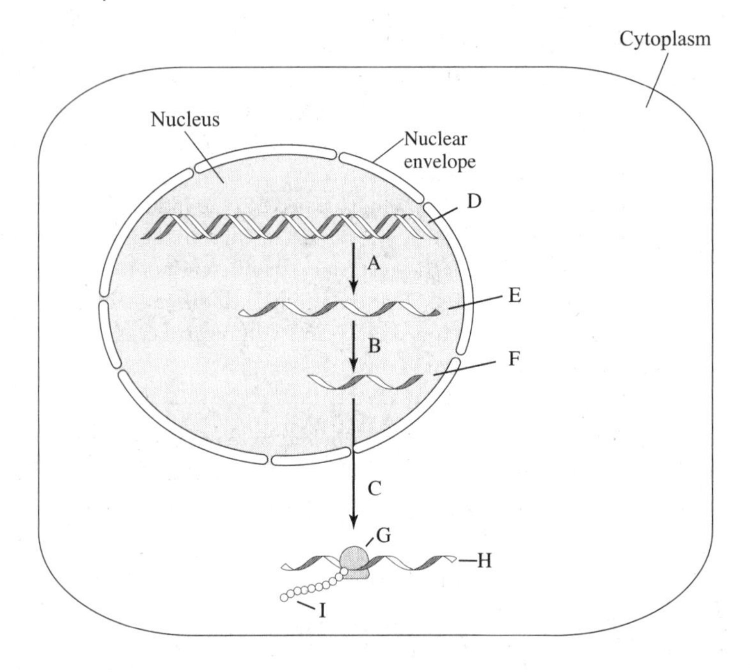

465. Which of the following is true of the process that molecules G, H, and I are performing?

(A) It involves the polymerization of nucleotide triphosphates in a sequence directed by the information contained in molecule I.
(B) It involves the hydrolysis of amino acids into RNA.
(C) It involves the polymerization of amino acids in a sequence directed by the information contained in molecule H.
(D) It involves the plus end polymerization of actin molecules, which elongates actin strands.

466. Changes in the sequence of monomers in which of the following molecules will have the greatest impact on the functioning of the cell?

(A) D
(B) E
(C) F
(D) I

467. Which of the following is true of the process labeled B?

(A) It involves removing some sequences of RNA from the molecule.
(B) It involves adding a long string of adenine nucleotides to the 3′ end.
(C) It involves adding a cap to the 5′ end.
(D) All of these are true of the process labeled B.

468. Which of the following is true of the process labeled A?

(A) It involves the synthesis of DNA using an RNA template.
(B) It involves adding amino acids together in long chains.
(C) It involves the synthesis of RNA using a DNA template.
(D) It involves the synthesis of DNA using a DNA template.

469. Which of the following is true of the process labeled C?

(A) It involves the synthesis of DNA using an RNA template.
(B) It does not occur in bacteria.
(C) It involves adding a long string of adenine nucleotides to the 3′ end.
(D) It does not occur in eukaryotes.

470. Which of the following does NOT affect the rate at which process A occurs?

(A) the presence or absence of start codons in molecule D
(B) the presence of transcription factors at the promoter region
(C) how tightly molecule D is bound to histones
(D) the degree to which molecule D is methylated

471. Two different varieties of a protein may be produced by the same gene sequence if different exons are used. During which process does the selection of exons occur?

(A) process A
(B) process B
(C) process C
(D) the process that molecules G, H, and I are performing

Questions 472–473

The diagram below shows crossing-over, a process that happens to homologous chromosomes during prophase I of meiosis.

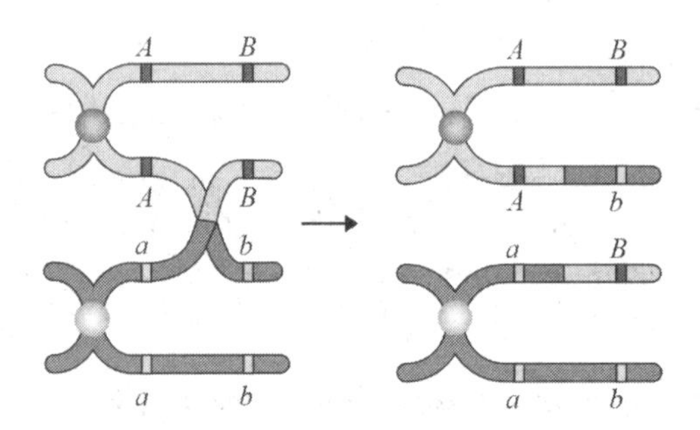

472. Which of the following best describes the importance of the process shown?

(A) The process contributes to genetic diversity by recombining alleles in new patterns.
(B) The process contributes to evolution by causing random point mutations.
(C) The process contributes to genetic drift by preventing heterozygosity.
(D) The process contributes to embryonic development by ensuring that all cells have the same DNA.

473. Which of the following is true of the segments labeled *B* and *b*, which represent different alleles?

(A) They likely differ in their entire sequence.
(B) Segment *B* is made of DNA, while segment *b* is composed of RNA.
(C) They likely differ in only one or a few nucleotides.
(D) They likely have the exact same sequence.

GRID-IN QUESTIONS

Questions 474–475

When designing primers for PCR, it is important to know the melting temperature (Tm) of the DNA sequences. This can be calculated using the following formula:

$$Tm = 2 \times (\text{number of A} + \text{number of T}) + 4 \times (\text{number of G} + \text{number of C})$$

474. **What is the melting temperature for the following sequence? Give your answer to the nearest whole number.**

ATGCGGCACC

475. **What is the melting temperature for the following sequence? Give your answer to the nearest whole number.**

ATATCGTACA

Questions 476–478

Blood type in humans is controlled by three alleles at a particular gene locus. Alleles A and B each produce a different polysaccharide marker that is expressed on the outside of red blood cells. The O allele does not express an antigen on red blood cells.

Blood Type	Genotype	Cell Diagram	Genotype	Cell Diagram
A	AA	A A	AO	A
B	BB	B B	BO	B
AB	AB	A B		
O	OO			

Note that an additional locus codes for the Rh factor, in which Rh+ is dominant over Rh–.

476. A woman with blood type O+, whose mother was Rh–, has a child with a man with blood type AB–. What are the odds that they will have a child who is blood type A+? Express your answer as a fraction or as a decimal value between 0 and 1.

477. A person with blood type B+ has a child with a person whose blood type is A–. Their first child has blood type O–. What are the odds that their second child will have blood type A–? Express your answer as a fraction or as a decimal value between 0 and 1.

478. Two people with blood type AB+ have a child that is A–. What are the odds that their second child will have blood type AB–? Express your answer as a fraction or as a decimal value between 0 and 1.

---

479. A diploid plant is heterozygous at three different gene loci and has a genotype of AaBbDd. If the plant is mated to itself, what fraction of the offspring will be homozygous recessive for all three loci? Express your answer as a fraction or as a decimal value between 0 and 1.

Questions 480–482

Coat color in Labrador Retrievers is primarily controlled by two genes. The gene for pigment production has a dominant allele that produces a black pigment (B) and a recessive allele that produces a brown pigment (b). A second gene locus controls the expression of pigment in the fur. The dominant allele (E) causes pigment expression in the fur, while the recessive allele (e) does not. Fur is yellow unless either a black or a brown pigment is present.

480. A breeder mates a black dog with a genotype of BbEe to a yellow dog with a genotype of Bbee. What is the probability of obtaining a brown puppy? Express your answer as a fraction or as a decimal value between 0 and 1.

481. A breeder mates a brown dog with a genotype of bbEe to a yellow dog with a genotype of Bbee. What is the probability of obtaining a yellow puppy? Express your answer as a fraction or as a decimal value between 0 and 1.

482. A breeder mates two brown dogs with genotypes of bbEe. What is the probability of obtaining a black puppy? Express your answer as a fraction or as a decimal value between 0 and 1.

Questions 483–485

Humans have 23 pairs of chromosomes. During the production of gametes, one of each pair is randomly allocated to each nucleus. As a result, there are 2^{23} = 8,388,608 possible combinations of chromosomes.

483. **Fruit flies have 4 pairs of chromosomes. How many different combinations of chromosomes are there, assuming that the chromosomes assort independently?**

484. **Peas have 7 pairs of chromosomes. How many different combinations of chromosomes are there, assuming that the chromosomes assort independently?**

485. **The genetic code is a triplet code with 4 possible nucleotides at each location. As a result, there are 64 different codons possible. Imagine that it was possible to add another 2 nucleotides at each location so that there were 6 possible nucleotides at each location. Assuming that there is still a triplet code, how many different codons would be possible?**

LONG FREE-RESPONSE QUESTIONS

486. **The vertebrate nervous system detects, transmits, and integrates signals from both the internal and the external environment. The brain is the primary organ that collects information from various sources and directs appropriate responses.**

   (a) Specialized nerve cells called receptor cells detect and transmit signals from the environment. Identify an external signal that humans can detect. Draw a diagram to illustrate how a receptor cell might detect and transmit that signal. Explain your diagram.

   (b) Identify an example in which the brain must integrate two sources of information to produce a response. Draw an illustration that shows the two sources of information being integrated in the brain in order to produce a response. Describe how that process occurs.

   (c) Animals often respond to stimuli with muscle movement. Draw a diagram that shows at least THREE of the steps by which motor neurons stimulate a muscle cell to contract. Describe the process.

487. *Bremia latucae* is a fungus that causes lettuce downy mildew. It infects both cultivated lettuce (*Lactuca sativa*) and a wild species called prickly lettuce (*Lactuca serriola*). A researcher crossed a mildew-susceptible cultivated lettuce plant (C) with a wild prickly lettuce plant (W) that was resistant to mildew. She tested the resulting hybrid plants for susceptibility to downy mildew. These results are shown in the table below in the row labeled C × W. She then crossed the hybrids with themselves. These are labeled (C × W) × (C × W) in the table. She also backcrossed the hybrids with cultivated lettuce (labeled (C × W) × C) and with wild lettuce (labeled (C × W) × W).

Cross or Species	Type of Cross or Strain	Resistant	Susceptible
C	Parent	0	1
W	Parent	1	0
C × W	Hybrid	531	0
(C × W) × (C × W)	Self-cross	388	144
(C × W) × C	Backcross	264	255
(C × W) × W	Backcross	497	0

(a) **Explain** how resistance to downy mildew is inherited. **Justify** your explanation with data from the backcrosses.

(b) **State** a null hypothesis for the self-cross data. **Calculate** a chi-square value, and **interpret** your result.

(c) Resistance to downy mildew has been shown to be the result of a gene-for-gene system. In this system, resistance occurs when a strain of a pathogen has an avirulence gene that matches a corresponding resistance gene in the host plant. Recognition results in the death of plant cells that the fungus needs in order to live. **Describe** a molecular mechanism by which this gene-for-gene system might work.

SHORT FREE-RESPONSE QUESTIONS

488. The MAP kinase pathway shown in the diagram below is an example of a signal transduction pathway.

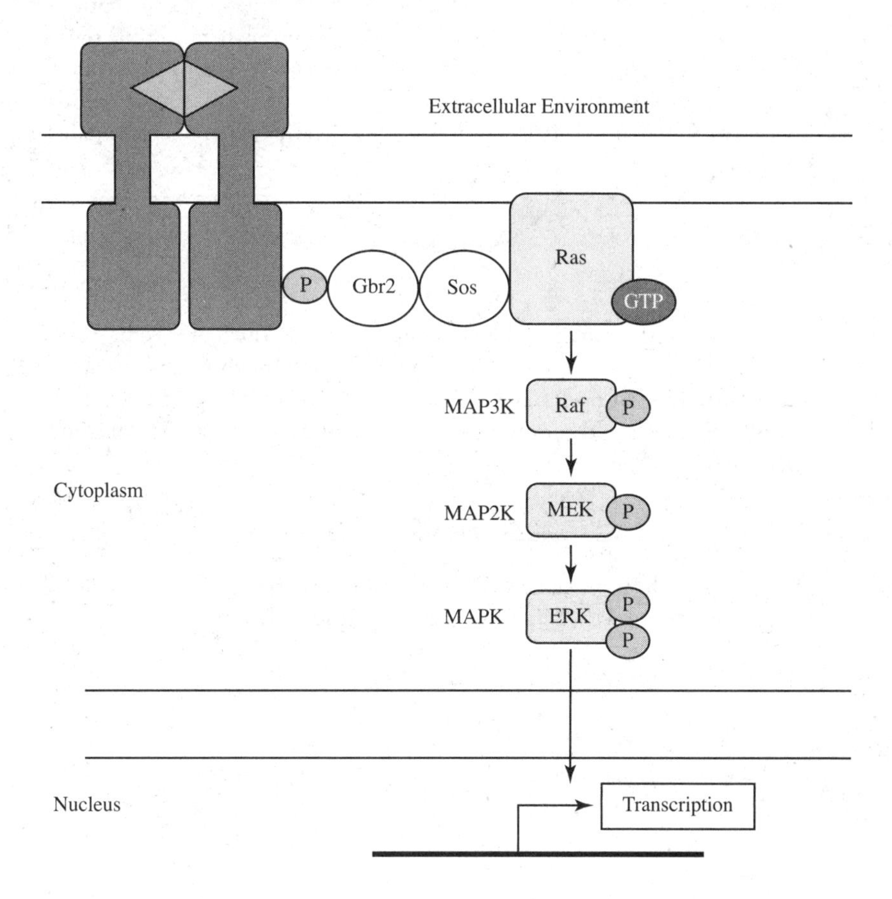

Briefly explain how processes that occur in each of the following contribute to controlling gene expression:

(a) The plasma membrane

(b) The cytoplasm

(c) The nucleus

489. The diagram below shows a restriction enzyme map of a plasmid that is 100 kb long.

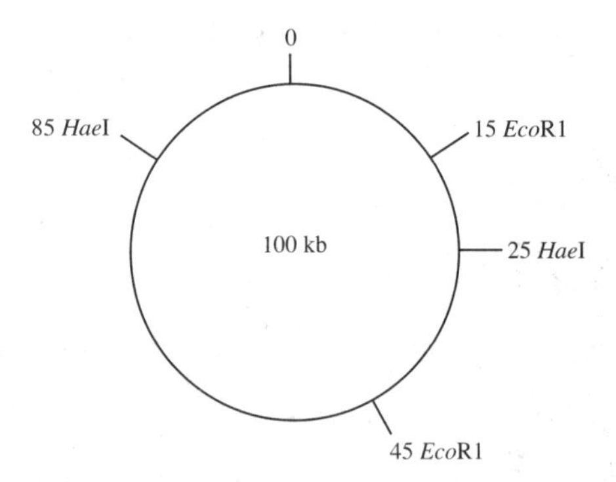

Three different samples of this plasmid were used. The first was digested with *Eco*R1. The second was digested with *Hae*I. The third underwent a double digest of *Eco*R1 and *Hae*I. Use the diagram of a gel below to show the results of this experiment. DNA will be loaded where the labels are shown.

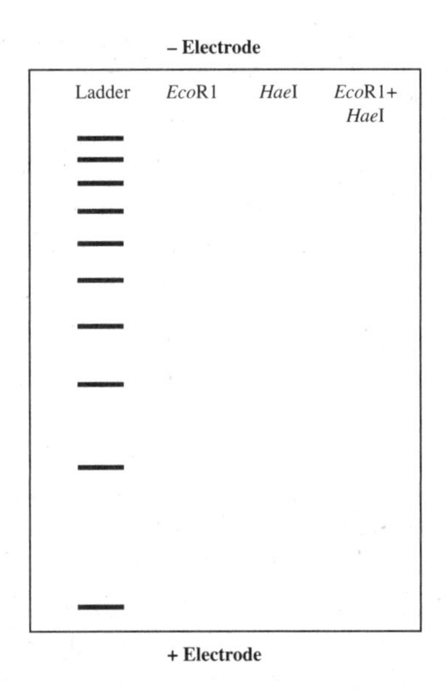

(a) Label the ladder fragments to show the pieces being 10, 20, 30, 40, 50, 60, 70, 80, 90, and 100 kb long.

(b) Draw bands on the gel to indicate the sizes of the DNA pieces for each of the digests listed.

Answers and explanations for all of the questions in this chapter can be found on pages 347–387.

Big Idea 4—Interactions

CHAPTER 4

Answers and explanations can be found on page 388.

BIG IDEA 4

Biological systems interact, and these systems and their **interactions** possess complex properties.

For the complete list of big ideas, learning objectives, enduring understandings, essential knowledge, and science practices, refer to the "AP Biology Course and Exam Description" from the College Board:

https://secure-media.collegeboard.org/digitalServices/pdf/ap/ap-biology-course-and-exam-description.pdf

MULTIPLE-CHOICE QUESTIONS

Questions 490–496

Adult herring gulls have a red spot on their beaks. When herring gull chicks peck at their parents' beaks, their parents feed them. A researcher had 50 newly hatched chicks that had been hatched in an incubator and had not yet been fed. She presented each chick with one of five randomly selected Popsicle sticks for one minute. Four of the sticks had a different colored spot on them (red, black, blue, or green). A fifth stick was left unpainted. Each of the five sticks was presented to 10 different chicks. The researcher then counted the number of times the chick touched the Popsicle stick with its beak in one minute (its pecking behavior). The results of this experiment are summarized in Table I.

Table I The Response of Herring Gull Chicks to Simulated Parent Beaks (Popsicle Sticks)

	Red Spot	Black Spot	Blue Spot	Green Spot	Unpainted
Pecks per Minute	27	15	6	4	2

In a second experiment, the researcher attempted to train chicks to peck at different colored spots. Using a new set of chicks, the researcher randomly assigned ten newly hatched chicks to each of the five training groups: red, black, blue, green, and unpainted (50 chicks total). In each training group, chicks were randomly exposed to different sticks with different colored spots on each one, but were only rewarded with food after pecking at the stick that had a colored spot on it (or a lack thereof for the unpainted group) that corresponded to their training group. For example, chicks in the green group were not fed when they pecked at sticks that were unpainted or that were painted with red, black, or blue spots. They were fed only when they pecked at a stick with a green spot on it. After 3 days of training, the chicks were tested to determine how frequently they pecked at each type of stick. The test was conducted by presenting each chick with one of the five different sticks and counting the number of times they pecked at each stick in one minute. All of the chicks were tested with all 5 test sticks, which were presented in a random order. The mean number of pecks per minute for each training group and test stick combination are summarized in Table II.

Table II The Effect of Training on Pecking Behavior in Herring Gull Chicks

	Pecking Frequency (the Number of Pecks/Min) When Tested After 3 Days of Training				
Training Group	**Unpainted Stick**	**Red Spot Stick**	**Black Spot Stick**	**Blue Spot Stick**	**Green Spot Stick**
Unpainted	15	1	1	0	0
Red Spot	0.5	42	3	1.5	1
Black Spot	0.6	2	39	2	0.5
Blue Spot	0.1	1	1	37	0.2
Green Spot	0	0.9	1.1	2	38

490. **Which of the following questions was being tested in the first experiment?**

(A) Are herring gull chicks born with an innate instinct to peck at red spots?
(B) Do herring gull chicks peck at red spots in order to obtain food?
(C) Can herring gulls see the color red?
(D) Do herring gull chicks learn to select red spots because their parents each have a red spot on their beaks?

491. **What was the function of the unpainted stick in the first experiment?**

(A) It was the independent variable.
(B) It was the dependent variable.
(C) It was the control against which the spotted sticks could be compared.
(D) It provided replication.

492. Which experimental question was being tested in the second experiment?

(A) What is the evolutionary reason for the spot-pecking behavior in herring gull chicks?
(B) How is pecking behavior controlled by genetics?
(C) Can the innate pecking behavior of herring gull chicks be modified by learning?
(D) What role does epigenetics play in controlling pecking behavior in herring gull chicks?

493. Which of the following is a reasonable conclusion from the data presented in Table I?

(A) Herring gull chicks are born with an innate tendency to peck at spots, and red spots produce the strongest response.
(B) Herring gull chicks peck at red spots because their parents have red spots on their beaks.
(C) Herring gull chicks peck at red spots because they are fed after doing so.
(D) Herring gull chicks do not discriminate among different colored spots.

494. Why did the researcher choose to work with newly hatched chicks?

(A) Newly hatched chicks are very hungry and produce stronger responses than older chicks.
(B) Newly hatched chicks had not yet been exposed to their parents, so any response must have been innate rather than learned.
(C) Older chicks stop pecking at spots.
(D) Newly hatched chicks are neotenous.

495. Which of the following is a reasonable conclusion from the data presented in Table II?

(A) Pecking behavior is completely innate.
(B) Pecking behavior is entirely learned.
(C) Herring gull chicks learned to peck at whatever stimulus was rewarded with food.
(D) Unpainted sticks were as strong of a stimulus as colored spots after training.

496. Which of the following would be helpful in determining whether the color of the spot made a difference in the second experiment?

(A) No extra information would be helpful. The numbers clearly show differences.
(B) A control group other than the unpainted stick would be helpful.
(C) A measure of variability, such as the standard error of the mean, would be helpful.
(D) A measure of central tendency, such as the mode, would be helpful.

Questions 497–502

Endophytic fungi live in healthy tissues of plants without causing disease, much as intestinal bacteria do in the human intestine. A student was interested in measuring the diversity of endophytic fungi in Douglas fir trees (*Pseudotsuga menziesii*) that were grown in a forest as compared to the endophytic fungi in Douglas fir trees that were grown in a nursery. To do this, he collected 1-year-old seedlings from three different nurseries and three forests located adjacent to the nurseries (a total of 6 locations). He soaked the seedlings in 20% bleach solution to kill surface fungi and plated pieces of tissue on water agar media. The student then identified fungi that grew. Table I shows the number of times each of 8 fungal species was isolated from the tissue of Douglas fir trees grown in the nursery and in the forest. Since the results from the three nurseries were similar and since the results from the three forests were similar, the results were combined in Table I.

Table I Fungal Species Isolated from the Tissue of Douglas Fir Seedlings Planted in Three Nurseries and in Three Adjacent Forests*

	Number of Times the Fungus Species Isolated	
Species	**Nursery**	**Forest**
A	522	86
B	127	55
C	17	72
D	5	24
E	0	18
F	0	65
G	0	17
H	0	7

*The numbers represent the totals from the three nurseries and the three forests.

In a second experiment, the student grew Douglas fir seedlings for 1 year in a nursery. He randomly selected six 10 meter by 10 meter plots. He sprayed three of the plots with fungicide every two weeks, and he sprayed the other three with water only. Each plot contained 1,000 seedlings. The number of seedlings that survived after 1 year of each treatment is presented in Figure I.

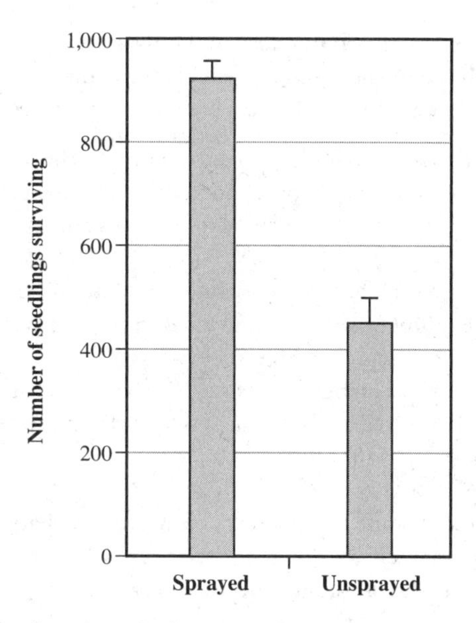

Figure I The survival of Douglas fir seedlings with regular pesticide application (sprayed) and without pesticide application (unsprayed)

The student then isolated fungi from the seedlings grown in the nursery plots described in the second experiment. The data are presented in Table II.

Table II Fungal Species Isolated from the Tissue of Pesticide Treated (Sprayed) and Untreated (Unsprayed) Douglas Fir Seedlings Planted in a Nursery

	Number of Times the Fungus Species Isolated	
Species	**Sprayed**	**Unsprayed**
A	519	91
B	132	47
C	19	68
D	4	33
E	0	12
F	0	6
G	0	0
H	0	0

497. Species richness is one component of biodiversity that expresses the number of different species in a particular environment. Which of the following is true of species richness based on the data in Table I?

(A) Species richness cannot be evaluated based on the information in Table I.
(B) There is no difference between the two environments because the same number of species is represented in both environments according to Table I.
(C) The forest has higher species richness because, according to Table I, there are 8 fungus species represented in the forest compared to 4 fungus species represented in the nursery.
(D) The nursery has higher species richness because fungus species *A* is more frequently isolated from the nursery than from the forest according to Table I.

498. Species evenness is one component of biodiversity that measures how similar the frequencies of different species are in an environment. Which of the following is true of fungus species evenness based on the data presented in Table I?

(A) Species evenness is higher in the nursery than in the forest because there are more of species *A* in the nursery than in the forest.
(B) Species evenness is the same in the forest and in the nursery because there is the same number of species in each environment.
(C) Species evenness is lower in the nursery than in the forest because the difference in frequencies of isolation was much greater in the nursery than in the forest.
(D) Species evenness is higher in the forest than in the nursery because there are more even numbers than odd numbers.

499. Based on Figure I, which of the following is an accurate statement regarding the effect of fungicide on plant survival?

(A) Fungicide application approximately doubled the number of seedlings that survived.
(B) Fungicide had no statistically significant effect on seedling survival.
(C) Fewer seedlings survived in the fungicide-treated plots.
(D) The experiment was not designed to answer this question.

500. Species richness and species evenness are two components of biodiversity. Species richness is high when the number of different species in a particular environment is high. Species evenness is high when many species are present in equal abundance. Which of the following is true about the data in Table II?

(A) Compared to the sprayed plots, biodiversity is high in the unsprayed plots because both species richness and species evenness are high.
(B) Compared to the sprayed plots, biodiversity is low in the unsprayed plots because both species richness and species evenness are low.
(C) Biodiversity is about the same in the sprayed and unsprayed plots because species richness is high and species evenness is low in the unsprayed plots.
(D) Biodiversity is about the same in the sprayed and unsprayed plots because species richness is high and species evenness is low in the sprayed plots.

501. Which of the following would provide the strongest evidence for a causal link between fungicide application and a decrease in endophyte biodiversity? (Refer to Table II.)

(A) finding that species *A* and species *B* inhabited tissues that were deeper in the plant and that species *E–H* were in tissues closer to the surface
(B) finding that species *A* and species *B* were fairly resistant to the fungicide used but that species *E–H* were highly susceptible
(C) finding that species *A* and species *B* rapidly colonized plants from airborne spores while species *E–H* were slower growing
(D) finding that the genetic diversity of the seedlings was lower in the unsprayed plots than in the sprayed plots

502. Endophytic diversity is consistently found to be low in 10-year-old Douglas fir trees that were attacked by a fungal pathogen that causes stem lesions called cankers. Based on this information, which of the following statements is most reasonable to make?

(A) A diverse endophytic community protects against canker-producing diseases.
(B) In response to a fungal attack, the plant produces compounds that kill endophytic fungi and reduce diversity.
(C) Endophytic fungi cause cankers.
(D) Endophytic diversity and pathogen infections are inversely correlated.

Questions 503–507

A particular fungus infects carpenter ants that live in the canopy of the rain forest. Fungal spores land on the ant and begin to secrete chitinases. The fungus eventually bores through the ant's exoskeleton and releases cells into the circulatory system. The fungus produces neurotransmitters that cause muscle contractions that cause the ant to fall to the forest floor where humidity is high. The fungus proliferates in the ant's body and causes it to climb vegetation and bite a leaf vein using its mandibles. The ant hangs from the leaf by its mandibles until the fungus consumes the ant and produces spores that travel on the wind.

503. **What is the effect of the chitinases in the situation described?**

(A) Fungal cell walls are composed of chitin. Fungi secrete chitinase to build cell walls.
(B) The chitinases interact with neurons in the ant to cause the ant's mandibles to attach irreversibly to a leaf vein.
(C) Chitinases are enzymes that break down chitin. Since the ant's exoskeleton is composed of chitin, this enzyme is important in allowing the fungus to penetrate into the ant's circulatory system.
(D) The chitinases are used by the ant as a defense against fungal intrusion since fungal cell walls are composed of chitin.

504. **Which of the following mechanisms best explains the muscle contractions?**

(A) The neurotoxin creates holes in the membranes of sensory neurons, disrupting action potentials.
(B) The neurotoxin increases blood filtration.
(C) The neurotoxin causes an increase in the level of glucose in the blood.
(D) The neurotoxin produced by the fungus has a similar shape and charge to the neurotransmitter that the ant uses to trigger muscle contraction.

505. **Researchers measured the average height above the forest floor of 100 ants attached to vegetation. The average was 25 cm. Which of the following would be the best experimental design to test the hypothesis that 25 cm above the forest floor provides the optimum conditions for spore production?**

(A) Inoculate 100 ants with the fungus in the laboratory, and provide them with artificial plants to climb. Measure the average height to which the ants climb.
(B) Collect data about the humidity, temperature, carbon dioxide, and light available at heights in the forest ranging from 0 cm to 100 cm.
(C) Collect 100 ants that have recently attached to leaves in the forest. Suspend the leaves at different heights ranging from 0 cm to 50 cm at 5 cm intervals. Measure the amount of spores produced from each ant.
(D) Sequence the genomes of the fungus and the ant species.

506. Why does the ant bite a leaf vein?

(A) The ant is hungry from the exertion caused by the disease and is attempting to obtain nourishment to fight the pathogen.
(B) The fungus causes the ant to perform this behavior so that the fungus can use the resources the ant obtains from the plant sap to grow.
(C) The fungus causes the ant to perform this behavior because biting into a leaf vein securely anchors the ant to the leaf while the fungus digests the ant's body.
(D) This is a vestigial response left over from the coevolution of the host and the parasite that no longer has a function.

507. Which of the following experiments would best test the hypothesis that the infected ant increases the survival of the hive by not returning to the hive to die?

(A) Select 20 ant hives. Place newly dead, infected ants into half of them and newly dead, uninfected ants into the other half. Measure the rates of infection in both halves.
(B) Observe 100 hives to determine what ants do with the bodies of dead ants.
(C) Measure the genetic relatedness of ants in a hive by sequencing polymorphic DNA in 100 individuals in each of 20 hives.
(D) Collect different strains of the fungus from infected ants from a broad geographical area. Infect other ants with these different strains in the laboratory to see if the newly infected ants display behavioral variability depending on the strain of the fungus.

Questions 508–512

Students placed 10 plant seeds in moist soil in chambers lit only with differently colored light-emitting diodes (LEDs). The LEDs were adjusted until the intensity of light was the same in each chamber. Seeds were permitted to germinate and grow for 7 days. Students measured the height, mass, number of leaves, and leaf size after 7 days of growth. They also noted the color of the leaves. The data, presented in the table below, are averages of the 10 seeds.

LED Color	Green	Blue
Height (cm)	12	4.3
Dry Mass (g)	0.09	0.41
Number of Leaves	2	6
Leaf Size (cm)	0.1	1.4
Leaf Color	Yellowish white	Green

508. Why did the students adjust the intensity of the LEDs?

(A) Plants grown at lower light intensities grow more slowly.
(B) LEDs vary widely in the intensity of light produced.
(C) They were adjusted to simulate sunlight.
(D) Since the students wanted to measure the effect of the wavelength of light, it was necessary to control other variables.

509. Which of the following is a reasonable interpretation of the height data?

(A) Plants are green because they absorb green light to use in photosynthesis. The increased photosynthesis led to taller plants.
(B) The differences in height were due to random chance.
(C) Plants grown under the green light were not able to absorb the light and use it for photosynthesis. The tall growth suggests that the plants were elongating in order to seek good-quality light.
(D) The differences in height were due to different mutation rates under different colors of light.

510. Which of the following represents a positive control that the students should have included in their experiment?

(A) adding full-spectrum light at the same intensity as the LEDs
(B) growing plants in complete darkness
(C) adding a red light at the same intensity as the LEDs
(D) adding an orange light at the same intensity as the LEDs

511. Which of the following represents a negative control that the students should have included in their experiment?

(A) adding full-spectrum light at the same intensity as the LEDs
(B) growing plants in complete darkness
(C) adding a red light at the same intensity as the LEDs
(D) adding an orange light at the same intensity as the LEDs

512. Which of the following would be the best measure of the amount of carbon fixed by the plants?

(A) height, because carbon fixation can only make plants grow taller
(B) leaf color, because the color of the leaf is the same as the color of light that the plant absorbs and light drives carbon fixation
(C) dry mass, because all of the carbon in a plant comes from the air
(D) the change in the level of carbon dioxide in the room

Questions 513–515

The graph below shows three different types of survivorship curves. Different species of organisms have different patterns of survival over their lifetimes.

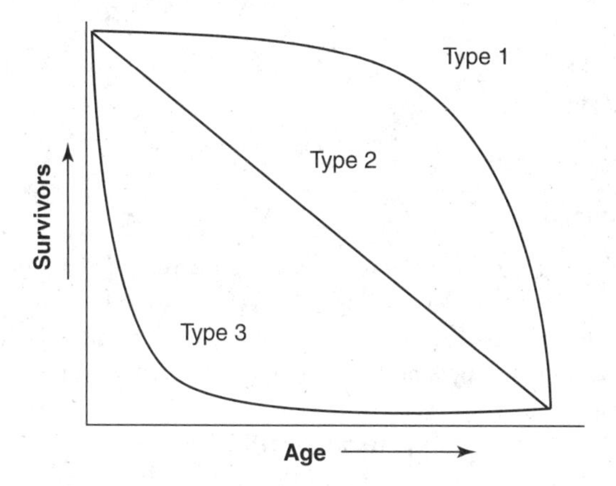

513. **Approximately 1 out of every 1,000 green sea turtle eggs that are laid will reach adulthood. Adults can live for 100 years. According to the graph, which type of survivorship curve do green sea turtles best match?**

(A) Type 1
(B) Type 2
(C) Type 3
(D) Green sea turtles do not follow any of the survivorship curves shown on this graph.

514. **Giant sequoia (*Sequoiadendron giganteum*) trees produce millions of very small seeds. Very few of those seeds reach adulthood. According to the graph, which type of survivorship curve do giant sequoia trees best match?**

(A) Type 1
(B) Type 2
(C) Type 3
(D) Giant sequoia trees do not follow any of the survivorship curves shown on this graph.

515. Elephants have relatively few young, have a gestation period of approximately 2 years, and provide extensive care for their offspring. As a result, a large percentage of young elephants live to adulthood. The death rate becomes high in old elephants. According to the graph, which type of survivorship curve do elephants best match?

(A) Type 1
(B) Type 2
(C) Type 3
(D) Elephants do not follow any of the survivorship curves shown on this graph.

Questions 516–518

The exponential growth equation (the exponential model) and the logistic growth equation (the logistic model) are used to model population growth. These equations, which you can find on the Reference Tables at the end of this book, are also reproduced below.

Exponential Growth Equation

$$\frac{dN}{dt} = r_{max}N$$

Logistic Growth Equation

$$\frac{dN}{dt} = r_{max}N\frac{K - N}{K}$$

Note that, in these equations, N is the population size, t is time, r_{max} is the maximum per capita growth rate of the population, and K is the carrying capacity.

The term $\frac{dN}{dt}$ is the population growth rate.

516. Which of the following is accurate with respect to the exponential growth equation?

(A) As the population size (N) reaches the carrying capacity (K), the population growth rate approaches zero.
(B) As the population size (N) increases, the population growth rate increases indefinitely.
(C) The intrinsic rate of increase (r_{max}) gets smaller as the population size (N) increases.
(D) The carrying capacity (K) decreases as the population size (N) increases.

517. Which of the following is accurate with respect to the logistic growth equation?

(A) As the population size (*N*) reaches the carrying capacity (*K*), the population growth rate approaches zero.
(B) As the population size (*N*) increases, the population growth rate increases indefinitely.
(C) The intrinsic rate of increase (r_{max}) gets smaller as the population size (*N*) increases.
(D) The carrying capacity (*K*) decreases as the population size (*N*) increases.

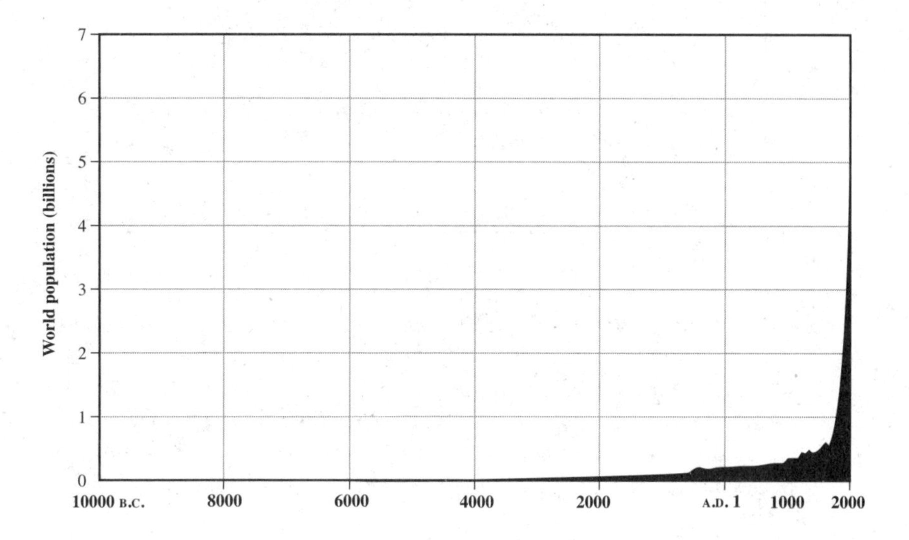

518. The graph above depicts the global human population size since 10000 B.C. Which of the growth models presented most accurately represents human population growth?

(A) the exponential model
(B) the logistic model
(C) The data presented match both the exponential model and the early part of the logistic model.
(D) The data presented do not match the exponential model or the logistic model.

Questions 519–524

A group of students grew cultures of *Paramecium aurelia* and *Paramecium caudatum* in separate cultures and in mixed cultures. The students' results are presented in Figure I.

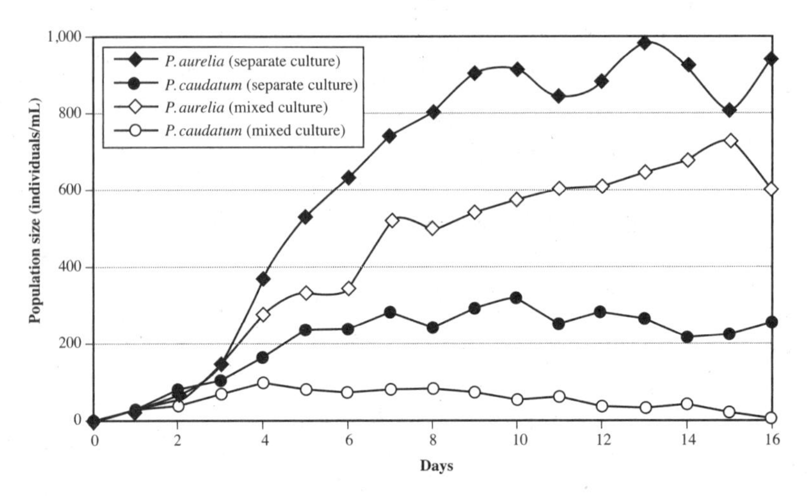

Figure I The growth of *P. aurelia* and *P. caudatum* in separate and mixed cultures

A second group of students performed a similar experiment, comparing separate and mixed cultures of paramecia. Like the first group, they also used *P. aurelia*, but they used *Paramecium burseria* as their second species instead of *P. caudatum*. They generated the results shown in Figure II.

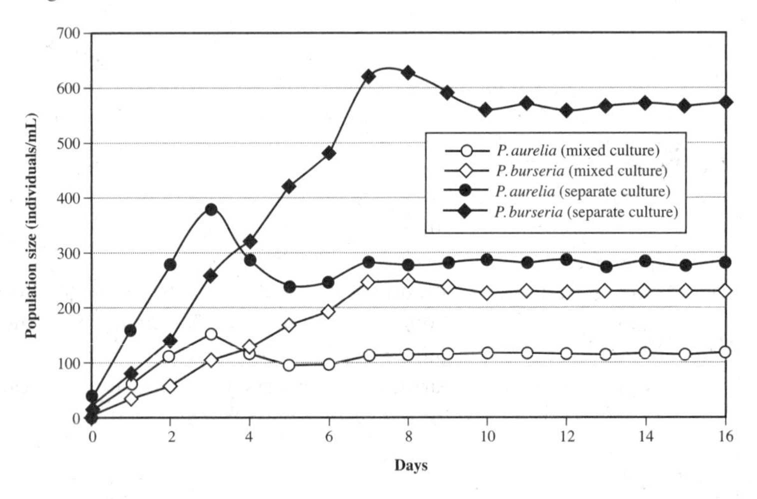

Figure II The growth of *P. aurelia* and *P. burseria* in separate and mixed cultures

519. Based on the data in Figure I, which of the following is true concerning mixed and separate cultures of paramecia?

(A) Paramecia grow better in separate cultures than in mixed cultures.
(B) Paramecia grow better in mixed cultures than in separate cultures.
(C) Paramecia grow equally well in separate cultures and in mixed cultures.
(D) There is not enough information to determine the answer.

520. Which of the following were the students in both groups likely investigating?

(A) the effect of parasitism on population growth
(B) the effect of competition on population growth
(C) the effect of predation on population growth
(D) the evolution of species of paramecia

521. Although the *Paramecium* species were fed the same amount of bacteria in separate cultures, *P. aurelia* had a larger maximum population size than *P. caudatum* (see Figure I). Which of the following might explain this observation?

(A) *P. aurelia* individuals are larger and consume more resources per individual than do *P. caudatum* individuals.
(B) *P. caudatum* individuals are larger and consume more resources per individual than do *P. aurelia* individuals.
(C) *P. aurelia* individuals have a higher reproductive rate and compete better for food than do *P. caudatum* individuals under the conditions in the test tube environment.
(D) *P. aurelia* individuals are more evolutionarily advanced than *P. caudatum* individuals.

522. In mixed cultures, both species of paramecia had access to the same food, yet *P. caudatum* eventually disappeared from the mixed culture when its population size reached 0 on day 16 (see Figure I). Which of the following does this illustrate?

(A) the idea of a keystone species, where the removal of one species reduces diversity in the community
(B) the concept that energy is lost as the trophic level increases
(C) the competitive exclusion principle, which states that if two species inhabit the same niche, one will go extinct due to differences in the ability to access the resources
(D) the principle of niche partitioning, which states that two species can coexist with the same resource if they have different strategies for exploiting it

523. Which of the following is an accurate interpretation of the data in Figure II?

(A) Both *P. aurelia* and *P. burseria* grew larger populations in separate cultures than in mixed cultures.
(B) Although *P. burseria* grew better in separate culture than it did in mixed culture, *P. aurelia* grew better in mixed culture than it did in separate culture.
(C) There was almost no difference between the growth of the two species in mixed cultures and the growth of the two species in separate cultures.
(D) *P. aurelia* was the dominant species in mixed culture, but it did not perform as well in separate culture.

524. In both experiments shown in Figures I and II, two species of paramecia had access to the same food. In Figure I, *P. aurelia* and *P. caudatum* were not able to coexist in mixed cultures. *P. caudatum* population levels in mixed cultures fell to 0 after 16 days. By contrast, in Figure II, *P. burseria* and *P. aurelia* were able to coexist in mixed cultures. When students carefully examined the test tubes from the experiment shown in Figure II, they discovered that, in mixed cultures, *P. burseria* were found more often at the bottom of the test tube and *P. aurelia* were found more often at the top of the test tube. In mixed cultures of *P. aurelia* and *P. caudatum* from the experiment shown in Figure I, however, both species were found throughout the tube during the first three days of the experiment. Which of the following does the coexistence of *P. burseria* and *P. aurelia* from the experiment shown in Figure II illustrate?

(A) the idea of a keystone species, where the removal of one species reduces diversity in the community
(B) the concept that energy is lost as the trophic level increases
(C) the competitive exclusion principle, which states that if two species inhabit the same niche, one will go extinct due to differences in the ability to access the resources
(D) the principle of niche partitioning, which states that two species can coexist with the same resource if they have different strategies for exploiting it

Questions 525–527

The graph below shows carbon dioxide concentrations measured at the Mauna Loa Observatory in Hawaii. The vertical lines represent January 15 of the year indicated. The value for the 15th of every month after January is also plotted.

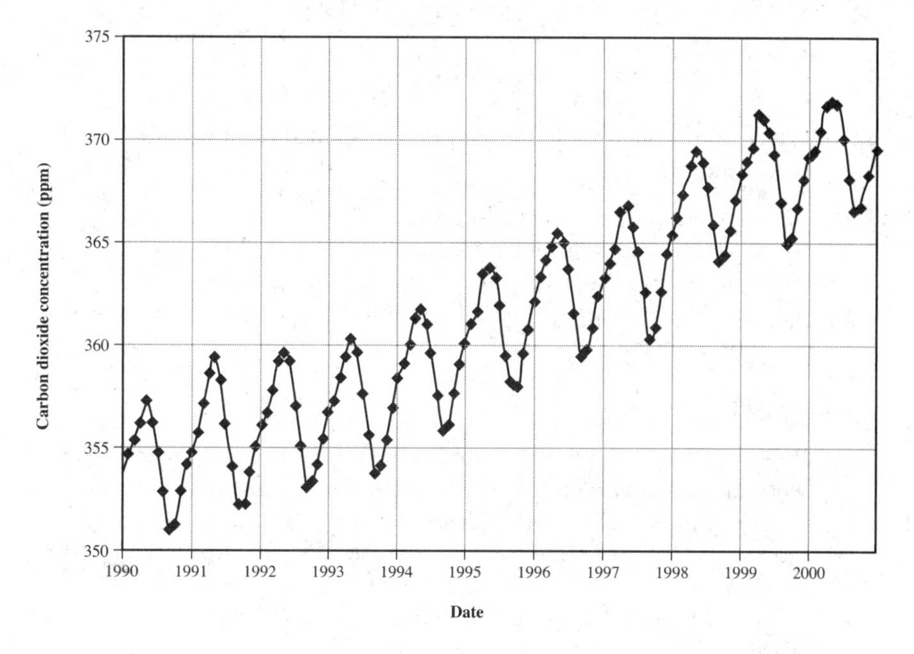

525. **Which of the following best explains the annual rise and fall of the carbon dioxide concentration?**

 (A) The carbon dioxide concentration rises in the summer months because people drive cars more in the summer. This additional fossil fuel consumption releases more carbon dioxide in the summer.
 (B) The carbon dioxide concentration rises in the summer months because people burn fuel to run air conditioners. When they turn their air conditioners off in the autumn, the carbon dioxide concentration decreases.
 (C) The carbon dioxide concentration rises through the winter months because there is more respiration occurring than photosynthesis. In the summer, photosynthetic carbon fixation decreases the carbon dioxide concentration.
 (D) The carbon dioxide concentration decreases in the summer because the ocean warms and is capable of dissolving more gas than in the winter when the water is cold.

526. Which of the following statements can be made based on the data presented in the graph?

(A) Earth's temperature increased during the 1990s.
(B) The carbon dioxide concentration measured at Mauna Loa Observatory increased by approximately 15 parts per million (ppm) during the 1990s.
(C) Earth's temperature decreased during the 1990s.
(D) The carbon dioxide concentration doubled during the 1990s due to the increased use of nuclear power.

527. Which of the following would slow the rate of increase in atmospheric carbon dioxide?

(A) an increase in the rate of photosynthesis
(B) a decrease in the rate of respiration
(C) an increase in the carbon dioxide dissolved in the ocean
(D) All of these would decrease the rate at which atmospheric carbon dioxide increases.

Questions 528–530

Rubisco is the primary enzyme that is responsible for turning inorganic carbon dioxide into organic molecules that are useful to life (carbon fixation). This enzyme can also use oxygen as a substrate. In that case, it leads to a metabolic pathway called photorespiration in which carbon dioxide is released instead of fixed. One group of plants performs a slightly different form of metabolism in which they use an enzyme called PEP carboxylase that is much more selective for carbon dioxide than rubisco is. These plants perform C-4 metabolism, which costs more energy per atom of carbon fixed than does C-3 metabolism, which relies on only rubisco.

Plants exchange gases with the environment through stomata. When water is abundant, stomata are generally kept wide open. However, they are closed when water is scarce. This closure prevents the exchange of oxygen and carbon dioxide in addition to slowing transpirational water loss.

528. Which of the following would be the best environment for C-3 plants that use only rubisco to fix carbon?

(A) a low carbon dioxide, high oxygen atmosphere with limited water in the soil
(B) a high carbon dioxide, low oxygen atmosphere with abundant water in the soil
(C) a low carbon dioxide, high oxygen atmosphere with abundant water in the soil
(D) a high carbon dioxide, low oxygen atmosphere with limited water in the soil

529. **Under what conditions do C-4 plants have a selective advantage over C-3 plants?**

(A) a low carbon dioxide, high oxygen atmosphere with limited water in the soil
(B) a high carbon dioxide, low oxygen atmosphere with abundant water in the soil
(C) a low carbon dioxide, high oxygen atmosphere with abundant water in the soil
(D) a high carbon dioxide, low oxygen atmosphere with limited water in the soil

530. **Which of the following is likely to be a consequence of higher global carbon dioxide levels?**

(A) C-3 plants will have a selective advantage over C-4 plants because there will be less need for the extra ATP cost per carbon atom fixed.
(B) C-4 plants will have a selective advantage over C-3 plants because C-4 plants are more efficient at fixing carbon.
(C) C-4 plants will be selected against because of the toxic nature of carbon dioxide.
(D) Both types of plants will show decreased growth due to the degradation of the environment.

Questions 531–534

The figure shown below depicts a portion of a sequence alignment of the rubisco large subunit from two vascular plants (*Arabidopsis thaliana* and *Glycine max*) and from two cyanobacteria (*Synechocystis* sp. PCC 6803 and *Microcystis aeruginosa*).

					90										100										110					
Arabidopsis/1-479	h	i	e	p	v	p	g	e	e	t	q	f	i	a	y	v	a	y	p	l	d	l	f	e	e	g	s	v	t	n
Glycine max/1-475	g	l	e	p	v	a	g	e	e	n	q	y	i	a	y	v	a	y	p	l	d	l	f	e	e	g	s	v	t	n
Synechocystis/1-470	d	l	e	a	v	p	n	e	d	n	q	y	f	a	f	i	a	y	p	l	d	l	f	e	e	g	s	v	t	n
Microcystis/1-471	d	i	e	p	v	p	n	e	d	n	q	f	f	c	f	v	a	y	p	l	d	l	f	e	e	g	s	v	t	n
	A																B													

531. **Which type of macromolecule is made of the type of monomers indicated by each letter in the sequences shown?**

(A) proteins, which are composed of amino acids
(B) nucleic acids, which are composed of nucleotides
(C) carbohydrates, which are composed of monosaccharides
(D) lipids, which are composed of ribosomes

532. **Which of the following statements is the most accurate comparison of the likely functions of the two regions indicated by the letters A and B?**

(A) Region A is more highly conserved than region B. Therefore, region A is more likely to have an important role in the function of rubisco.
(B) Region B is more highly conserved than region A. Therefore, region B is more likely to have an important role in the function of rubisco.
(C) Region A is more variable than region B. Therefore, region A is likely to comprise the binding site for carbon dioxide.
(D) Region B is less variable than region A. Therefore, region B is likely to function as a transmembrane domain for rubisco.

533. **Which of the following is the most accurate statement concerning the evolution of the two regions?**

(A) Region A has a higher number of differences because it has had longer to evolve than region B.
(B) Although these four sequences evolved from a single sequence in a common ancestor, region A changed more because it experienced less selective pressure against mutations than did region B.
(C) Region B was likely shared by the common ancestor of the four species shown, but region A is probably the result of horizontal gene flow because the similarity among species is so low.
(D) Since region B is more similar for the four organisms, it represents a more recent addition to rubisco than region A.

534. **Which of the following would be the best approach to infer the phylogenetic history of these four species using the sequences presented?**

(A) Compare the number of letters in each region. The fewer the number of letters, the older the sequence is.
(B) Compare the number of similarities in region B. The more similarities there are, the more distantly related the four species are.
(C) Compare the number of differences in region A. The fewer the number of differences, the more closely related the four species are.
(D) There is no way to infer phylogenetic relatedness using this information.

---

535. **Some people have attempted to link military use of sonar with whale beachings. Which of the following experiments would provide the strongest evidence that sonar causes changes in the behavior of whales?**

(A) discovering that military sonar uses the same wavelengths that killer whales use for echolocation
(B) observing military ships near the sites of whale beachings
(C) discovering high levels of nitrogen gas in the blood of beached whales
(D) recording the diving behavior of whales in the presence of and in the absence of sonar

Questions 536–541

In an experiment on the effect of sonar on the behavior of blue whales, researchers attached a sensor to individual whales and measured the depth to which the whales dove. After 150 minutes, the experimenters turned sonar on for 50 minutes. The results below are typical of the responses seen.

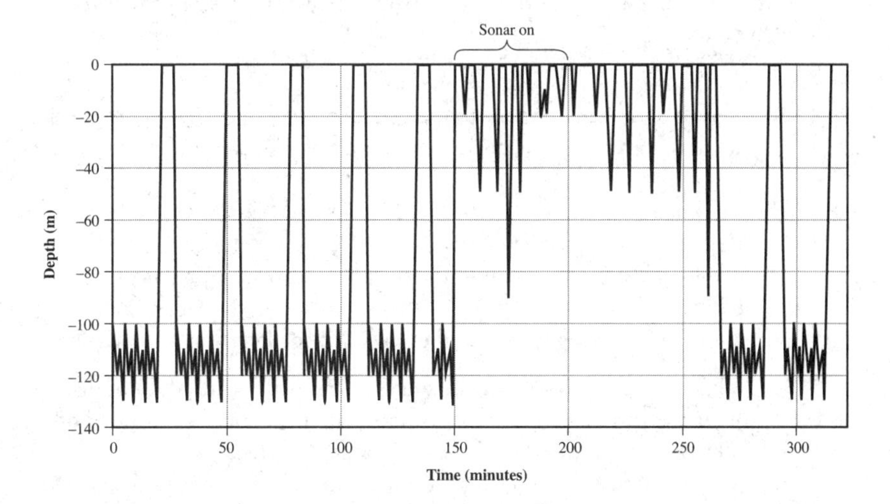

536. **Why did the researchers record data for 150 minutes before turning on the sonar?**

(A) They wanted to let the whales get used to having the data recording device on.
(B) They wanted to make sure that the sonar equipment was working properly.
(C) They wanted to observe normal behavior so that they would have a baseline against which to compare the data.
(D) They wanted to wait for the whales to surface so that the whales could hear the sonar.

537. **Blue whales dive to obtain food. Which of the following best describes the location of their food?**

(A) at the surface of the water
(B) from 0 to 100 meters deep
(C) between 100 and 130 meters deep
(D) below 130 meters deep

538. Which of the following best matches the duration of a typical dive?

(A) approximately 5 minutes
(B) approximately 10 minutes
(C) approximately 25 minutes
(D) more than 35 minutes

539. What was the purpose of the whales periodically returning to the surface?

(A) Since whales are mammals, they have lungs and need to surface to obtain oxygen for cellular respiration.
(B) Whales periodically return to the surface to avoid getting decompression sickness, known as "the bends," as a result of dissolved nitrogen coming out of solution.
(C) The whales are chasing krill, small crustaceans that need to surface periodically to ventilate their gills.
(D) Whales periodically return to the surface to check for predators.

540. Which of the following best describes the effect of sonar on blue whales?

(A) Whales surfaced soon after the sonar was turned on and did not dive to the same depth again for about an hour after the sonar was turned off.
(B) Whales thought they heard orcas, a dangerous predator.
(C) Whales confused the sonar with the sounds of their prey, krill, and surfaced to chase the source of the sound.
(D) The whales dove deep immediately after hearing the sonar.

541. Based on this experiment, what negative impact would sonar have on blue whales in nature?

(A) Sonar would prevent blue whales from diving deep to obtain food, which the animals need in order to stay alive, grow, and reproduce.
(B) Sonar would be a minor inconvenience for blue whales since they can feed at other depths.
(C) Sonar would cause whales to beach themselves.
(D) It is impossible to draw conclusions from this experiment.

Questions 542–550

Dwindling supplies of fossil fuels are driving the search for renewable sources of energy. Biofuels are a promising area of research. Hybrids among poplar trees (trees in the genus *Populus*) are one group of organisms that are being considered as potential biofuels because of their rapid growth rate (up to 3 meters per year under ideal conditions). Insect pests and fungal pathogens represent a serious challenge to realizing the potential of hybrid poplar trees as viable biofuels. *Septoria musiva* is a fungal pathogen that grows on poplar trees, causing a leaf spot and canker disease. A canker is a sunken area on the stem of a tree. Observations by growers have suggested that the disease is more severe during extended periods of dry weather (drought). Researchers interested in the effects of drought stress on the ability of poplar trees to resist the fungus grew two clones of poplar. One clone was grown in drought (stressed) conditions. The other was grown in well-watered conditions. Researchers then placed either sterile fungal growth media or growth media containing the fungus into wounds made by removing a leaf from the trees. After 80 days of growth, the researchers measured the extent of the infections in terms of the length of the lesion (canker) produced and the estimated percentage of the stem circumference that was affected. Figure I shows the canker length and the percentage of stem girdling for NM6 and NE308 trees. Each tree clone was exposed to sterile media and media with the fungus. Each was also grown under both stressed and watered conditions. This resulted in 8 different groups (represented by the 8 different bars shown in Figure I).

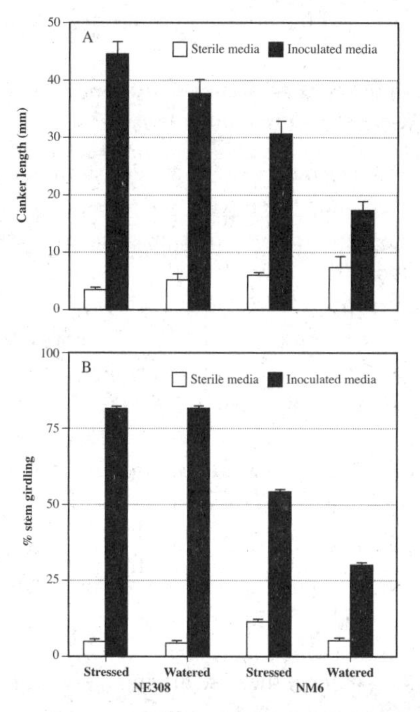

Figure I The effect of sterile media and inoculated media on canker length (mm) and the percentage of stem girdling for stressed and watered hybrid poplar trees

After 80 days of growth, the researchers examined each tree for the presence or absence of fungal spores. There were no spores on the trees exposed to sterile media, so they were omitted from the analysis. The data are presented in Table I.

Table I The Absence and Presence of Spores Observed on Cankers of Poplar Trees Inoculated with *Septoria musiva*

		Conidia (Spores)	
Water Treatment	**Clone**	**Absent**	**Present**
Watered	NE308	15	30
Watered	NM6	41	4
Stressed	NE308	20	25
Stressed	NM6	33	12

542. What was the purpose of applying sterile fungal growth media to wounds?

(A) This was to encourage the growth of other fungi.
(B) This was the control treatment. Any effect seen in inoculated wounds but not in the control could be attributed to *S. musiva*.
(C) This was to help wound healing.
(D) This was to prevent the growth of *S. musiva*.

543. Which of the following is an accurate interpretation of the effect of drought stress with respect to canker length?

(A) Drought stress led to decreased canker length in both poplar clones when they were inoculated with *S. musiva*.
(B) Drought stress had no effect on canker length for either poplar clone when infected with *S. musiva*.
(C) Drought stress led to increased canker length for NM6 but not for NE308 when they were inoculated with *S. musiva*.
(D) Drought stress led to increased canker length in both poplar clones when they were inoculated with *S. musiva*.

544. Which of the following is an accurate interpretation of the effect of drought stress with respect to the percentage of stem girdling?

(A) Drought stress led to an increased percentage of stem girdling in NE308 but not in NM6 when both clones were inoculated with *S. musiva*.
(B) Drought stress led to an increased percentage of stem girdling in NM6 but not in NE308 when both clones were inoculated with *S. musiva*.
(C) Drought stress had no effect on the percentage of stem girdling for either clone when inoculated with *S. musiva*.
(D) Drought stress led to a decreased percentage of stem girdling in both clones when they were inoculated with *S. musiva*.

545. Table I only shows data from inoculated plants because fungal spores were never observed on trees exposed to sterile media, as stated in the passage. What is the significance of the fact that there were no fungal spores seen on trees exposed to sterile media?

(A) It suggests that trees exposed to sterile media were highly resistant to *S. musiva.*
(B) It demonstrates that these trees were under significant drought stress.
(C) If spores were on these trees, it would indicate that there was some other source of the fungus other than the treatment applied.
(D) If there were no spores on these trees, it would indicate that each tree's immune system killed the fungus.

546. Which of the following is an accurate interpretation of the results of the fungal spore observations seen in Table I?

(A) The fungus was less likely to produce spores on NM6 than on NE308.
(B) The fungus was more likely to produce spores on NM6 than on NE308.
(C) There was no difference between NM6 and NE308 with respect to the number of fungal spores produced.
(D) It is not possible to accurately interpret the results seen in Table I.

547. After reviewing the passage as well as the data presented in Figure I and Table I, which of the following conclusions about the two different clones do the data support?

(A) NM6 is more resistant to the fungus in terms of canker growth, but NE308 is more resistant in terms of spore production.
(B) There are no significant differences between the clones in terms of canker length, percentage of stem girdling, and the production of fungal spores.
(C) NM6 is more resistant to the fungus than NE308 in terms of canker length, percentage of stem girdling, and the production of fungal spores.
(D) NE308 is more resistant to the fungus than NM6 in terms of canker length, percentage of stem girdling, and the production of fungal spores.

548. After reviewing the passage as well as the data presented in Figure I and Table I, which of the following conclusions about drought stress do the data support?

(A) Drought stress had no effect on fungal resistance in either clone.
(B) Drought stress had no effect on the ability of NE308 to resist the fungus, but it increased resistance in NM6.
(C) Drought stress increased the ability of both clones to resist the fungus in terms of canker length and the incidence of spores. This effect was also seen for NM6 in terms of the percentage of stem girdling.
(D) Drought stress decreased the ability of both clones to resist the fungus in terms of canker length. This effect was also seen for NM6 in terms of the percentage of stem girdling and the incidence of spores.

549. **Which of the following is the most likely explanation for why there was no apparent effect of stress on the percentage of stem girdling for NE308 plants that were inoculated?**

(A) As the stems became more colonized, the fungus had more resources. Therefore, colonizing the plant became easier.
(B) The trees were overwhelmed by fungal colonization and no longer produced compounds that inhibited fungal growth.
(C) After completely colonizing the circumference of the stem, the fungus could no longer grow in that dimension and the canker could only grow in length.
(D) None of these explanations are consistent with the data.

550. **Based on the research presented, which of the following would be the best advice the researchers could give to hybrid poplar growers in order to limit the damage caused by *S. musiva*?**

(A) Plant clone NE308, and do not irrigate during a drought.
(B) Plant a mixture of the two clones, and irrigate during a drought.
(C) Plant clone NM6, and do not irrigate during a drought.
(D) Plant clone NM6, and irrigate during a drought.

Questions 551–552

Psoriasis is a dermatological condition in which patches of dead cells remain attached to the skin and cause itching. A spa in the mountains of Turkey has achieved success in treating psoriasis with a treatment called ichthyotherapy. Patients spend time lounging in natural ponds that contain "doctor fish" (*Garra rufa*), which eat dead skin cells. Afterward, the patients lie in the sun.

551. **Researchers attempting to re-create this effect in a clinic exposed 65 patients with psoriasis to baths containing doctor fish for 30 minutes followed by 30 minutes in a UVB sunbed. They reported a 75% reduction in psoriasis symptoms. Which of the following is a valid criticism of the experimental design?**

(A) The sample size was not large enough.
(B) Baths and sunbeds are not natural.
(C) There was no control treatment.
(D) No dependent variable was measured.

552. **An ichthyotherapy spa that opened in Wisconsin was shut down due to concerns about the potential for the transmission of skin pathogens from patient to patient. Which of the following would be the most convincing evidence that the transmission of a pathogen had occurred?**

(A) patients leaving the spa with uncured psoriasis
(B) isolating algae from the spa water
(C) patients developing new skin diseases that other patients had
(D) discovering that the same fish were used on multiple patients

Questions 553–556

Cleaner fish participate in a mutualistic relationship with larger fish by removing parasites from larger fish of a different species. Ecologists interested in the importance of this relationship to reef communities undertook a study of 18 patch reefs in which they removed cleaner fish from 9 randomly selected reefs and left the other 9 patch reefs as controls. The ecologists measured the abundance of different sizes of fish in the presence and absence of cleaner fish (see Figure I). They also measured species abundance and species richness in the presence and absence of cleaner fish (see Figure II).

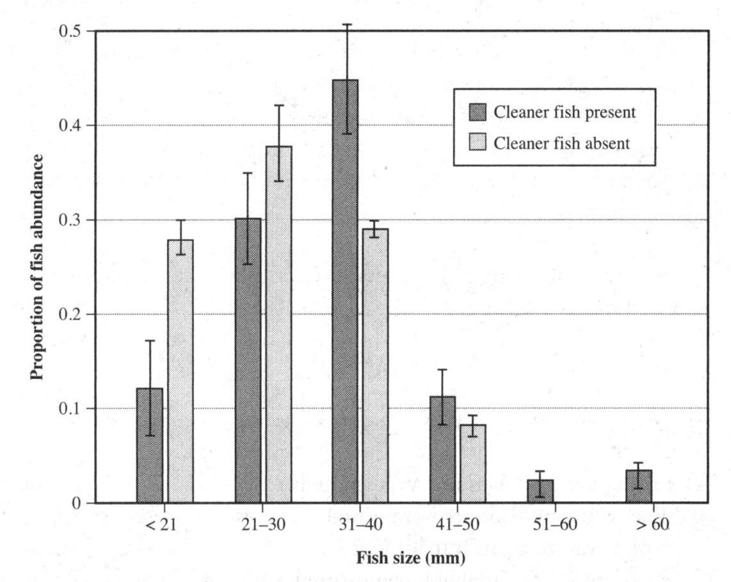

Figure I Fish size distribution in the presence and absence of cleaner fish

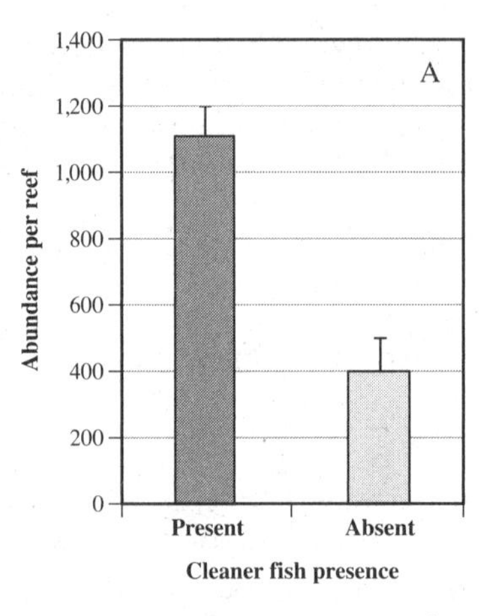

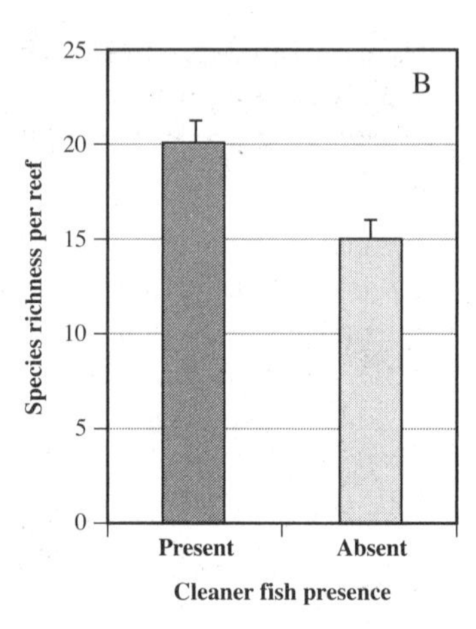

Figure II Species abundance and species richness in the presence and absence of cleaner fish

553. **Which of the following is an accurate interpretation of the data presented in Figure I?**

(A) The reefs with the cleaner fish present had a larger proportion of bigger fish.
(B) There was no significant difference between the reefs with and without cleaner fish.
(C) The reefs with the cleaner fish absent had a larger proportion of bigger fish.
(D) The reefs with the cleaner fish present had a greater abundance of all sizes of fish.

554. **Which of the following is a reasonable hypothesis that might explain the results shown in Figure I?**

(A) Fish inhabiting the reefs with the cleaner fish present lived longer and therefore attained greater sizes.
(B) Fish inhabiting the reefs with the cleaner fish present grew faster because the parasites were not an additional resource sink.
(C) Fish inhabiting the reefs with the cleaner fish absent produced more offspring than those with the cleaner fish present, resulting in a greater proportion of small fish.
(D) All of these are reasonable hypotheses that explain the data presented.

555. Based on Figure II, which of the following accurately compares the abundance of fish in the presence and in the absence of the cleaner fish?

(A) There was no difference in fish abundance among these treatments.
(B) The reefs with the cleaner fish absent did not support any other fish species.
(C) The reefs with the cleaner fish present supported less than half as many fish as those with the cleaner fish absent.
(D) The reefs with the cleaner fish present supported more than twice as many fish as those with the cleaner fish absent.

556. Which of the following is an accurate interpretation of the graphs in Figure II?

(A) Both the abundance of fish and the species richness were significantly lower in the presence of the cleaner fish.
(B) Both the abundance of fish and the species richness were significantly higher in the presence of the cleaner fish.
(C) The abundance of fish was higher in the presence of the cleaner fish, but the species richness was lower in the presence of the cleaner fish.
(D) There was no difference in the abundance of fish or in the species richness for either treatment.

Questions 557–561

Students inoculated 5 mL of bacterial growth medium with the bacterium *Pseudomonas fluorescens* in test tube microcosms. They added the predator *Tetrahymena thermophila* to half of the tubes. The students noted the appearance of the microcosm, the number of different types of colonies that formed when plated on agar media, and the density of the cultures after 10 days. Their data are presented in Table I.

Table I The Comparison of Microcosms With and Without Predators After 10 Days

	Bacteria-Only	**Bacteria + Predator**
Description of the Microcosm	Cloudy suspension	Film on the surface and cottony spread on the bottom of the culture
Number of the Types of Bacteria Present Based on the Colony Shape	1	3
Bacterial Density	1.3×10^9 cells/mL	2.7×10^8 cells/mL

Students identified three different types of bacterial colonies: smooth, fuzzy edge, and wrinkled. Bacteria-only microcosms produced only smooth colonies. All three types of bacterial colonies were found in the bacteria + predator microcosms.

Students then conducted a second experiment in which they inoculated fresh media with the bacteria isolated from the bacteria + predator microcosms. None of these new microcosms contained predators. The students observed the growth in the media over a period of 10 days and then reisolated bacteria from the cultures. The data are summarized in Table II.

Table II The Comparison of Simple Microcosms Inoculated with Different Types of Bacteria After 10 Days

	Inoculated with Smooth Bacteria	**Inoculated with Fuzzy Edge Bacteria**	**Inoculated with Wrinkled Bacteria**
Description of the Microcosm	Cloudy	Top growth at first, then cloudy	Bottom growth at first, then cloudy
Smooth Colonies Reisolated	35	28	32
Fuzzy Colonies Reisolated	0	2	0
Wrinkled Colonies Reisolated	0	0	1

557. **Based on the data in Table I, which of the following explains the difference in the number of the types of bacteria present in each of the microcosms?**

(A) In microcosms without predators, mutant bacteria that were able to escape predation were selected for.
(B) In microcosms with predators, mutant bacteria that were able to escape predation were selected for.
(C) In microcosms with predators, predation caused mutations to form.
(D) In microcosms without predators, bacteria were free to diverge and did so readily.

558. Based on the data in Table I, which of the following explains the difference in the bacterial density in the two different microcosms?

(A) Fewer bacteria were in the microcosms with predators because the predators consumed some of the bacteria.
(B) Fewer bacteria were in the microcosms with predators because there was more competition for resources in the microcosms with predators.
(C) More bacteria were in the microcosms with predators because the bacteria were parasitic on the predators.
(D) The bacterial density in each microcosm was the same.

559. Which of the following best explains the source of the genetic material for the bacteria displaying different colony morphologies in the first experiment described?

(A) horizontal gene transfer from the predator
(B) contamination from the students' hands
(C) random mutation that led to new phenotypes
(D) natural selection that introduced new genetic material

560. In the second experiment, all three cultures wound up with cloudy growth similar to the growth of the wild-type bacteria without the predator in the first experiment. Which of the following best explains this?

(A) In an environment with no predator, there is a selection pressure against the wild-type.
(B) Without a predator, bacteria have no need to produce antiherbivory compounds.
(C) In an environment with no predator, there is a selection pressure against the mutant types.
(D) In the absence of a predator, all bacteria live. This leads to a crowded and therefore cloudy environment.

561. In the second experiment, cultures inoculated with fuzzy edge and wrinkled bacterial colonies produced smooth colonies. Which of the following best explains why this happened?

(A) Without a predator present, different genes in the bacteria were turned on, leading to the smooth phenotype.
(B) In the absence of selective pressure from a predator, mutants that reverted to wild-type had an advantage and proliferated.
(C) The bacteria took up genetic information from dead, smooth bacteria and expressed those genes.
(D) None of these are reasonable explanations for this phenomenon.

Questions 562–564

A biofilm is a community of microorganisms that grow in a particular environment. Plaque is a bacterial biofilm that forms on teeth. Among the bacteria that form plaque are *Streptococcus mutans* and *S. sanguinus.* Enamel is a hard layer on the surface of teeth that is primarily composed of calcium phosphate. It can be broken down under acidic conditions. *S. mutans* grows best in small depressions in the teeth, toward the tooth side of the biofilm where oxygen is scarce. *S. sanguinus* grows best near the top of the biofilm where oxygen levels are high.

562. **Which of the following best explains why *S. mutans* is able to dissolve enamel and cause dental caries (cavities) while *S. sanguinus* is not?**

(A) *S. mutans* populations grow larger than *S. sanguinus* populations in the presence of oxygen. High populations of *S. mutans* lead to apoptosis, which releases oxidases that degrade enamel.
(B) *S. mutans* produces hydrochloric acid as a result of the hydrolysis of water, while *S. sanguinus* performs lactic acid fermentation. Hydrochloric acid dissolves enamel, but lactic acid does not.
(C) *S. mutans* produces carbon dioxide as a result of aerobic respiration of sugars, while *S. sanguinus* performs anaerobic fermentation. Carbon dioxide dissolves enamel.
(D) *S. mutans* produces lactic acid as a result of anaerobic fermentation of sugars, while *S. sanguinus* performs aerobic respiration. Lactic acid dissolves enamel.

563. **Brushing teeth without toothpaste has been shown to be effective in preventing dental caries. Which of the following most likely explains the protective benefit of brushing alone?**

(A) Brushing moves the bacteria in the biofilm to new places in the mouth.
(B) The toothbrush breaks DNA strands so that the bacteria are not able to replicate properly.
(C) Brushing leads to a lower pH, which kills bacteria.
(D) Brushing decreases the thickness of the biofilm so that more oxygen penetrates to the teeth, causing less lactic acid production.

564. **Toothpastes typically contain an alkaline component that raises the pH. Which of the following is the most likely effect of this on the bacterial populations?**

(A) The lower pH prevents bacterial growth.
(B) The higher pH favors bacteria like *S. mutans* that grow better than *S. sanguinus* under these conditions.
(C) The higher pH favors bacteria like *S. sanguinus* that grow better than *S. mutans* under these conditions.
(D) The higher pH prevents all bacteria from growing.

Questions 565–568

Students interested in developing methods to sequester carbon dioxide (CO_2) set up ten bioreactors containing microalgae using two sources of air input. Five of the reactors used exhaust air from the school's natural gas burning power plant. This exhaust air from the power plant has high levels of carbon dioxide. The other five used air from the environment. Air was bubbled through each bioreactor using an airstone. The students measured the algae concentration with a spectrophotometer and expressed it as a percent decrease in transmittance. The data are presented in the graph below.

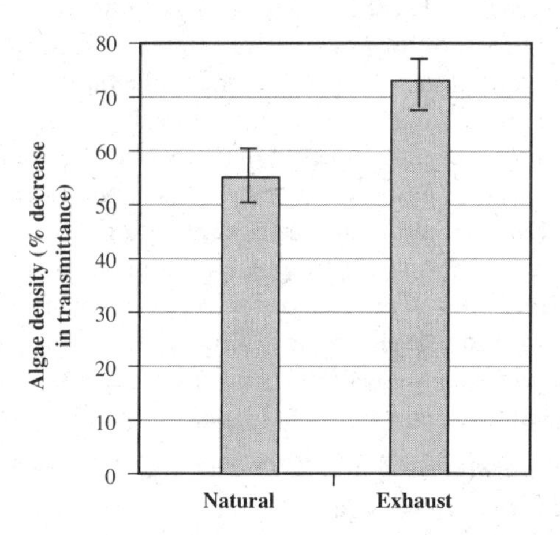

565. **Which of the following is an assumption that the students made concerning the algae concentration?**

(A) Exposure to exhaust increased the number of bacteria.
(B) The algae performed photosynthesis.
(C) The percent decrease in transmittance was due to the presence of algae.
(D) The algae used CO_2 as a source of carbon.

566. **Which of the following would be a method by which students could use the transmittance data they collected to determine an actual concentration of algae?**

(A) Compare these transmittance values to a "blank" with no algae in it but with all other factors held the same, such as the amount of fertilizer.
(B) Compare the transmittance values to values from samples with known concentrations of algae.
(C) Calculate the absorbance values, and use those instead because they directly relate to algae density.
(D) Use a pipette to transfer a drop of algae to a microscope slide, and count all of the cells on the slide.

567. Which of the following is a reasonable conclusion based on the data?

(A) Algae fed natural air grew faster because the air was purer.
(B) Algae fed air from the school power plant exhaust grew faster because there was a higher concentration of CO_2.
(C) Algae fed natural air and exhaust air grew at approximately the same rate.
(D) Algae fed air from the school power plant exhaust grew too fast and released more CO_2 than they fixed.

568. The students presented a poster of their findings at a science fair. In their poster, they claimed that the increase in algae growth was due to increased CO_2 in the exhaust of the power plant. A poster judge asked the students the following question: "How do you know the increased algae growth was due to carbon dioxide and not due to some other component of power plant exhaust?" The students could not answer the judge's question, so they went home and read up more on power plants. They learned that power plants, like the one their school used, contain high levels of CO_2 as well as smaller amounts of other compounds, including carbon monoxide (CO), nitrogen oxides (NO_x), nitrous oxides (N_xO), methane (CH_4), sulfur dioxide (SO_2), and volatile organic compounds (VOC). The students then decided to conduct a follow-up experiment to address the judge's question. They considered several additional bioreactor setups, including:

I. Environmental air with no CO_2 at 1 atmosphere of pressure
II. Exhaust air with enough CO_2 removed to match the level of CO_2 in environmental air
III. Environmental air with additional pure CO_2 to raise the concentration of CO_2 to the same level as that of the exhaust air
IV. Pure CO_2 at 1 atmosphere of pressure

Which of the following experiments would best test the hypothesis that the increased growth observed was due to carbon dioxide and not due to some other component of exhaust air?

(A) Repeat the experiment they conducted initially, and add bioreactor I.
(B) Repeat the experiment they conducted initially, and add bioreactors II and III.
(C) Repeat the experiment they conducted initially, and add bioreactors II and IV.
(D) Repeat the experiment they conducted initially three more times to see if the results are consistent.

Questions 569–571

AP Biology students investigated the effects of amending soil with used coffee grounds. To do this, they placed 100 mL of soil into each of 16 identical pots. In 4 pots, there were 0 g of coffee grounds. Into another 4 pots, they placed 8 g of coffee grounds. Into the third set of 4 pots, they placed 16 g of coffee grounds. Into the fourth set of 4 pots, they placed 24 g of coffee grounds. The students then planted lettuce seeds in each pot. They measured the plant height on days 7, 14, 21, and 35. Their data are presented in the graph below.

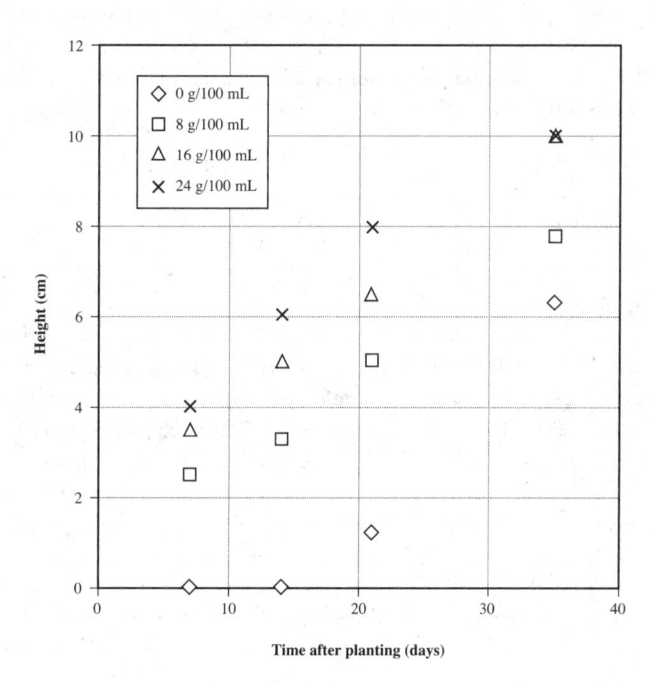

569. **Which of the following is an accurate interpretation of the data?**

(A) There was no difference in the growth rate among the different treatments.
(B) Lettuce grew most rapidly over 5 weeks in soil with the lowest concentration of coffee grounds.
(C) Lettuce grew most rapidly over 5 weeks in soil with the highest concentration of coffee grounds.
(D) It is impossible to determine the answer from this graph.

570. Which of the following would NOT be a testable hypothesis that the students could evaluate in a future experiment?

(A) Lettuce grown with coffee grounds is healthier to eat than lettuce grown without coffee grounds.
(B) Caffeinated and decaffeinated coffee grounds have different effects on plant height and mass.
(C) The growth rates shown in this graph are constant. Therefore, lettuce plants grown with more coffee grounds will yield more lettuce at an earlier harvest date.
(D) Growing plants in pure coffee grounds produces the fastest-growing lettuce.

571. Which of the following treatments had the fastest rate of growth between day 21 and day 35?

(A) 0 g/100 mL
(B) 8 g/100 mL
(C) 16 g/100 mL
(D) 24 g/100 mL

Questions 572–574

The standard method for testing the safety of mascara involves spreading a sample of the cosmetic on the cornea of the eyes of live rats and measuring the degree of irritation. Researchers interested in developing a humane method of testing the toxicity of mascara developed a procedure in which they exposed paramecia (a unicellular eukaryote) to different brands of mascara and measured the growth rate of the paramecia. The data are presented in the graph below.

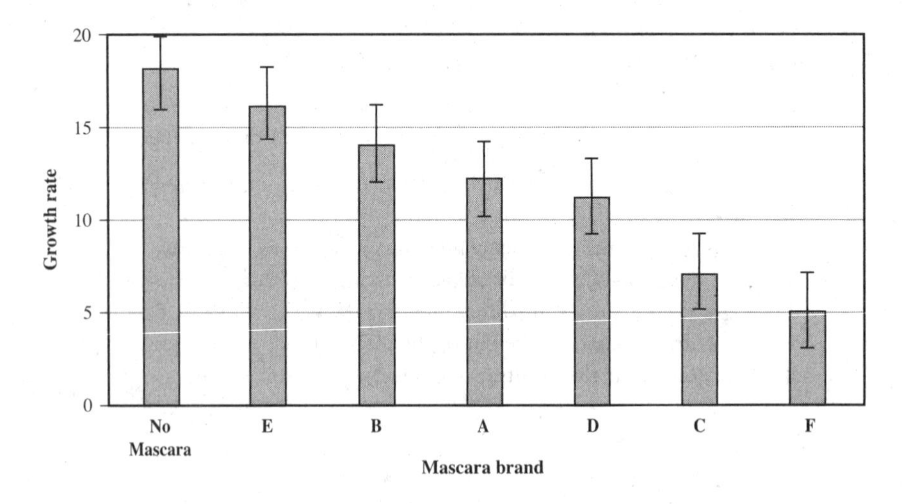

572. Which of the following is an accurate interpretation of the data shown?

(A) Some brands of mascara significantly promoted the growth of paramecia, while others showed no significant difference from the control.
(B) Some brands of mascara significantly inhibited the growth rate of paramecia, while other brands showed no significant difference from the control.
(C) None of the brands significantly decreased the growth of paramecia.
(D) Mascara brand F is probably the least toxic brand.

573. Which of the following would best support this assay as a replacement for the standard method in which products are applied to the cornea of the eyes of live rodents?

(A) if the standard method produced the same values for growth as the paramecium assay
(B) if the reaction was consistent across several species of paramecia
(C) if the standard method ranked the different brands in the same order as the paramecium assay with respect to toxicity
(D) if the paramecium assay produced results that were consistent with consumer feedback

574. Which of the following is an underlying assumption of the paramecium assay?

(A) Human and rat eyes contain paramecia.
(B) Paramecia are able to metabolize mascara.
(C) The same factors that affect the growth rate in paramecia will cause irritation in the rat eye used in the standard method as well as in the human eye.
(D) An increased growth rate of paramecia would indicate an increase in the toxicity of the mascara.

Questions 575–578

Students conducted an experiment to investigate the importance of soil organisms on plant health. They collected soil from the school garden. They autoclaved approximately half of the soil twice in order to kill all soil organisms present. They mixed this sterile soil with different amounts of unsterilized soil and planted Wisconsin Fast Plants (*Brassica rapa*) in the different soil mixtures. After three weeks, the students measured the height of the plants that grew. The graph on page 260 shows the results.

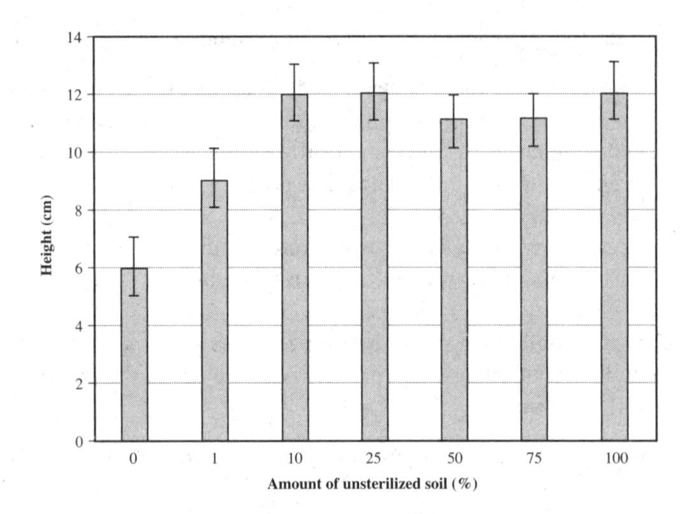

575. **Which of the following is the broadest question that the students were attempting to answer with this experiment?**

(A) Does the mineral content of soil change after autoclaving?
(B) Does the soil contain pathogens that affect the growth of *Brassica rapa*?
(C) What type of soil provides the optimum conditions for growing *Brassica rapa*?
(D) Does the presence of soil organisms have an effect on plant growth?

576. **Which of the following is an accurate interpretation of the results shown?**

(A) Soil microorganisms attack plant roots and represent a significant energy drain.
(B) Sterilizing the soil led to increased plant growth.
(C) Plants grown in sterile soil were significantly shorter than any soil containing at least some unsterilized soil.
(D) Nitrogen-fixing bacteria led to a decrease in the growth rate of infected plants.

577. **Which of the following is a reasonable conclusion to draw from the results shown?**

(A) The presence of one or more of the organisms found in unsterilized soil was beneficial to the health of the plants.
(B) Unsterilized soil was detrimental to the health of the plants due to the microorganisms it contained.
(C) Soilborne pathogens represent a threat to both laboratory and wild populations of *Brassica rapa*.
(D) The energy used by nitrogen-fixing bacteria in the soil was made available to plants in sterilized soil.

578. **Which of the following could the students do to determine if the soil contained microorganisms after autoclaving?**

(A) transfer samples of the autoclaved soil to sterile media that support the growth of microorganisms and see if any grow
(B) perform a DNA extraction and PCR to see if any DNA is present
(C) examine the autoclaved soil under a microscope and look for microorganisms
(D) spread samples of the soil onto moist paper towels in the laboratory, provide a source of nutrition such as oatmeal, add tap water to prevent dehydration, and examine for growth after a week

Questions 579–580

The graph below shows lynx and snowshoe hare population levels as estimated by pelt harvest records. The snowshoe hare is an herbivore and is the predominant food source for the lynx.

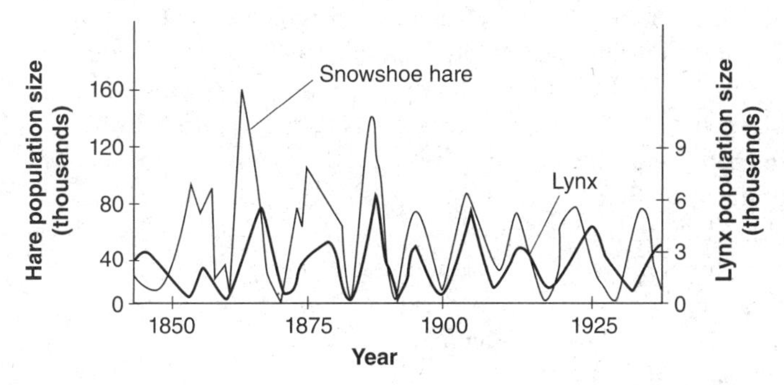

579. **One hypothesis to explain the cycling is that the predator population size drives the population size of the prey. When the predator population increases beyond a certain point, the lynx eat more snowshoe hares than the snowshoe hares can replace through reproduction. The snowshoe hare population crashes as a result of predation by the overabundant lynx. The lower availability of snowshoe hares to serve as food for the lynx causes many lynx to die from starvation. Which of the following would be a reasonable experiment to test the hypothesis that the lynx population levels would fall due to a lack of food?**

(A) Maintain the lynx and snowshoe hare populations in an enclosure to see if the same cycles occur.
(B) Supply the lynx with alternative food sources when the snowshoe hare population falls.
(C) Remove the lynx, and see if the snowshoe hare population grows beyond previously established maximum levels.
(D) Provide the snowshoe hares with alternative food sources when the lynx population falls.

580. **One hypothesis suggests that as the snowshoe hare population increases, plants produce more antiherbivore toxins. This leads to poorer nutrition and increased mortality in the snowshoe hares. This drop in the snowshoe hare population causes a decrease in the lynx population. Which of the following experiments would test this hypothesis?**

(A) Remove the lynx from the environment.
(B) Provide the snowshoe hares with food sources that do not contain these toxins.
(C) Sample the plants for toxins in the years after the snowshoe hare population crashes.
(D) Trap snowshoe hares, and remove them from this environment to keep their population levels low.

Questions 581–584

Toxoplasma gondii is a single-celled eukaryotic organism that infects both cats and mice. Cats get infected when they eat infected mice. *T. gondii* performs sexual reproduction in the cat, and the resulting propagules (called oocysts) are deposited in the cat's feces. Mice that come into contact with cat feces can become infected with the pathogen, which invades the brain, liver, and muscle tissue of the mice. Researchers did an experiment in which they placed either bobcat urine or rabbit urine on one side of a box. They then introduced mice with and without the *T. gondii* infection and recorded the time each mouse spent near or away from the urine. The results are shown in Figure I.

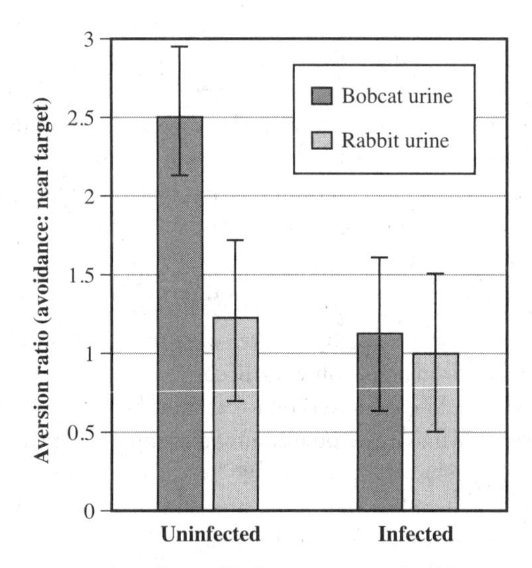

Figure I The effect of bobcat urine and rabbit urine on the avoidance behavior of mice

Researchers interested in the effect of the *T. gondii* infection on humans tested blood from 35 cat hoarders (defined by having more than 9 household cats), 35 individuals who kept 1–2 household cats, and 35 individuals who did not keep cats. Table I shows the results of the study.

Table I The Number of People Infected with *Toxoplasma gondii* Based on the Number of Cats They Own

	No Cats	1–2 Cats	More Than 9 Cats
Number of People Infected	5	15	35

581. **What was the purpose of the rabbit urine?**

(A) It was an experimental group to determine whether the mice were attracted to urine from a fellow prey species.
(B) Rabbit urine was used to mask the scent of the bobcat urine.
(C) It was present to provide the mice with a choice between the rabbit urine and the bobcat urine.
(D) It was the control for the bobcat urine. Without this treatment, it would be impossible to determine whether the mice were simply avoiding the urine.

582. **Which of the following is a reasonable conclusion to make based on the data shown in Figure I?**

(A) The mice that were infected with *T. gondii* lost their aversion to bobcat urine.
(B) The mice that were infected with *T. gondii* lost their aversion to rabbit urine.
(C) The uninfected mice preferred bobcat urine to rabbit urine.
(D) *T. gondii* causes mice to be attracted to cat urine so that the mice will become infected and be eaten by the cats.

583. **Which of the following is an evolutionary explanation for the data shown in Figure I?**

(A) Mice without an aversion to cat urine are more likely to obtain food despite the presence of cats. *T. gondii* makes this adaptive behavior more likely. Even though these mice are more likely to be eaten, their increased reproduction offsets any cost.
(B) Infected mice lose their aversion to cat urine because *T. gondii* secretes oxytocin, which is a neurotransmitter that stimulates neurons in the reward center of the mouse's brain, causing the mice to experience pleasure.
(C) Mice without an aversion to cat urine are more likely to come into contact with cats, making it more likely for the mice to be eaten. This would allow *T. gondii* to infect the cat and complete its life cycle.
(D) *T. gondii* secretes pheromones that are excreted in cat urine. These pheromones are highly attractive to infected mice and balance the aversive effects of cat urine.

584. Which of the following is the strongest statement that can be reliably made based solely on the results shown in Table I?

(A) There is a strong correlation between an infection with *T. gondii* and the number of cats owned.
(B) *T. gondii* infection influences people to keep many cats.
(C) People who own cats are more likely to become infected with *T. gondii*.
(D) All of these statements are equally reliable statements.

GRID-IN QUESTIONS

Questions 585–586

Septoria musiva is a fungal pathogen that grows on poplar trees (trees in the genus *Populus*). Researchers interested in the effects of drought stress on the ability of poplar trees to resist the fungus grew two clones of poplar. One clone was grown in drought (stressed) conditions. The other was grown in well-watered conditions. Researchers then placed either sterile fungal growth media or growth media containing the fungus into wounds made by removing a leaf from the trees. After 80 days of growth, the researchers examined each tree for the presence or absence of fungal spores. There were no spores on trees exposed to sterile media, so they were omitted from the analysis. The data are presented in the table below.

		Conidia (Spores)	
Water Treatment	**Clone**	**Absent**	**Present**
Watered	NE308	15	30
Watered	NM6	41	4
Stressed	NE308	20	25
Stressed	NM6	33	12

585. Perform a chi-square analysis to test the hypothesis that there was no effect of water treatment on the presence of spores in clone NM6. Report your chi-square value to the nearest tenth.

586. Perform a chi-square analysis to test the hypothesis that there was no difference between NM6 and NE308 with respect to the absence of spores in the stressed trees. Report your chi-square value to the nearest tenth.

Questions 587–588

The mark-recapture method is one way to estimate the size of an animal population. Researchers capture a sample of animals, mark them, and then release them. After permitting the animals to mix for some period of time, the researchers then resample the population. The proportion of marked individuals that are captured in the second sample should be the same as the ratio of the total number of marked animals in the area (which were marked in the first sample) to the total population size.

A group of students used the mark-recapture method to determine the size of the populations of sunfish in two lakes on the school campus. One of the lakes was surrounded by a forest, and the other was surrounded by athletic fields. Using nets at random locations around the lakes, the students caught and marked fish on one day. In the lake surrounded by the forest, the students caught 254 fish, marked them, and released them back into the same lake where they were originally found. As a result, the students knew that there were 254 marked fish in the lake surrounded by the forest. This is shown in the following table where the number of marked fish column intersects with the total in the lake surrounded by the forest. They used the same procedure for the lake surrounded by the athletic fields. The students caught 163 fish, marked them, and then released them back into the same lake where they were originally found. As a result, the students knew that there were 163 marked fish in the lake surrounded by the athletic fields. The following week, the students caught 261 fish from the lake surrounded by the forest and noticed that 16 of these fish caught in the second sample had been marked the previous week. In the lake surrounded by the athletic fields, they caught 152 fish, of which 21 had been marked the previous week. The data are presented in the table below.

	Number of Marked Fish	Number of Fish
Total in the Lake Surrounded by the Forest (Marked During the First Week)	254	?
Second Sample from the Lake Surrounded by the Forest	16	261
Total in the Lake Surrounded by the Athletic Fields (Marked During the First Week)	163	?
Second Sample from the Lake Surrounded by the Athletic Fields	21	152

587. Calculate the population size for the lake surrounded by the forest to the nearest whole number of fish.

588. Calculate the population size for the lake surrounded by the athletic fields to the nearest whole number of fish.

Questions 589–590

The graph below shows carbon dioxide concentrations measured at the Mauna Loa Observatory in Hawaii. The vertical lines represent January 15 of the year indicated. The value for the 15th of each month after January is also plotted.

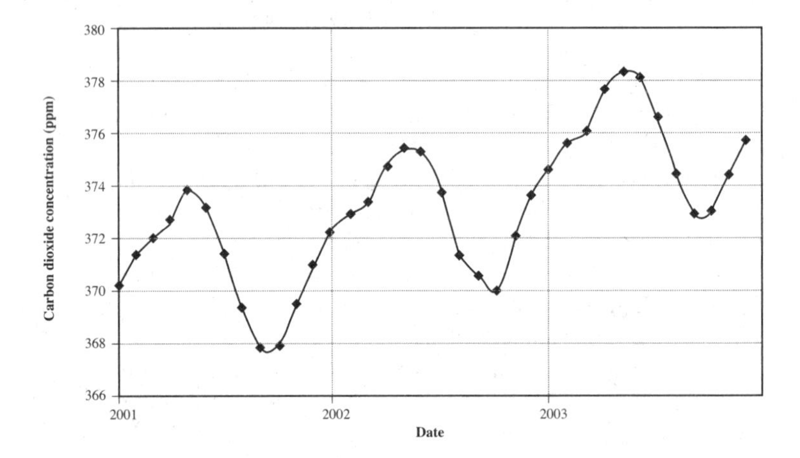

589. Calculate the change in the carbon dioxide concentration from the minimum value in 2002 to the maximum value in 2003. Give your answer to the nearest part per million (ppm).

590. Calculate the average change in carbon dioxide per year from January 2001 to January 2003. Give your answer to the nearest part per million per year (ppm/yr).

---

591. The graph on page 267 shows the survival of Douglas fir seedlings over 1 year in a nursery under typical management practices (sprayed with fungicides) and without spraying fungicides. What percentage of the survival of seedlings in sprayed plots was achieved by seedlings in the unsprayed plots? Express your answer to the nearest whole percent.

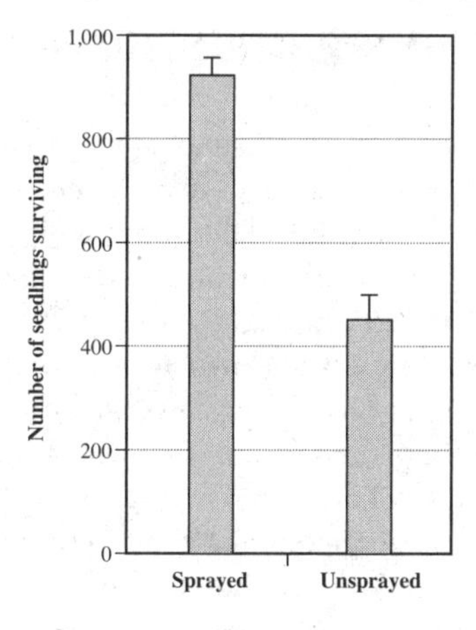

Questions 592–595

The graph below is from an experiment that compared the growth of different species of *Paramecium* in mixed and in separate cultures.

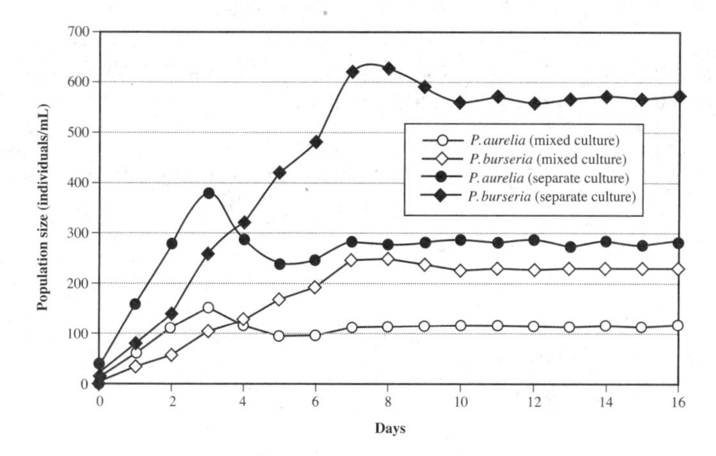

592. Calculate the growth rate for *P. aurelia* in separate culture from day 1 to day 3. Express your answer to the nearest whole number of individuals per mL per day.

593. Calculate the growth rate for *P. burseria* in mixed culture from day 1 to day 7. Express your answer to the nearest whole number of individuals per mL per day.

594. Determine the carrying capacity for *P. aurelia* in separate culture. Give your answer to the nearest whole number of individuals per mL.

595. Determine the carrying capacity for *P. burseria* in separate culture. Give your answer to the nearest whole number of individuals per mL.

LONG FREE-RESPONSE QUESTIONS

596. An AP Biology class performed a study of species diversity in suburban lawns. Lawn management practices were divided into three treatment groups: professionally managed, high-input do-it-yourself (DIY), and low-input DIY. Professional management included regular mowing as well as fertilizer and pesticide treatment. High-input DIY management included the same practices as professional management but were performed by the homeowner. Low-input DIY management included mowing as the only management technique (no fertilizer or pesticide treatment). The students also took photographs of a 3 m × 3 m segment of each lawn and asked their parents to rate each picture for its attractiveness on a scale of 1 to 10, with 1 being the least attractive and 10 being the most attractive. These photographs were of only vegetation and did not include identifying traits that would indicate which of the three treatment groups it was from or whose house it was from. The photographs were all taken at approximately the same time of the day using a single camera. The results of the study are displayed in the table below. Note that the species richness is the number of different species found. The Shannon diversity index is a common measure of diversity in which high numbers represent diverse populations and low numbers represent low diversity.

	Low-Input DIY	High-Input DIY	Professionally Managed
Species Richness	21	12	7
Shannon Diversity Index	1.7	1.1	0.9
Attractiveness Rating	4	7	9

(a) **Compare** the species richness and the diversity among the three different management techniques. **Explain** how the choice of management techniques might have led to the differences.

(b) **Identify** ONE factor that might have led to the observed differences in the species richness and diversity. **Describe** a controlled experiment to test that factor.

(c) **Discuss** the impact that social views on aesthetics might have on the diversity and stability of an ecosystem.

597. Researchers interested in the effects of rising carbon dioxide levels on crop growth measured the effects of different concentrations of atmospheric carbon dioxide on the amount of photosynthesis performed by corn and by wheat. The table below summarizes their data.

	Photosynthesis ($mg\ CO_2$ assimilated $dm^{-2}\ h^{-1}$)	
Carbon Dioxide Concentration (ppm)	**Corn**	**Wheat**
50	12	6
100	21	15
200	49	31
400	53	53
600	54	66
800	54	68
900	54	68

(a) Graph the data using the grid provided.

(b) Describe the shape of the curve for corn. Explain the physiological reason for the shape of the curve for corn between 0 ppm and 200 ppm and between 400 ppm and 900 ppm.

(c) Compare the responses of the corn and the wheat to increased carbon dioxide.

(d) Global atmospheric carbon dioxide is currently at approximately 410 ppm. Use the data from the graph to predict the effects on corn growth and wheat growth if the carbon dioxide levels continue to increase.

SHORT FREE-RESPONSE QUESTIONS

598. An AP Biology student wanted to test the efficacy of catnip extract as a natural insect repellent. She treated banana slices with either catnip extract or water. The student then exposed the banana slices to a population of fruit flies and counted the number of times that flies landed on the banana slices for a period of 5 minutes. The results are shown in the table below.

	Banana Slices Treated with Catnip Extract	Banana Slices Treated with Water
Number of Fruit Fly Landings	27	44

(a) Perform a chi-square test on the data. Specify the null hypothesis that you are testing, and enter the values from your calculations in the table below.

Null Hypothesis:			
	Observed (o)	Expected (e)	$\frac{(o-e)^2}{e}$
Catnip Extract			
Water			
Total			

(b) Explain whether the null hypothesis is supported by the chi-square test at the 0.05 level.

599. The codling moth (*Cydia pomonella*) is an insect pest of apple and pear trees. Adult codling moths lay eggs on developing fruits just after the petals fall off of the flower. After hatching, larvae burrow into the fruit and eat for approximately 3 weeks before leaving the fruit to overwinter and pupate the following year. *Cydia pomonella granulovirus* (CpGV) is a virus that causes disease in the codling moth and is used as a biological control agent. The virus is sprayed onto trees shortly after the petals fall.

 (a) Briefly explain the rationale for the timing of the virus application.

 (b) Identify ONE potential environmental risk of this practice, and explain what could be done to manage the risk you identified.

600. Biology students exposed zebrafish (*Danio rerio*) to alcohol (ethanol) by placing them in tanks that contained 0%, 0.5%, and 1% ethanol for one hour. After an hour of exposure to ethanol, these fish were moved to a test tank that contained regular water. The fish were video recorded for an hour. The students viewed the video and recorded how much time each fish spent in three zones of the tank (top, middle, and bottom). This is a standard test for anxiety used by researchers. Work by other researchers has established that anxious fish avoid the top of the tank. Each bar in the graph below represents the average of 7 fish. Error bars depict the standard error of the mean.

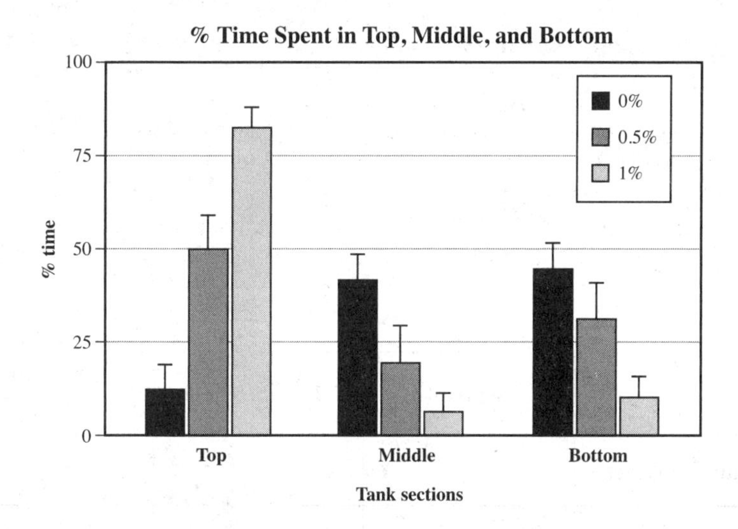

 (a) State a hypothesis that the students were testing.

 (b) Interpret the data in terms of the hypothesis.

Answers and explanations for all of the questions in this chapter can be found on pages 388–415.

ANSWERS

BIG IDEA 1—EVOLUTION

Multiple-Choice Questions (pages 3–54)

1. **(B)** Natural selection acts on preexisting variability in populations. In the organic field, there will be some fraction of plants that have resistance to glyphosate, even though the plants have never been exposed to it. LD_{50} is the level of a compound that kills 50% of the organisms. Choice (A) is incorrect because it relies on the misconception that evolution works on acquired traits. Natural selection can select only what is already present in the gene pool. Choice (C) is incorrect because there is no evidence of interbreeding. Choice (D) relies on the misconception that organisms have mechanisms to change their genomes in response to the environment. (LO 1.5)*

2. **(C)** All of the plants that grew to maturity were resistant to 2,820 g/ha glyphosate because the ones that weren't resistant died. The offspring of those plants that survived had a high frequency of resistance alleles, regardless of how resistance was inherited. Therefore, choices (A) and (B) are incorrect. It is also unlikely that all of the susceptible alleles were removed from the population. Therefore, it is very unlikely that all of the offspring would be resistant, making choice (D) incorrect as well. (LO 1.7)

3. **(A)** Although some plants might be able to tolerate a nearly 100-fold increase in pesticide concentration, it is unlikely that many would be able to tolerate this increase. Choices (B), (C), and (D) are incorrect because they project that at least half to all of the plants would be resistant to this much higher concentration of pesticide. (LO 1.7)

4. **(D)** Seeds and fruits are structures that allow plants to move from place to place, so choice (D) is the correct answer. Random mutation is possible but unlikely, especially in large numbers without any selection pressure. Therefore, choice (A) is not correct. Choice (B) is not correct because there is a biological explanation for this occurrence. Choice (C) would require several events to occur that are less likely than seed spread. (LO 1.2)

5. **(D)** In incomplete dominance, hybrids between pure breeding types show an intermediate phenotype. Choice (A) is incorrect because no data on sex were discussed. If resistance was controlled by either a recessive alelle or a dominant allele as choices (B) and (C) suggest, one phenotype would appear as 100% and the other would appear as 0% in the F1s. (LO 1.4, LO 3.14)

6. **(A)** Traits that are controlled by many gene loci, such as a person's height, show a normal distribution. Choice (B) should produce a 3:1 ratio, and choice (C) should produce a 1:2:1 ratio, neither of which are accurate. The

*Following each answer explanation in this book, you will find, in parentheses, the Learning Objective(s) (LO), from the AP Biology Curriculum Framework, that each question corresponds to.

distribution of traits does not appear to be caused by the sex of the plant, so choice (D) is incorrect. (LO 1.4, LO 3.14)

7. **(D)** Genetic drift is a result of random fluctuations in allele frequency. Since each individual in a small population is a larger proportion of the total, single changes have a greater effect on the allele frequency. In small populations, alleles at any given gene locus tend to go to either fixation (100% frequency) or extinction (0% frequency). Choice (A) is incorrect because there is no evidence of southern bears migrating to the north. Increased natural selection against highway-crossing bears should lead to decreased gene flow, not increased gene flow as stated in choice (B). Additionally, if there were increased gene flow, diversity should increase, not decrease. Choice (C) is incorrect because a higher mutation rate should lead to *increased* diversity. There is also no evidence that pollution had an effect. (LO 1.4)

8. **(C)** By connecting the northern and southern populations, the bears were able to move between the two populations, increasing gene flow and genetic diversity. Wildlife crossings are pathways over or under roadways. The bears do not need to be able to read signs in order to use the crossings, as suggested in choice (A). Although the crossings likely resulted in different populations mating, mutation rates are not affected by the extent of inbreeding. Therefore, choice (B) is incorrect. In addition, a decreased mutation rate would lead to decreased diversity, not an increase as is seen in the graph. Sexual selection decreases diversity because some individuals do not mate and their alleles are removed from the population. Therefore, choice (D) is incorrect. (LO 1.6)

9. **(D)** Introducing bears from another population will bring whatever alleles the new bears carry into the existing population, increasing diversity. Choices (A) and (C) would increase the population size, but these actions would not necessarily have an effect on diversity. Installing bear-proof garbage cans, as choice (B) suggests, should help prevent the bears from becoming reliant on humans for food. However, this action should have no effect on genetic diversity. (LO 1.8)

10. **(B)** Warm winters would not kill as many flies as cold winters. In cold winters, being able to enter diapause is much more important. Since winters are variable, the selection pressure on the diapause minus allele is variable. Choice (A) is incorrect because inbreeding would lead to less diversity, not greater diversity. Additionally, because fruit flies are typically in diapause or dead during the winter, no mating occurs. Nonrandom mating decreases variability, so choice (C) is incorrect. Choice (D) is incorrect because there is no evidence to suggest that genetic drift was higher in May. (LO 1.2)

11. **(D)** Genetic drift is random, and random change does not explain the nonrandom trends in the graph. Since the question asked which of the answer choices does NOT account for the trend shown in the graph, choice (D) is

the correct answer. The ability to enter diapause is selected for in the winter because it allows flies to survive cold weather. This selection for diapause is the same as selecting against the diapause minus allele. Since the statement in choice (A) is true, this is not the correct answer. Southern fruit flies should have a higher frequency of the diapause minus allele because diapause is less valuable in warm climates. Immigration from the south could have increased the allele frequency in the population being studied. Once again, since the statement in choice (B) is true, this is also not the correct answer. The cost of diapause is that insects don't mature as quickly. As a result, diapause is selected against in the summer months. This therefore selects for the diapause minus allele. This makes choice (C) a true statement and therefore not the right answer. (LO 1.2)

12. **(C)** Males are less frequent in deep water in both lakes. Therefore, there must be some selective force acting against them. If random chance was the explanation, the same pattern should not occur in both lakes. Therefore, choice (A) is incorrect. If males were selected *for* in deep water, there would be more males, not less. Therefore, choice (B) is not correct. Males are not more abundant in deep water, which makes choice (D) incorrect. (LO 1.5)

13. **(D)** When Lake A snails were challenged with parasites from Lake A, the frequency of infection in shallow water was approximately 0.48 (48%). When Lake B snails were challenged with parasites from Lake B, the frequency of infection in shallow water was approximately 0.35 (35%). When Lake A snails were challenged with parasites from Lake B, and when Lake B snails were challenged with parasites from Lake A, the frequency of infection was much lower. Therefore, the parasites must be adapted to infect the snails that inhabit the lake in which the parasites live. Snails were challenged with parasites in the laboratory and therefore could not move to deep water, so choice (A) is not correct. Snail immunity was not tested in this experiment since immune responses require prior exposure to the parasite, and there is no evidence of this in the experiment. Therefore, choice (B) is incorrect. Resistance in this case is likely caused by the fact that the parasite lacks genes to cause an infection. The fact that not all of Lake A's snails are infected suggests that some of the snails do have resistance genes. Therefore, choice (C) is incorrect. (LO 1.4)

14. **(C)** Diploid organisms make haploid gametes, which contain 50% of the parents' genes. Therefore, the offspring will contain 50% of each parent's genes. Asexual reproduction is the only type of reproduction that passes 100% of an organism's genes to its offspring, so choice (A) is incorrect. In a haplo-diploid genetic system, such as that used by honeybees, worker bee sisters are 75% related. Of course, this doesn't apply to diploid snails. In addition, the question asked about offspring, not sisters. Therefore, choice (B) is incorrect. Grandchildren share 25% of their genes with grandparents

in this system. However, this question asked about offspring, so choice (D) is incorrect. (LO 1.10)

15. **(C)** Evolutionary fitness is high when an individual contributes a large proportion of its genes to the future gene pool. Choice (A) is incorrect because it conflates *athletic* fitness with *evolutionary* fitness. If athleticism improves an individual's ability to contribute to the gene pool, it contributes to high evolutionary fitness. However, this is not necessarily the case. For example, in turtles a thicker shell and a quicker startle response are likely more important than speed and agility. Longer-lived, disease-free organisms are likely to leave more offspring, but this is not necessarily the case. For example, salmon die immediately after spawning. Salmon spare no reserve energy for life after spawning. Therefore, choice (B) is incorrect. Obtaining food and controlling territory may help an individual attract mates and produce offspring, but these are a means to an end (offspring); they do not make an individual fit. For example, a male bird with no territory might have high fitness if he is sneaky and mates with many females in other birds' territories. This makes choice (D) incorrect. (LO 1.9)

16. **(A)** Sexually reproducing organisms pass on only half of their genes to their offspring. To match the asexual snail that passes on all of its genes, the sexually reproducing snail would need to have twice as many offspring as that of the asexual snail. Choice (B) suggests that sexual and asexual reproduction pass on the same proportion of genes. This is not true. You may have arrived at choice (C) by dividing 100 by 2 instead of multiplying 100 by 2. Organisms that reproduce sexually need twice as many offspring, not half as many, to match the fitness of organisms that reproduce asexually. If you guessed choice (D), then you likely divided 100 by 4 instead of multiplying 100 by 2. (LO 1.9)

17. **(B)** In the presence of parasites, the cost of sexual reproduction is outweighed by the benefit of increased variability. Sexual reproduction provides a selective advantage in some situations but not in others, so choice (A) is incorrect. The low diversity in an asexual population would make it possible for a single genotype of a parasite to decimate the population. As a result, asexual reproduction is less likely in the presence of parasites, so choice (C) is incorrect. Organisms that reproduce asexually may have higher fitness in some environments. However, if the environment changes, they may have lower fitness. Therefore, choice (D) is incorrect. (LO 1.5)

18. **(C)** Evolutionary fitness is dependent on environmental conditions. What is fit in one environment may be unfit under a different set of circumstances. In this experiment, humans are the selective agent. Although obtaining food and surviving are important for fitness, a rat that never reproduces has a fitness of zero because it does not contribute genetically to the next generation. Therefore, choice (A) is incorrect. In this experiment, the only

selective pressure was due to the experimenter choosing to breed aggressive rats in one group and choosing to breed tame rats in the other group. There is no indication that the rats had differing access to food. Additionally, it is reproduction, not survival alone, that determines fitness. Thus, choice (B) is not correct. The rats were not permitted to compete for mates in this experiment. This makes choice (D) incorrect. In the wild, competition may play a role in fitness. However, selecting for aggression should make the rats in the treatment group more aggressive than rats in nature. The naturally occurring level of aggression is most likely to be the optimum level for the environment. Therefore, higher levels of aggression than those found in the natural population should be maladaptive. (LO 1.2)

19. **(B)** The genes involved in tameness likely controlled the levels of hormones and neurotransmitters, such as adrenaline, noradrenaline, dopamine, testosterone, and estrogen. These genes also controlled a myriad of other phenotypes. Additionally, selecting a particular gene inadvertently selects for any genes that are physically near the selected genes. The description of the experiment only states that the researchers selected for tameness. It gives no indication that the researchers selected for variable coat colors, droopy ears, or a decreased brain size, as choice (A) suggests. Choice (C) is incorrect because artificial selection has no effect on mutation rates. The question asks about *tame* rats, so choice (D) is irrelevant to the question. In this experiment, female rats were only given access to mates chosen by experimenters, so mate choice was not permitted. (LO 1.4)

20. **(D)** The table shows higher corticosteroid levels for the aggressive rats before and after handling them. There seems to be a correlation between high levels of corticosteroids and aggressiveness. However, there is no basis for concluding a causal relationship, so choice (A) is wrong. Choice (B) is wrong because domestication probably selects for *low* corticosteroid levels. The tame rats did not have higher baseline corticosteroid levels, so choice (C) is wrong. (LO 1.4)

21. **(C)** Environmental conditions largely determine the selective forces on populations. These highly aggressive rats are probably more aggressive than necessary for the environment, and they would likely incur a cost for such high aggressiveness. It is possible to have a high frequency of one allele or another. However, genes cannot be concentrated, so choice (A) is incorrect. Evolution happens without human intervention, so choice (B) is incorrect. Natural selection is not random. The variation it selects from may be random, but natural selection depends on the conditions in the environment. Therefore, choice (D) is incorrect. (LO 1.8)

22. **(B)** Before and after handling is a time-linked independent variable that belongs on the *x*-axis. Corticosteroid concentration is the dependent variable that belongs on the *y*-axis. Each treatment group (tame vs. aggressive) is a separate bar. Standard errors of the mean are displayed as error bars. Pie

charts are useful to depict portions of a whole. Since the corticosteroid concentration of each treatment is independent of the other treatments, a pie chart, as shown in choice (A), is not an appropriate display. There are very few applications in science where pie charts are the best choice for data display. Line graphs are used for continuous variables. Tameness (tame vs. aggressive) is a discrete variable and should not be depicted as shown in choice (C). Choice (D) is incorrect because the corticosteroid concentration is the dependent variable and should be plotted on the *y*-axis. Also, SEMs should not be displayed as separate bars. (LO 1.4)

23. **(A)** The dugong is a species that is known to be outside the elephantid group. Any differences between the dugong and the members of the elephantid group must have occurred in the time since the evolutionary lineages separated. As such, it provides a "root" for the cladistic tree and serves as a reference point for comparisons among the elephantids. Choice (B) is incorrect because the main goal of the study was to resolve differences among the elephantids, not between elephantids and a closely related nonelephantid. An animal species that is too distantly related would not be helpful in resolving differences among species that are closely related, so choice (C) is incorrect. Choice (D) is incorrect because there is a solid rationale for including the dugong, as explained in choice (A). (LO 1.19)

24. **(C)** The Asian elephant's closest ancestor is the mammoth because there are the fewest differences between them. To compare two species, find a cell where a row and a column intersect. For example, where the row for the mammoth and the column for the Asian elephant intersect, you will see that there are only 2 amino acid differences between them. All of the other species have more differences from the Asian elephant: 7 for the African elephant (choice (A)), 12 for the mastodon (choice (B)), and 25 for the dugong (choice (D)). (LO 1.19)

25. **(C)** The African elephant's closest ancestor is the mammoth because there are the fewest differences between them. To compare two species, find a cell where a row and a column intersect. For example, where the row for the mammoth and the column for the African elephant intersect, you will see that there are only 6 amino acid differences between them. The Asian elephant is a close second with 7 amino acid differences (choice (A)). The mastodon is more distant with 11 amino acid differences (choice (B)), and the dugong is most distant with 24 amino acid differences (choice (D)). (LO 1.19)

26. **(B)** Species with fewer differences in their sequences are more closely related and have a more recent common ancestor. The longer that species are separate, the more time there is for mutations to accumulate. To find the number of differences between any two animals, find the cell where the row for one species and the column for the other species intersect. The mammoth and Asian elephant only have 2 differences in their amino acid sequences and

should be placed on the same branch, very closely together. The next most closely related are the mammoth and the African elephant (6 differences) followed by the African elephant and the Asian elephant (7 differences). Therefore, the branch connecting the African elephant to the mammoth/ Asian elephant branch should be further to the left (back in time). The next most closely related species is the mastodon, which is between 11 and 12 differences from the other elephantids. Finally, the dugong is the most distantly related with between 23 and 26 amino acid differences from the rest of the species shown. Choice (A) is incorrect because it shows the dugong and the mastodon being closely related. The mastodon is more closely related to the mammoth than to the dugong. Choice (C) is incorrect because it shows the Asian elephant being more closely related to the African elephant than to the mammoth. The Asian elephant has only 2 differences with the mammoth, whereas it has 7 differences with the African elephant. Choice (D) is incorrect because it shows the mammoth and the mastodon being most closely related when in fact the Asian elephant is clearly more closely related to the mammoth. (LO 1.17)

27. **(C)** The molecular evidence suggests that there are undiscovered fossil angiosperms that are older than 130 million years old but younger than 215 million years old. Looking at rocks that span that range would be the best strategy for testing the hypothesis that angiosperms diverged earlier than 130 million years ago. Choice (A) is incorrect and unethical. Scientists need to rely on data to drive conclusions rather than manipulate data to get the answer they want. Choice (B) is incorrect because there is no reason to believe there would be any angiosperm fossils older than 215 million years old. Although scientists would likely find fossil angiosperms in rock strata younger than 125 million years old, this discovery would do nothing to resolve the difference between the fossil record and the molecular data. Therefore, choice (D) is not correct. (LO 1.32)

28. **(A)** Since the conquering female has not yet mated with any of the males in the territory, none of the eggs or chicks are related to her. Once these unrelated chicks are killed by the conquering female, the males are immediately able to mate with the conquering female, begin caring for her eggs, and begin feeding her chicks once they hatch. Killing the unrelated chicks, although gruesome, increases the number of offspring the conquering female would have and therefore increases her contribution to the gene pool. The conquering female owns an entire territory with plenty of food, aside from baby jacanas. There must be an adaptive function beyond nutrition to drive this behavior. Therefore, choice (B) is wrong. As long as there is a next generation, there will be competition, whether it is from her own chicks or from another bird's chicks. Therefore, choice (C) is incorrect. Although choice (D) may be true, it is more of an indirect advantage of the behavior than the more direct selective advantage described in choice (A). Therefore, choice (D) is not the correct answer. (LO 1.2)

29. **(D)** RNA viruses rely on their error-prone replicases (RNA-dependent RNA polymerases) to generate variety. This variability allows for rapid evolution. A stable genome, as suggested in choice (A), is useful in *unchanging* environments. Adaptation to changing environments requires variability. Viral genomes are small (on the order of 10,000 base pairs) and typically do not encode proteins beyond the absolute necessities. Therefore, choice (B) is incorrect. Choice (C) is wrong since DNA-editing mechanisms would not work on RNA because DNA and RNA have significant structural differences. (LO 1.4)

30. **(B)** RNA is capable of acting as both an information molecule, like DNA, and a catalyst, like protein. Therefore, it seems reasonable to believe that RNA performed these functions in early living things. DNA and protein evolved later as specialized molecules that were better than RNA at their functions. However, neither DNA nor protein could carry out both functions. Choice (A) should be rejected because the phrase "proved beyond any doubt" is an unscientific statement. Although this experiment demonstrated that it was possible to engineer self-replicating RNA molecules, and we may speculate that such a molecular system could have worked in early life, that's the extent of what we can conclude. Spontaneous generation, as mentioned in choice (C), is the formation of living organisms without descent from similar organisms (such as mice spontaneously forming from rags or maggots spontaneously forming from meat). An RNA molecule that can self-replicate in a test tube lacks many of the fundamental properties of life, such as a membrane. Therefore, this is not a correct choice. Choice (D) is incorrect because the experiment *supported* the RNA world hypothesis. It showed that RNA could act as an information molecule while simultaneously acting as an enzyme capable of copying that information. (LO 1.31)

31. **(A)** Errors in DNA replication randomly generate mutations in the genetic code. The Petri plate was initially sterile. There was no mention of plasmids present, as suggested by choice (B). Choice (C) is incorrect because there was no mention that any viruses were present. The plate was inoculated with a single strain of bacteria. Conjugation between identical organisms is unlikely to produce variability, so choice (D) is not correct. (LO 1.13)

32. **(B)** Once a section is full, bacteria begin getting pushed into the next section. Individuals that cannot survive in the higher concentration of antibiotic do not survive. Eventually, a bacterium that had a mutation from replication gets pushed over and is able to multiply in the higher concentration of antibiotic. Mutation occurs when bacteria replicate their DNA. Mutation is intrinsically tied to reproduction. If bacteria stopped growing, they would stop dividing and therefore stop replicating DNA, so choice (A) is incorrect. Evolution does not work by organisms gradually acquiring resistance, so choice (C) is incorrect. Mutations arise and are either adaptive or not

adaptive. Choice (D) confuses developmental changes with evolution and is therefore incorrect. (LO 1.13)

33. **(C)** The bacteria in the initial culture were susceptible to 3 units of antibiotic. They would certainly not be resistant to 3,000 units of antibiotic. Therefore, there would be no growth. Choice (A) is incorrect because only mutations present at the time of exposure can be selected for. Exposure to a new environment does not induce mutations. Choice (B) is incorrect because if there were mutants present in the initial population that could grow in the presence of 3,000 units of the antibiotic, the bacteria would be able to colonize the entire plate without mutants having to arise. Choice (D) can also be ruled out because the passage did not mention the presence of plasmids. (LO 1.5)

34. **(A)** There are usually evolutionary trade-offs for adaptations. In this case, the energy used to provide antibiotic resistance could have been used for more rapid growth. Choice (B) does not explain biologically how the bacteria were stunted and is logically circular. Choice (C) is incorrect because bacteria must be near each other in order to compete. The two types of bacteria (resistant and susceptible) were grown in separate Petri plates. Therefore, they would not have been able to compete with one another. The question does not mention the presence of viruses. If viruses were present, it is unlikely that selection for antibiotic resistance would have affected virus resistance. Therefore, choice (D) is incorrect. (LO 1.5)

35. **(B)** Although it is likely easier to evolve resistance to a higher concentration of the same drug, the same processes that make that possible also make adaptation to a new drug possible. Although most of the bacteria likely died because they didn't have resistance to the new drug, only a single bacterium was needed to colonize the new environment. Even a 1 in a billion chance is a certainty if you have enough bacteria, so choice (A) is incorrect. Choice (C) is incorrect because bacteria do not choose genes. Since a different chemical likely has a different target in the bacterium, selecting for resistance to a new drug is likely to be harder than adaptation to a higher concentration of the same drug. Therefore, choice (D) is not correct. (LO 1.5)

36. **(B)** On the fatty acid, the carboxylic acid group (COOH) can hydrogen bond with water molecules. In addition, the hydrocarbon chain is hydrophobic. The phospholipid has both a hydrophobic end and a hydrophilic end. The phosphate group (PO_4), the amino group (NH3), and the oxygen atoms on the fatty acid tails are all hydrophilic, while the hydrocarbon chains are hydrophobic. As a result, both of these molecules are capable of forming bilayers when placed into water. Choice (A) is incorrect because fatty acids *do* have a hydrophilic end. Choice (C) is incorrect because fatty acids are *not* completely hydrophilic and *can* form bilayers in water. Choice (D) conflates fatty acids with proteins. (LO 1.28)

37. **(D)** Living organisms must have a boundary, such as a membrane, that defines the limits of the organism. There must have been a stepwise pathway, from nonliving things to the complex organisms of today, that involved simpler molecules performing the functions of more complex ones. Neither fatty acids nor phospholipids are capable of catalyzing reactions. Therefore, choice (A) is incorrect. Choice (B) reverses the likely evolutionary sequence. Fatty acids were likely to be major components of early membranes and were replaced by phospholipids later. Choice (C) is incorrect because fatty acids and phospholipids are *not* capable of storing genetic information. (LO 1.28)

38. **(B)** A mechanism that explains how fatty acids could be produced from inorganic carbon without the action of a cell would provide a link between nonliving chemicals and protocells. Fatty acids are key components of phospholipids and are incorporated in nearly all modern membranes; no discovery is necessary. Therefore, choice (A) is incorrect. The synthesis of fatty acids using acetyl CoA is a well-studied, complex phenomenon that happens in modern cells. This process is much too complex to be helpful in explaining how early protocells arose, so choice (C) is incorrect. Ribosomes are RNA/protein complexes that currently produce proteins. Lipids are not involved. As a result, it would be unlikely that a mechanism by which fatty acids interact with nucleic acids to produce protein would be discovered. If it were, it would not help to support the idea that early protocells had fatty acid-based membranes. Therefore, choice (D) is incorrect. (LO 1.28)

39. **(A)** In choice (A) we see a spherical bilayer growing into a cylinder, a shape with a higher surface area to volume ratio than a sphere. The cylinder then fragments into several small spheres. Choice (B) shows a sphere growing in size. A large sphere has a lower surface area to volume ratio than a small sphere, which is not described in the passage. Choice (C) shows a sphere directly forming several spheres and then combining into a cylinder. This is not described in the passage since several spheres of the same size would have the same surface area to volume ratio as the original sphere. Choice (D) shows a large sphere fragmenting into several smaller ones. Although this increases the surface area to volume ratio, the fragmenting depicted in this image does not happen in the order described in the passage. (LO 1.28)

40. **(D)** The passage describes a spherical protocell consuming fatty acids from the environment (eating) and becoming larger (growing). Fragmentation into multiple spheres is analogous to reproduction by division. Death by apoptosis is not described, so choice (A) is incorrect. For competition to occur, there would have to be more than one type of vesicle. Since there is no mention of different types of vesicles, choice (B) is incorrect. This process does not show variability or natural selection, so choice (C) is incorrect. (LO 1.30)

41. **(D)** Resprouting trees are genetically identical to the previous trees that died. Repeated exposure to the same genetic material cannot cause the trees to

become resistant. It might be possible for a somatic mutation to lead to resistance, but this is not what choice (D) states. Since the statement in choice (D) is incorrect, this is the correct answer. If the American chestnut did have any resistance, we would have known about it only by seeing that trees survived. Harvesting trees before infection preserved the value of the wood but prevented potentially resistant individuals from reproducing. Therefore, since the statement in choice (A) is true, this is not the correct answer. Resprouting produces trees that are genetically identical to the original trees. Sexual reproduction is necessary to generate variability on which natural selection can act. Thus, the statement in choice (B) is true, so it is not the correct answer. Since the fungus was imported on Japanese chestnut trees, it is reasonable to think that there would be resistance genes in the environment where the pathogen originated. If there weren't, the disease would have wiped out those Japanese chestnut trees as well. Therefore, the statement in choice (C) is correct, making this an incorrect answer as well. (LO 1.25)

42. **(D)** Imported species have a tendency to be invasive, especially if there are no natural enemies and if there is a ready food source. In the case of the chestnut blight, all three of the conditions mentioned in choices (A), (B), and (C) were met. (LO 1.2, LO 4.22)

43. **(A)** The goal of the breeding program was to move resistance genes into American chestnut. As a result, the foundation wanted only a small portion of the genomes of the Asian species. Backcrossing and screening for resistance isolated individuals with the desired genes while retaining the diversity present in American chestnut. Backcrossing to a susceptible type would *decrease* the amount of disease resistance, so choice (B) is incorrect. The fact that crosses were made ensures sexual compatibility, so choice (C) is incorrect. The taste of the fruit was not mentioned, so choice (D) is incorrect. (LO 1.8)

44. **(D)** The amount of necrosis produced on the Transgenic American chestnut leaves was much lower than that of the "resistant" Chinese chestnut variety. The error bars do not overlap, so there are statistically significant differences. Since the Transgenic American chestnut plants were more resistant, choices (A), (B), and (C) are incorrect. (LO 1.8)

45. **(C)** Planting only this clone of American chestnut would mean that any other pathogen that could attack this clone could attack all of the individual trees. Since people generally don't consume bark, which is where the gene is expressed, allergic reactions are highly unlikely. Therefore, choice (A) is incorrect. Hybridizing with wild chestnut would probably be good because it would spread resistance genes in a more diverse population. This result, however, is highly unlikely because existing trees generally don't survive to maturity, so choice (B) is incorrect. Choice (D) is a moral stance, not an evolutionary argument. (LO 1.8)

46. **(D)** The misunderstanding demonstrated by this criticism is that evolution relies *entirely* on random processes. Natural selection is nonrandom. Choice (A) is incorrect because evolution is not goal directed. Organisms are not trying to evolve into something else. Organisms vary, and some varieties are advantageous in a particular environment. Choice (B) maintains that there is no misunderstanding. However, this is not true as explained in the information provided above. Choice (C) accepts the false premise that evolution relies on random chance alone and is therefore incorrect. The amount of time for a modern eukaryotic cell to arise by random chance is likely more than the time that the universe has existed (although this is difficult to estimate). (LO 1.30)

47. **(B)** Viruses are entirely dependent on their hosts for reproduction. As a result, they must coevolve with their hosts. Choice (A) is incorrect because the branching patterns differ very little. Viruses do not choose hosts, so choice (C) is incorrect. Since the branching patterns are not dissimilar and because evolution is not completely random, choice (D) is wrong. Evolution relies on random variation, but natural selection is nonrandom. (LO 1.17)

48. **(A)** The branch that leads to host species A and host species B comes off of the same node as the branch that leads to host species C. Additionally, the branch that leads to virus c shares a node with the branch that leads to viruses d, e, f, and g. Choice (B) is incorrect because host species C is not most closely related to host species D and because virus c is more distantly related to viruses a and b than it is to the remaining viruses. Choice (C) describes host species and virus relationships that are the reverse of what the trees show. Choice (D) is incorrect because host species are more closely related to each other than they are to the viruses that infect them. (LO 1.17)

49. **(D)** The most likely explanation is that host species C and host species D were similar enough that the ancestor of virus c and virus d was able to jump to a new host. Choice (A) is incorrect because ancestral species are not shown on these trees. Only currently existing species are shown. Choice (B) is incorrect because, while the host species that virus c relied on may have gone extinct, this would not have caused a mutation. Choice (C) is not correct because host species C and virus c are clearly exceptions to the pattern. They arise off of different branches. Host species C is most closely related to host species A and host species B, while virus c is most closely related to virus d. All of the other branching patterns match well. (LO 1.18)

50. **(B)** Alleles are passed through the populations that overlap, except for populations A and E. For any allele to get from population A to population D, it must first be passed to population B and then to population C. Therefore, populations A and D are further apart from each other genetically than populations B and C are from each other. Choices (A) and (C) are incorrect because there really is no way to compare two adjacent populations to two

other adjacent populations. Choice (D) states the opposite of what is likely to be true. Populations A and E are likely more *distantly* related to each other than populations A and D are to each other. (LO 1.8)

51. **(D)** When birds from population A mate with birds from population E, there is a large investment of energy without any return because no offspring are produced. This is a strong selection for individuals to mate within their own population. Populations A and E are not geographically isolated, as suggested by choice (A). Since they are sympatric (live in the same area), populations A and E would be exposed to the same predators, making choice (B) incorrect. Bird songs are important in species recognition. Birds that do not sing a song correctly will not have a chance to mate, so choice (C) is incorrect. (LO 1.20)

52. **(B)** Populations B, C, and D allow populations A and E to exchange genetic information, even though direct hybridization is not possible. Without these bridge populations, there would be no way for populations A and E to exchange genetic information. Organisms that are genetically isolated are considered to be two different species. Choice (A) is incorrect because the background information states that A + E hybrid eggs do not hatch. Therefore, hybridization would not be able to prevent speciation from occurring. Although small populations are more at risk of extinction, the question provides no information to believe that populations A and E are not sufficiently large enough to be sustainable. Therefore, choice (C) is not correct. Choice (D) is incorrect because the recolonization of vacant habitats would have no effect on hybrid viability. (LO 1.20)

53. **(A)** The bone structures shown are examples of homologous structures. The underlying pattern of bones suggests common ancestry. Choice (B) is incorrect because convergent evolution produces structures that have a similar form but often different underlying structures. It is unlikely that separate evolutionary paths would all lead to bones, let alone the pattern of bones shown. Therefore, choice (C) is not correct. Choice (D) is incorrect because there is no way to determine which of the species was the ancestral species from the figure shown. In fact, these species all shared a common ancestor that was none of the animals shown. (LO 1.16)

54. **(A)** HA1 and HB1 appear to have diverged from one another at node 9. All of the species on the top branch after node 9 have HA1 and all of the species on the bottom branch after node 9 have HB1. This is before the species diverged from one another starting at nodes 7 and 8. For choice (B) to be correct, the divergence of HA1 and HB1 would have had to follow the same pattern in multiple lines. This is unlikely. If choice (C) were correct, one would expect the mouse HA1 and HB1 to be grouped together rather than with the other mammals. If *Hox* genes did not arise by gene duplication, we would not expect any sequence homology. HA1 and HB1 must have

sequence homology because they are grouped more closely together than they are with HD8. Therefore, they likely evolved from the same ancestral sequence, so choice (D) is incorrect. (LO 1.15)

55. **(C)** Node 7 includes all of the HA1 sequences, and node 8 includes all of the HB1 sequences. Therefore, all of the HA1 sequences are more similar to each other than any HA1 sequence is to an HB1 sequence. Choice (A) is incorrect because the divergence between the lines that gave rise to humans and gorillas happened before the modern species arose. Choice (B) is incorrect because node 1 includes *Homo* and *Gorilla*, suggesting that these sequences are more similar to each other than *Gorilla* is to *Pan*. Choice (D) is incorrect because nodes 7 and 8 suggest a common ancestor among these organisms, not that primates evolved from mice. (LO 1.18)

56. **(C)** If all the organisms in a phylogenetic tree share a trait, the simplest explanation is that the common ancestor also shared that trait. Choice (A) has this reasoning backward. Choice (B) relies on the misunderstanding that the only type of egg is a chicken egg. If choice (B) had read "the chicken and the chicken egg evolved at the same time," it would have been correct. Choice (D) is wrong because a conclusion can be drawn. (LO 1.17)

57. **(C)** This explanation requires the fewest changes to occur in the evolutionary history of all of the species shown and is therefore the best choice. Choices (A) and (B) each require more changes to have occurred during the evolutionary history of all of the species shown in order to result in the pattern shown. Therefore, these choices are incorrect. Choice (D) is incorrect because a conclusion can be drawn from the data presented. (LO 1.18)

58. **(D)** Since the two phylogenetic trees don't match, we know that there are differences. Additional studies might reveal changes that could resolve the differences so that the trees would match. The fact that the trees don't correspond to each other makes choice (A) incorrect. Since the actual phylogenetic relatedness of the firefly species is not known, it is impossible to know whether the molecular analysis or the morphological analysis is closer to reality. Therefore, choice (B) and choice (C) are incorrect. (LO 1.18)

59. **(B)** The students were applying an artificial selection pressure to the plants. The students did not consider the sex of the plants, so choice (A) is not correct. Although some genetic drift may have occurred, this was not the primary effect. Therefore, choice (C) is wrong. No new plants were added to the population. Therefore, immigration was not occurring, and choice (D) is incorrect. (LO 1.7)

60. **(A)** A good control group undergoes everything the experimental group does except the thing being tested, which is selection in this case. Hybridizing the different selected groups would likely produce an intermediate phenotype. However, this would not be useful as a control group, so choice (B) is

incorrect. Exposure to the same amount of light and using the same microscope at the same magnification are examples of controlled variables. A control group is not the same as a controlled variable, so choices (C) and (D) are incorrect. (LO 1.7)

61. **(B)** Comparing similar leaves from different plants ensures consistency and is an example of a controlled variable. Selecting only bald plants is an example of an experimental treatment, not a controlled variable. Therefore, choice (A) is incorrect. Choices (C) and (D) are examples of controlled variables, but these were not in the description of the experiment. (LO 1.7)

62. **(D)** Since the differences are great enough that the hairiest bald-selected plant has fewer trichomes than the baldest hairy-selected plant, the two average trichome numbers are clearly different. The percentages presented in choices (A) and (B) are arbitrary distinctions. Choice (C) is incorrect because the ranges do not overlap. (LO 1.7)

63. **(D)** A quick way to determine whether significant differences are likely to exist is to compare the averages of the treatments to the standard error of the mean. For example, if we compare the effect of the fan on the hairy plants, the average percent changes in mass are 21 with a standard error of 4.1 for the hairy plants with the fan and 7 with a standard error of 2.5 for the hairy plants without the fan (see the table provided for this question). The standard error describes a range in which the true average value is likely to exist. Therefore, if we add the standard error to the low number ($7 + 2.5 = 9.5$), and subtract the standard error from the high number ($21 - 4.1 = 16.9$), we can then compare the highest possible value for the no fan–hairy group with the lowest possible value for the fan–hairy group. Since 9.5 is still much lower than 16.9, it seems likely that a more complex statistical test (such as an analysis of variance) will indicate significant differences. If we drew a graph with the averages and the standard errors as error bars, the error bars would not overlap. If the error bars were to overlap, choice (A) would be correct. However, they would not overlap. Comparing the fan–hairy group to the no fan–bald group does not isolate the effect of the fan. Therefore, choice (B) is incorrect. In most cases where error bars do not overlap, a statistical test will show significant differences. For choice (C), we would compare the average of the fan–bald group, which had an average percent change in mass of 39 with a standard error of 2.1, to the average of the no fan–bald group, which had an average percent change in mass of 24 with a standard error of 3.9. For the fan–bald group, subtracting the standard error from the average would give 36.9 (since $39 - 2.1 = 36.9$), while for the no fan–bald group, adding the average and the standard error would give 27.9 (since $24 + 3.9 = 27.9$). Since the low value would still be lower than the adjusted high value, we would expect these results to be significantly different. Since choice (C) says that these results are not likely to be significantly different, this choice is incorrect. (LO 1.7)

64. **(C)** If you were to graph the results using the averages and if you were to add the standard errors as the error bars, you would be able to visually determine whether or not the error bars overlap. This would help you develop a reasonable idea of whether or not the differences were statistically significant. Since the error bars for these treatments would not overlap, a more complex statistical test (which is beyond the scope of AP Biology) would probably yield statistically significant results. Choice (A) is incorrect because the standard error was actually quite small. Choice (B) does not isolate the effect of the fan, so it cannot be correct. In order to test the effect of the fan, you must compare treatments that are only different in the presence of or in the absence of the fan. This choice compares hairy plants in the presence of the fan with bald plants in the absence of the fan. No direct comparison is possible. Choice (D) is incorrect because the standard errors were similar for both plants exposed to the fan and those that were not exposed to the fan. (LO 1.5)

65. **(A)** Air movement moves water vapor away from the leaf, decreasing water potential. Since the air is dryer than the leaf, water evaporates more rapidly from the leaf. The leaf might experience a temperature drop from water evaporating. However, a higher water potential will decrease, not increase, the likelihood of evaporation occurring. Therefore, choice (B) is incorrect. Water is not actively transported out of the xylem since the xylem does not have plasma membranes. Therefore, choice (C) is incorrect. The table clearly shows that the fan has an effect on transpiration, so choice (D) is not correct. (LO 1.5)

66. **(A)** Trichomes (hairs) trap gases and slow their movement. In this case, trichomes slow the movement of water vapor. This makes the environment near the leaf have higher water potential and a less steep gradient for evaporation. If significant water were lost through trichomes, choice (B) might be correct. Most water lost from a leaf exits through the stomata. The data show that trichomes decrease transpiration rates, so choices (C) and (D) are incorrect. (LO 1.5)

67. **(C)** In small populations, the effect of random events is large. The frequency of sexual selection is not dependent on population size, so choice (A) is incorrect. Although inbreeding is more common in small populations, inbreeding does not affect the mutation rate. Therefore, choice (B) is incorrect. In an infinitely large population, the effect of randomness disappears. Therefore, choice (D) is not correct. (LO 1.6)

68. **(A)** Since random events affect allele frequencies more in a small population, alleles usually become fixed (100% of alleles) or extinct (0% of alleles) over time. Large populations are *less* likely to lose genetic variation than small populations, so choice (B) is incorrect. Mutation rates do not depend on population size, so choice (C) is incorrect. All kinds of selection, including

sexual selection, lead to *decreased* variability. Therefore, choice (D) is not correct. (LO 1.6)

69. **(B)** Alleles from the larger population flowing into the small population periodically refresh alleles that the small population loses due to genetic drift. Choice (A) has this backward. Choice (C) is incorrect because immigration does have an effect on the smaller population. Immigration does not remove alleles from the gene pool, so choice (D) is not correct. (LO 1.6)

70. **(C)** A change in allele frequencies indicates that evolution is happening on the population level. Provided that the forward mutation rate and the backward mutation rate do not change, there will always be an equilibrium point where the two balance. Therefore, choice (A) is incorrect. When allele frequencies stay the same, no evolution is occurring and the population is in Hardy-Weinberg equilibrium. Therefore, choices (B) and (D) are incorrect. (LO 1.6)

71. **(D)** A larger number of *A1* alleles with a lower mutation rate balance a smaller number of *A2* alleles with a higher mutation rate. There was no selection pressure in this simulation, so no deleterious alleles were removed, making choice (A) incorrect. Choice (B) is incorrect because there never was any selection pressure in this simulation; thus natural selection could not have stopped. There are fewer copies of allele *A2* in the population, so choice (C) is incorrect. (LO 1.13)

72. **(A)** An invasive species spreading uncontrollably would have a large impact on the native forest. Choices (B) and (C) would lead to a smaller impact on the forest. Interbreeding with native species, while unlikely, would only lead to more diversity in the adelgid population. There is no clear reason for this to have a large impact on the native forest, so choice (D) is incorrect. (LO 1.2, LO 4.22)

73. **(A)** In a small population, a single person with a rare allele will skew the frequency of that allele more so than in a large population. Choice (B) is incorrect because no information is given about the diet of the OOM population. Additionally, individuals with MSUD who eat diets rich in valine, leucine, and isoleucine would die. No information is given about mosquito-borne illnesses. Therefore, there is no reason to believe this is an example of heterozygote advantage as choice (C) asserts. There is no link between the number of offspring an individual has and the rate at which mutation occurs. Therefore, choice (D) is incorrect. (LO 1.7)

74. **(B)** In order to show symptoms of MSUD, the man's mother would have had to be homozygous. Therefore, she must have passed on one of her alleles to her son. The fact that the man does not have MSUD means that he is a heterozygote for the allele. Therefore, he has a 50% chance of passing the MSUD allele to his child. The information provided states that the disorder

is autosomal, so choices (A) and (D) are incorrect because they both rely on a discussion of the X chromosome, which is not involved in autosomal inheritance. The fact that the man's mother had the disorder gives us definite information about his genotype. Additionally, people who are not part of the OOM population can also have MSUD, so choice (C) is incorrect. (LO 1.4, LO 3.26)

75. **(B)** The woman is part of the world population that has a frequency of homozygous recessives of $\frac{1}{185{,}000}$. Using the Hardy-Weinberg equations, this is the term q^2. From this, it is possible to calculate the frequency of heterozygotes ($2pq$) as 0.4%. To reach this calculation, first take the square root of $\frac{1}{185{,}000}$ to find that $q = 0.00232$. To find p, rearrange the equation $p + q = 1$ so that it reads as $p = 1 - q$, which equals $1 - 0.00232 = 0.99768$. Next, multiply $2 \times p \times q = 2 \times 0.99768 \times 0.00232 = 0.004$, which equals 0.4%. This also means that the chance is higher than 0%, so choice (A) is incorrect. Choices (C) and (D) rely on the misunderstanding that marriage somehow changes a person's genetic makeup. (LO 1.4, LO 3.26)

76. **(B)** In order for the parents to have a child with the disease, the mother has to be a heterozygote. Having a first child without the disease gives us no information in addition to what we had before. Therefore, the odds of the second child having the disease do not change. Choices (A), (C), and (D) all rely on the mistaken assumption that the odds change for subsequent children. (LO 1.4, LO 3.26)

77. **(C)** Since the couple had a child with the disorder, the woman must be a heterozygote. It has already been established that the man is a heterozygote because his mother had the disease. The odds of two heterozygotes producing a homozygous recessive child are 25%. Therefore, choices (A), (B), and (D) are incorrect. (LO 1.4, LO 3.26)

78. **(A)** Since *Pax-6* turned on all three types of eyes, it is most likely a regulatory gene that controls the genes that make eyes. Choice (B) is incorrect because, if the genes were completely different sequences in each species, they wouldn't be able to be transferred and show the same function. If choice (C) were correct, we would expect it to be able to turn on different types of eyes in a single organism but it can't. Ribosomes do not produce mRNA; they read it. Therefore, choice (D) is incorrect. (LO 1.16)

79. **(B)** *Pax-6* likely evolved before different types of eyes evolved from a common ancestor. The fact that a single sequence can turn on different types of eyes suggests that the gene is ancestral to different eye types. If eyes evolved independently, they would be unlikely to use the same regulatory sequence. Therefore, choice (A) is incorrect. The fact that the same genetic sequence is used to turn on eyes in different species supports the idea that the eye

evolved from common sources. Therefore, choice (C) is not correct. The fact that the *Pax-6* gene doesn't induce fruit fly eyes in mice suggests that the gene is missing information for making a compound eye. Therefore, choice (D) is not correct. (LO 1.16)

80. **(B)** The null hypothesis should state that there is no mate selection occurring. If there is no mate selection, it should be equally likely for any pairing to be successful. This would produce a 1:1:1:1 ratio. Choices (A) and (D) give ratios not seen in the table. Choice (C) gives the correct alternative hypothesis and is therefore incorrect. (LO 1.20)

81. **(C)** The degrees of freedom are found by the number of options minus 1 (4 – 1 = 3). If you look at the chi square table (see the Reference Tables at the end of this book) under 3 degrees of freedom for the 0.01 cutoff, the value is 11.34. Since the χ^2 value is less than 11.34, the data are not different enough from what would be expected by random chance to reject the null hypothesis. Choice (A) is incorrect because we should accept, not reject, the null hypothesis. Choices (B) and (D) have an incorrect number of degrees of freedom and the wrong cutoff value. (LO 1.20)

82. **(B)** Since all of the numbers in Table I are close in value, there is no evidence of mating preference. Table II also shows no evidence of mating preference for the same reason. If flies on the same media preferentially hybridized, the numbers in the cells in Table I where M1 and M2 intersect would be significantly higher than in the cells where M1 and M1 intersect. The same pattern would be seen in Table II. Therefore, choice (A) is incorrect. Choice (C) is incorrect because neither Table I nor Table II depicts matings between males and females grown on different media. Table III does depict matings between males and females grown on different media, but the question does not ask about Table III. For choice (D) to be correct, Table II would need to show significantly higher numbers in the cells where S1 and S2 intersect and significantly lower numbers in the cells where S1 and S1 intersect and in the cells where S2 and S2 intersect. Table I would also need to show significantly higher numbers in the cells where M1 and M1 intersect and in the cells where M2 and M2 intersect and significantly lower numbers in the cells where M1 and M2 intersect. This is not the case, so choice (D) is incorrect. (LO 1.21)

83. **(C)** Populations grown on different media had lower mating frequencies than those grown on the same media. Note that the value for maltose to maltose matings was 22 and the value for starch to starch matings was 21. The maltose to starch matings were 7 and 8. This also explains why choice (A) is incorrect, since the flies did not preferentially hybridize; they did the opposite. Choice (B) is incorrect because mate preference seems to depend on the media the fly was raised on, not on their sex alone. If all males preferred females grown on starch and all females preferred males grown on

maltose, we would expect a much higher number of successful matings when males grown on maltose (whom all females would prefer) were mated with females grown on starch (whom all males would prefer) as opposed to any other combination. Males grown on starch (whom no females would prefer) paired with females grown on maltose (whom no males would prefer) would be especially unsuccessful. Choice (D) is incorrect because there is a definite preference for mates grown on the same media. Random mating does not always take place. (LO 1.21)

84. **(C)** Maltose-selected flies preferred maltose-selected mates, and starch-selected flies preferred starch-selected mates. Isolation alone did not create preferences due to genetic drift, so choice (A) is wrong. There was no immigration or emigration of flies in the experiment. Therefore, choice (B) is incorrect. Mate choice controls reproductive isolation, which is central to speciation events. Therefore, choice (D) is incorrect. (LO 1.24)

85. **(A)** *Tetrahymena* use their contractile vacuole to expel excess water that enters when the organisms are hypertonic to their environment. As the environment becomes saltier, the organisms have less need to pump out water and have no need to do so when the environment is isotonic. The organisms were hypertonic to their environment at a salt concentration of 0 mM NaCl. They became less hypertonic the closer the solution came to 500 mM NaCl. Therefore, choices (B) and (C) are incorrect. There is no reason to believe that ATP production would be impossible at 500 mM NaCl, so choice (D) is wrong. (LO 1.12, LO 2.12, LO 4.5)

86. **(C)** Since the organisms stop pumping at 500 mM NaCl, they go from being hypertonic to the environment to being isotonic. A higher NaCl concentration than 500 mM NaCl would make the *Tetrahymena* hypotonic to the environment. Water would flow out, and the cells would shrink. Adding aquaporins to the membrane should make water flow out faster, not cause the cell to stay the same size. Therefore, choice (A) is incorrect. No information is provided that suggests that the contractile vacuole can secrete NaCl, so choice (B) is incorrect. A hypertonic environment would cause water to flow out, not in. Therefore, choice (D) is wrong. (LO 1.12, LO 2.12)

87. **(D)** Pumping water against its concentration gradient requires energy, so the process must be active transport. Since the process restores osmolarity levels, the mechanism must be negative feedback. Choices (A) and (B) are incorrect because passive transport would not move water out of the cell. In addition, choice (A) incorrectly states that this is a positive feedback loop. Choice (C) is incorrect because positive feedback loops amplify situations; they do not restore set points. (LO 1.12, LO 2.12)

88. **(C)** Osmolarity is the independent variable and should be on the *x*-axis. Contractile vacuole activity is the dependent variable and should be on the *y*-axis. Choice (A) is incorrect because pie graphs are used to show fractions

of a whole, and the data are not additive. Choice (B) has the dependent and independent variables graphed on the wrong axes. Among other errors, the graph in choice (D) implies that the standard error of the mean is dependent on the number of contractions per minute. It is hard to imagine a scenario where that would be the case. (LO 1.12, LO 2.9)

89. **(B)** Each step in the dilution series took 1 mL and diluted it to a total volume of 10 mL. Therefore, the concentration of each dilution changes by a factor of 10. Dilution 5 grew about 20 colonies, and dilution 4 grew about 200. Therefore, dilution 3 should have grown about 2,000 yeast colonies. Choice (A) is less than dilution 4 and is therefore wrong. Choices (C) and (D) are too high. (LO 1.3)

90. **(C)** The students plated 100 μL of solution, which is 0.1 mL. If there are 230 yeast in 0.1 mL, there ought to be 2,300 yeast in 1 mL. Choice (A) is incorrect because it divides the number of colonies listed in the table by 10 instead of multiplying that number by 10. Choice (B) is wrong because it ignores the dilution factor caused by plating 0.1 mL. Choice (D) is too high and involves multiplying the number of colonies by 100. (LO 1.3)

91. **(B)** As shown in the answer to question 90, there are 2,300 yeast/mL in dilution 4. Therefore, dilution 3 has 10 times more, dilution 2 has 100 times more, dilution 1 has 1,000 times more, and dilution 0 has 10,000 times more. Multiply $2{,}300 \times 10{,}000 = 23{,}000{,}000 = 2.3 \times 10^7$ yeast/mL. Choices (A) and (C) suggest that the more concentrated solution has fewer yeast than the dilution solutions, which doesn't make sense. Choice (D) doesn't have enough zeros. (LO 1.3)

92. **(D)** Lawns look smooth and green because they are made of lots of blades of grass, just like the overlapping colonies on the plate of bacteria. There is a limit to cell size, and bacteria do not regularly fuse their membranes. Therefore, choice (A) is incorrect. Only a single bacterium is needed to make a colony, so choice (B) is incorrect. There is no mention of viruses in the stem of the question. Therefore, there is no reason to believe they would be on the agar plates, making choice (C) wrong. (LO 1.3)

93. **(B)** Many of the lines end in period II. This indicates extinction. If choice (A) were true, there would be many new branches produced in period II. Choice (C) is incorrect. Two species might interbreed. However, the lines shown indicate higher taxonomic levels (orders) that contain many species. Choice (D) has the timeline backward. Period II came before period III. (LO 1.20)

94. **(C)** Changes that take place over long geological times do not lead to mass extinctions because the rate of extinction is balanced by the rate of speciation. The remaining choices, (A), (B), and (D), all occur over short periods of time in a geological sense and can cause mass extinctions as seen in period II. (LO 1.20)

95. **(B)** The large number of branches generated in period III indicates new species being produced. Choice (A) is incorrect because the number of orders *increased.* It did not decrease. Choice (C) is incorrect because the number of orders did not stay the same. Since the number of branches in period III is fewer than those that existed before period II, it is unlikely that there is more diversity. Therefore, choice (D) is incorrect. (LO 1.20)

96. **(D)** This pattern can be seen several times in Earth's history. The comet that is thought to have caused the extinction of the nonavian dinosaurs led to an adaptive radiation like the one shown in this diagram. Choice (A) is incorrect because different species cannot inbreed. Deleterious mutations are removed from populations but do not cause entire orders to go extinct. Therefore, choice (B) is incorrect. Choice (C) is incorrect because the other orders went extinct. They did not evolve faster. (LO 1.20)

97. **(B)** Punctuated equilibrium includes periods of rapid evolution (horizontal lines) between periods of stasis (vertical lines). In contrast, gradualism has a relatively constant evolutionary rate. Both models include natural selection as a major mechanism of evolution, so choice (A) is incorrect. Both models show the production of 4 different species. In the gradualism model shown, one lineage ends in a shaded circle, one ends in an unshaded square, one ends in a shaded square, and the final ends in a shaded rectangle. In the punctuated equilibrium model shown, one lineage ends in a shaded triangle, one ends in an unshaded triangle, one ends in a shaded circle, and the final ends in an unshaded square. Therefore, choice (C) is wrong. Punctuated equilibrium was proposed to explain the difference between what was observed in the fossil record and what Darwin described. It is substantially different from gradualism because gradualism predicts a constant rate of evolution producing small changes over long periods of time whereas punctuated equilibrium predicts a changeable rate of evolution producing rapid bursts of evolution and periods with no evolution. Therefore, choice (D) is not correct. (LO 1.13)

98. **(C)** The fossil record shows long periods where one form predominates and then is rapidly replaced by another. Punctuated equilibrium explains why this happens. One explanation for the apparent jerkiness of the fossil record is that there are gaps in it because not enough fossils have been collected. The gradualism model, on the other hand, suggests that many intermediate fossils existed that haven't yet been found. Choice (A) is true for gradualism. However, the choice is incorrect because the question asks which prediction would not be made by gradualism. Both models explain how diversity evolves, so choice (B) is incorrect. Choice (D) is incorrect because neither model suggests that organisms will eventually give rise to sentient beings. (LO 1.13)

99. **(C)** Stasis is maintained by a good fit between the niche and the species. Phenotypes that differ significantly are not well adapted and are selected against. Random fluctuations are destabilizing, so choice (A) is incorrect. Evolution does not occur because organisms strive to change, so choice (B) is incorrect. Diversifying selection selects against the average phenotype and for the extremes. This could lead to speciation, not stasis. Therefore, choice (D) is incorrect. (LO 1.18)

100. **(D)** In 2016, the largest galls were larger than in previous years, and the average gall size went up as well. Choice (A) is the opposite of what is shown. Choice (B) is incorrect because the distribution did change. For the gall size to fluctuate widely, one would expect the gall size to get large and then small and then large again. This did not happen, so choice (C) is incorrect. (LO 1.26)

101. **(B)** More parasitoids would cause a selection pressure away from small galls and toward larger galls as would fewer woodpeckers. Choice (A) is the opposite of the situation shown. Stabilizing selection would lead to more galls of average size and fewer extremes on both sides. Therefore, choice (C) is incorrect. Diversifying selection would occur if the middle-sized galls were selected against. Therefore, choice (D) is wrong. (LO 1.26)

102. **(A)** This experiment directly tests the wasps' choice of galls. Choice (B) answers the question of how frequently wasps attack small galls. However, it gives no information about choice of galls. The question asks about wasp choice, not woodpecker choice. Therefore, choice (C) is incorrect. The size of each wasp's ovipositor may explain the evolutionary advantage of a wasp choosing small galls. However, it does not test the hypothesis stated in the question, so choice (D) is incorrect. (LO 1.11)

103. **(A)** If the wasps are presented with only one gall each, they will be forced to attempt to lay their eggs on it. If the hypothesis is correct, larvae should only develop on the smaller galls. If the hypothesis is incorrect, the wasps might refuse to lay eggs on the larger galls, supporting the idea that they choose which galls to lay eggs on. Alternatively, they may lay eggs that do survive even on the larger galls. This would also disprove the hypothesis in the question. In either case, the experiment described in choice (A) is the best experiment to test the hypothesis in the question. If the effect seen is not choice but is simply a matter of ovipositor length, this experiment should reveal that fact. Choice (B) is incorrect because it does not provide a control group. Choice (C) is incorrect because dissecting large galls for wasp damage would not give any information that answers the question of the effects of wasps attempting to lay on larger galls. Choice (D) doesn't test wasp egg laying. (LO 1.11)

104. **(D)** The structural similarities between the chloroplast and cyanobacteria suggest similar evolutionary histories. These structural similarities also predict DNA sequence similarities, so choice (A) is incorrect. For choice (B) to be

correct, each one of the approximately 500,000 species of existing plants would have to incorporate cyanobacteria from the environment and use them for photosynthesis. The hypothesis that all species of plants evolved from earlier organisms that incorporated cyanobacteria is a much simpler explanation. Since choice (B) is a much more complex explanation, it is less likely to be correct. Choice (C) requires the evolution of smaller genomes from larger ones, circular chromosomes from linear ones, and 70S ribosomes from 80S ribosomes. It also requires these changes in genome size and shape and in ribosome size to have happened many times. Choice (C) is therefore a much more complex model than the one presented in choice (D). In choice (D), cyanobacteria that already had all of these properties were incorporated as part of ancient eukaryotic cells whose descendants are the currently existing species of plants. The more complex model described in choice (C) is less likely to be correct. (LO 1.14)

105. **(C)** Since all plants have chloroplasts with the same structure, the simplest explanation is that chloroplasts evolved before the divergence of plants. Choice (A) suggests multiple origins of the chloroplast, which is less likely than a single origin. Mitochondria produce carbon dioxide from respiration. This is a catabolic process, and it does not create starch. Therefore, choice (B) is incorrect. The endoplasmic reticulum does not contain DNA, but according to the table, the chloroplast does. Therefore, choice (D) is incorrect. (LO 1.15)

106. **(A)** The chloroplast and cyanobacteria are both photosynthetic and have prokaryotic chromosomes. Therefore, they should be most closely related. *E. coli* is a bacterium and has a circular chromosome. However, is not photosynthetic. Therefore, the *E. coli* should be more closely related to the chloroplast and cyanobacteria than to the plant nucleus. Choices (B) and (D) are incorrect because both trees show the chloroplast and the plant nucleus being the most closely related. Choice (C) shows *E. coli* as the nearest relative to cyanobacteria. However, the chloroplast is closer to cyanobacteria. (LO 1.18)

Grid-In Questions (pages 54–57)

When working through the grid-in, long free-response, and short free-response questions that involve math, remember that your answer may vary slightly from the answer provided due to rounding or due to reading a graph differently. See page ix of the Introduction of this book for more information regarding how to avoid rounding errors. As a general rule, if you are within 5% of the actual answer, then your answer will be marked as correct.

107. **0.0023** The frequency of the recessive phenotype in the world population is 1 in 185,000 (5.4×10^{-6}). In the Hardy-Weinberg equation, this is q^2. To find the recessive allele frequency, simply take the square root of q^2. Be sure to report the number of digits specified in the question. (LO 1.1)

108. **0.0513** The frequency of the recessive phenotype in the OOM population is 1 in 380 (2.6×10^{-3}). In the Hardy-Weinberg equation, this is q^2. To find the recessive allele frequency, simply take the square root of q^2. Be sure to report the number of digits specified in the question. (LO 1.1)

109. **0.45** From the graph, we find that the frequency of aa is 0.3 in June (month 6). Using the Hardy-Weinberg equation, $q^2 = 0.3$. The question asks for the frequency of the dominant allele, A, which is represented in the formula as p. First, take the square root of q^2 (0.3) to find q, which equals 0.55. Then, use the formula $p + q = 1$ to find p.

$$p + 0.55 = 1$$
$$p = 0.45$$

(LO 1.3)

110. **81** From the graph, the recessive allele frequency (q) at 30° N is 0.9. The question asks for the recessive phenotype, which is q^2. In this case, q^2 is 0.9^2, which equals 0.81. Multiply 0.81 by 100 to get the percentage. (LO 1.6)

111. **18** From the graph, the recessive allele frequency (q) at 30° N is 0.9. The frequency of the dominant allele is $1 - q = 1 - 0.9 = 0.1$. The question asks for the percentage of heterozygotes, or $2pq$, expressed as a percentage, which is $2(0.1)(0.9) = 0.18 \times 100 = 18\%$. (LO 1.6)

112. **0.1** From the graph, the recessive allele frequency (q) at 30° N is 0.9. The frequency of the dominant allele is $1 - q = 1 - 0.9 = 0.1$. (LO 1.6)

113. **36** At 34° N latitude, the recessive allele frequency (q) is 0.6. The question asks what percentage are homozygous recessive, which is q^2. In this case, $q^2 = 0.6^2 = 0.36$. Multiply 0.36 by 100 to get the percentage. (LO 1.6)

114. **48** From the graph, the recessive allele frequency (q) at 34° N is 0.6. The frequency of the dominant allele is $1 - q = 1 - 0.6 = 0.4$. The question asks

for the percentage of heterozygotes, or $2pq$, expressed as a percentage, which is $2(0.4)(0.6) = 0.48 \times 100 = 48\%$. (LO 1.6)

115. **0.4** From the graph, the recessive allele frequency (q) at 34° N is 0.6. The frequency of the dominant allele is $1 - q = 1 - 0.6 = 0.4$. (LO 1.6)

116. **50** Mating a pure breeding black fly (BB) with a pure breeding light brown fly (bb) would produce all heterozygous flies (Bb). When the heterozygotes are permitted to mate, 75% of their offspring should be black and 25% should be light brown. If all of the assumptions of the Hardy-Weinberg equilibrium are met, there should be no evolution, and the allele frequencies should remain the same. Therefore, 25% of the offspring should show the recessive (light brown) phenotype. If there are 200 flies, then 25% of the flies is 50 flies. (LO 1.3)

117. **0.41** Since we expect a 1:1:1:1 ratio and the average of the values is 22.75 $\left(\frac{23 + 25 + 21 + 22}{4} = 22.75\right)$, the expected values are all 22.75. Use the chi-square formula:

$$\chi^2 = \sum \frac{(o - e)^2}{e}$$

Find the expected values by taking the average of the observed values in Table II.

Table II Matings of S1 and S2 Males and Females (Expected Values)

	Females	
Males	**S1**	**S2**
S1	22.75	22.75
S2	22.75	22.75

Then subtract the expected value from the observed value for each box.

Table II Matings of S1 and S2 Males and Females
(Observed Value – Expected Value ($o - e$))

	Females	
Males	**S1**	**S2**
S1	23 – 22.75 = 0.25	25 – 22.75 = 2.25
S2	21 – 22.75 = –1.75	22 – 22.75 = –0.75

Next, square each result from the table above.

Table II Matings of S1 and S2 Males and Females (Observed Value – Expected Value $(o - e)^2$)

	Females	
Males	**S1**	**S2**
S1	$(0.25)^2 = 0.625$	$(2.25)^2 = 5.0625$
S2	$(-1.75)^2 = 3.0625$	$(-0.75)^2 = 0.5625$

Then divide each result from the table above by the expected value.

Table II Matings of S1 and S2 Males and Females $\left(\text{Observed Value} - \text{Expected Value}\left(\frac{(o-e)^2}{e}\right)\right)$

	Females	
Males	**S1**	**S2**
S1	$\frac{0.625}{22.75} = 0.0275$	$\frac{5.0625}{22.75} = 0.2225$
S2	$\frac{3.0625}{22.75} = 0.1346$	$\frac{0.5625}{22.75} = 0.0247$

Add all of these final values together to get 0.41:

$$0.0275 + 0.2225 + 0.1346 + 0.0247 = 0.41$$

(LO 1.20)

118. **13.59** Since we expect a 1:1:1:1 ratio and the average of the values is 14.5 $\left(\frac{22 + 8 + 7 + 21}{4} = 14.5\right)$, the expected values are all 14.5.

Use the chi-square formula:

$$\chi^2 = \sum \frac{(o-e)^2}{e}$$

Find the expected values by taking the average of the observed values in Table III.

Table III Matings of M1 and S1 Males and Females (Expected Values)

	Females	
Males	**M1**	**S1**
M1	14.5	14.5
S1	14.5	14.5

Then subtract the expected value from the observed value for each box.

Table III Matings of M1 and S1 Males and Females
(Observed Value – Expected Value ($o - e$))

	Females	
Males	**M1**	**S1**
M1	22 – 14.5 = 7.5	8 – 14.5 = –6.5
S1	7 – 14.5 = –7.5	21 – 14.5 = 6.5

Next, square each result from the table above.

Table III Matings of M1 and S1 Males and Females
(Observed Value – Expected Value ($(o - e)^2$))

	Females	
Males	**M1**	**S1**
M1	$(7.5)^2 = 56.25$	$(-6.5)^2 = 42.25$
S1	$(-7.5)^2 = 56.25$	$(6.5)^2 = 42.25$

Then divide each result from the table above by the expected value.

Table III Matings of M1 and S1 Males and Females
$\left(\text{Observed Value – Expected Value}\left(\frac{(o-e)^2}{e}\right)\right)$

	Females	
Males	**M1**	**S1**
M1	$\frac{56.25}{14.5} = 3.8793$	$\frac{42.25}{14.5} = 2.9138$
S1	$\frac{56.25}{14.5} = 3.8793$	$\frac{42.25}{14.5} = 2.9138$

Add all of these final values together to get 13.59:

$$3.8793 + 2.9138 + 3.8793 + 2.9138 = 13.59$$

(LO 1.20)

Long Free-Response Questions (pages 58–60)

When working through the grid-in, long free-response, and short free-response questions that involve math, remember that your answer may vary slightly from the answer provided due to rounding or due to reading a graph differently. See page ix of the Introduction of this book for more information regarding how to avoid rounding errors. As a general rule, if you are within 5% of the actual answer, then your answer will be marked as correct.

119. This question asks you to design an experiment to cause microevolutionary change and speciation artificially. The rubric below explains how to grade your response. (LO 1.11)

(a) Design a controlled experiment to cause microevolutionary change and speciation artificially in a sexually reproducing organism of your choice. (1 point for each item listed below—6 points maximum. Note that although 8 points are listed below, you will only be awarded 6 points maximum for any combination of the items, and you do not need to include all 8 items to receive full credit for your experimental design.)

Rationale	The organism should have a short life cycle so that many generations can be observed. Fruit flies, nematodes, Wisconsin Fast Plants, and C-ferns would all be acceptable choices for the organism. Note that the organism must also be sexually reproducing. Therefore, *E coli* would not be an acceptable choice.
Conditions	Manipulate the independent variable. Any reasonable change in environment, such as food source, timing of emergence, etc., is acceptable.
	Specify at least 2 variables to be controlled, such as temperature, humidity, day length, etc.
	Use a large sample size (3 in each treatment group minimum).
	Conduct the experiment over a large number of generations (10 minimum).
Assessing Microevolutionary Change	Look for a change in phenotype or genotype. Any reasonable test will be accepted.
Speciation	Describe mating tests and tests for gene flow.
Data Analysis	Describe statistical manipulations, such as chi-square, that apply to this experiment.

(b) Identify which of the five assumptions of the Hardy-Weinberg equilibrium will NOT be met in your experiment. Discuss the importance of TWO of these unmet assumptions with respect to microevolutionary change and speciation. (1 point awarded for each assumption and importance pair listed on the following page—4 points maximum. Only unmet assumptions count. Note that although 5 points are listed in the following table, you will only be awarded 4 points maximum for any combination of the items, and you do not need to include all 5 items to receive full credit for your response.)

Assumption	Importance
Large population size	A large population size increases the effect of genetic drift, which can lead to a change in allele frequencies, which is microevolutionary change.
No net mutation	Mutation from one allele to another can cause microevolutionary change. It is unlikely that this will be unmet.
No immigration or emigration	Immigration introduces alleles into a population, and emigration removes them. Either immigration or emigration can lead to a change in allele frequencies.
Random mating	Mate choice is one key factor that reinforces speciation because it contributes to the genetic isolation of different groups.
Natural selection	Natural selection results in changes in both phenotypes and allele frequencies in response to environmental conditions.

120. This question asks you to graph and interpret data. (LO 1.13)

(a) Graph the birth weights for 1990 and 2015 on the grid provided. (1 point for each of the following—3 points maximum)

- The labels for the title, the axes, and the legend are complete and correct.
- The independent variable is on the *x*-axis, and the dependent variable is on the *y*-axis.
- All data points are plotted correctly.

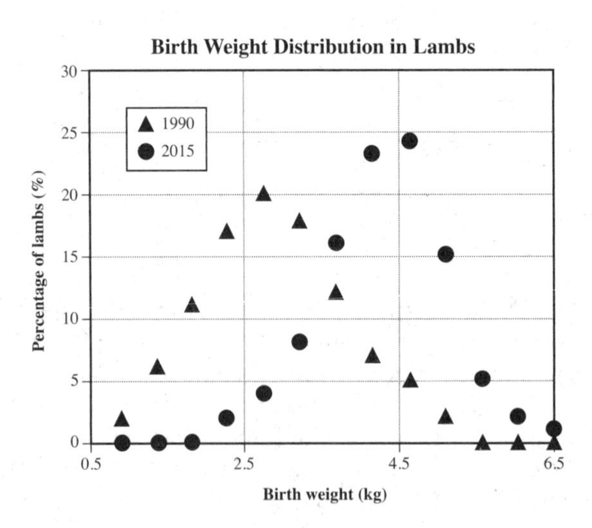

(b) Describe the selective forces that act on high birth weights and on low birth weights. (1 point for each of the following—3 points maximum)

- Low birth weight is selected against because lambs that are too small don't have enough energy to survive cold conditions.
- Farmers breed only sheep that produce large lambs.
- High birth weight is selected against because large lambs don't fit through the birth canal easily. This can kill the mother as well as the lamb.

(c) Compare the distributions in 1990 and 2015. Explain the differences in terms of the selective forces you identified in part (b). (1 point for each of the following—4 points maximum. Note that although 5 points are listed below, you will only be awarded 4 points maximum for any combination of the items, and you do not need to include all 5 items to receive full credit for your response.)

Difference (1 point each)	**Explanation (which must be paired with the correct difference listed to the left of it for 1 point each)**
The distribution is shifted to the right for 2015 compared to 1990 (or shifted to the left in 1990 compared to 2015).	One reason for this distribution was directional selection for larger lambs by farmers. The farmers chose to breed sheep that produced larger lambs.
	Another reason for this distribution was better access to veterinary help for difficult births. This reduced deaths during birth.
The distribution is taller and thinner for 2015 than for 1990 (or shorter and wider in 1990 compared to 2015).	One reason for this distribution was that farmers culled from the herd sheep that produced smaller lambs, but there is an upper limit to the size of a lamb that can be born. This lead to a larger fraction of sheep near the mean and fewer further away.

121. This question asks you to analyze some data collected by fishery scientists. (LO 1.2)

(a) Based on the phenotype data, calculate the expected number of each genotype. Record your answer in the table. Assume that the population is in Hardy-Weinberg equilibrium. (1 point for calculating all of the correct expected values)

The frequency of animals showing the recessive phenotype gives us the value of q^2.

$$q^2 = \frac{85}{1{,}000} = 0.085 \text{ (this is the expected aa)}$$

$$q = \sqrt{q^2} = \sqrt{0.085} = 0.2915$$

Use the equation $p + q = 1$ to determine p.

$$p + 0.2915 = 1$$
$$p = 0.7085$$
$$p^2 = 0.5019$$

The term p^2 is the frequency of homozygous dominant individuals. Multiply the frequency of the homozygous dominant genotype (0.5019) by 1,000 (number of fish sampled) to get the number of individuals that are expected to have that genotype.

$p^2 \times 1{,}000 = 0.5019 \times 1{,}000 = 502$ (This is the expected AA.)

The term $2pq$ is the frequency of heterozygous individuals. Multiply the frequency of the heterozygous genotype by 1,000 (the number of fish sampled) to get the number of individuals that are expected to have that genotype.

$2pq = 2 \times 0.7085 \times 0.2915 = 0.413 \times 1{,}000 = 413$ (This is the expected Aa.)

(b) State the null hypothesis, calculate the chi-square statistic, and interpret the test in terms of the null hypothesis.

The null hypothesis is that the observed values do not differ from the expected values. (1 point)

The chi-square should be calculated like this:

$$\chi^2 = \frac{(468 - 502)^2}{502} + \frac{(477 - 413)^2}{413} + \frac{(85 - 85)^2}{85} = 12.22$$

(1 point for correct calculations and 1 point for the correct answer)

Since there are 3 categories, there are 2 degrees of freedom. Use the chi-square table (see the Reference Tables at the end of this book) to look up the 0.05 cutoff on 2 degrees of freedom. The value you will find is 5.99. Since 12.22 > 5.99, reject the null hypothesis. There is significant evidence that the observed values differ from the expected values. (1 point for rejecting the null hypothesis and 1 point for the explanation for rejecting the null hypothesis)

(c) Identify the selective forces that may be affecting the population. Explain how these forces could lead to the data shown. (1 point for each selective force and 1 point for each matching explanation—4 points maximum. Note that although 6 points are listed in the following table, you will only be awarded 4 points maximum for any combination of the items, as long as the explanation matches the selective force, and you do not need to include all 6 items to receive full credit for your response.)

Selective Force (1 point each—maximum of 2 points)	Explanation (1 point each—maximum of 2 points)
Homozygous recessive is selected against.	Sterile fish do not contribute to the gene pool of the next generation.
Homozygous dominant is less frequent than expected.	Homozygous dominant fish are possibly more susceptible to disease than are heterozygotes.
Heterozygotes occurs more frequently than expected.	Heterozygous fish are possibly resistant to disease. Heterozygotes may be more attractive to mates.

Short Free-Response Questions (pages 61–62)

When working through the grid-in, long free-response, and short free-response questions that involve math, remember that your answer may vary slightly from the answer provided due to rounding or due to reading a graph differently. See page ix of the Introduction of this book for more information regarding how to avoid rounding errors. As a general rule, if you are within 5% of the actual answer, then your answer will be marked as correct.

122. Describe the relationship between the frequency of the recessive allele and the latitude at which the samples were collected. Identify ONE evolutionary cost and ONE benefit to diapause. Explain the pattern shown in the graph. (LO 1.2)

     A maximum of 4 points can be awarded for correctly answering this question. A complete answer includes the following:

     - The recessive allele decreases in frequency as the latitude north increases until the frequency reaches zero. (1 point)
     - One cost is that diapause suspends development and therefore lengthens the time to reach sexual maturity. (1 point)
     - One benefit to diapause is that it allows fruit flies to survive cold weather. (1 point)
     - The allele preventing diapause is selected against in northern climates because organisms that can't enter diapause die. In the south, there is a selective advantage to organisms that don't enter diapause because they mature earlier. In essence, if southern flies were able to enter diapause, they would waste time preparing for very cold weather that would never come. (1 point)

123. Use the information provided to construct a phylogenetic tree for the selected organisms using the tree provided on page 62. Indicate at which

point each of the derived characters arose. Describe the most recent common ancestor of the cat and the lizard in terms of the characters listed, and explain your description. (LO 1.19)

A maximum of 4 points can be awarded for correctly answering this question.

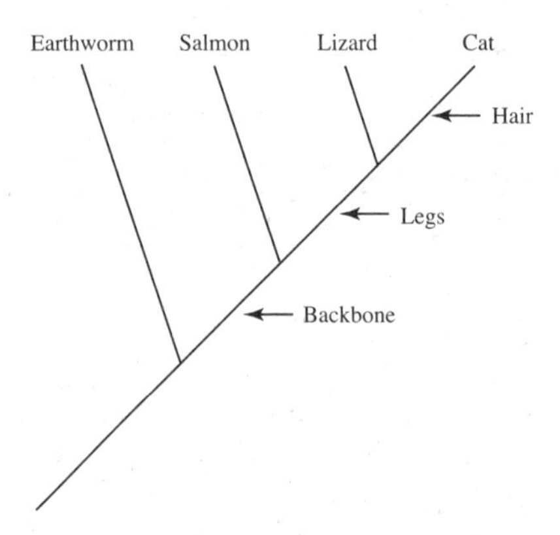

(1 point is awarded for correctly positioning the organisms on the tree. Note that it would be acceptable for the lizard and the cat to be switched if the arrow for hair pointed at the branch leading to the cat. 1 point is awarded for correctly placing the characteristics at the correct locations.)

The most recent common ancestor of the cat and the lizard should have a backbone and legs but no hair. (1 point)

This ancestor should have all of the shared ancestral characteristics but none of the derived characteristics. (1 point)

124. Identify TWO possible adaptive functions that might lead stepwise from shape 1 to shape 5. Describe how genetic variability and natural selection may have worked together in this case to cause adaptation. (LO 1.13)

A maximum of 4 points can be awarded for correctly answering this question.

(a) One point would be awarded for each of the following. (2 points maximum)

- A curved leaf can more easily hold nutrients from falling debris.
- A curved leaf can hold more water.
- A curved leaf has less exposed surface area, so there is less evaporation of stored water.
- The higher the walls are, the harder it is for anything to fall out or escape.

(b) One point would be awarded for each of the following. (2 points maximum)

- The leaf shape is variable and is controlled by genetics (or it is heritable).
- Plants that are better able to collect water and/or nutrients are more likely to have more offspring.

BIG IDEA 2—ENERGY

Multiple-Choice Questions (pages 63–126)

125. **(C)** Species III's osmolarity rose from about 385 mOsmol/L to about 465 mOsmol/L over 6 days (about 13 mOsmol/L per day). This is a steeper slope than seen in the other choices. Species I fell from about 475 mOsmol/L to about 450 mOsmol/L. This is a change of about 4 mOsmol/L per day, so choice (A) is incorrect. Species II fell from about 365 mOsmol/L to about 345 mOsmol/L, or about 3 mOsmol/L per day, so choice (B) is incorrect. Choice (D) is incorrect because the slopes are not identical. (LO 2.12)

126. **(B)** In freshwater, Species II had a pump activity of about 0.5 µmol/h mg protein. This value is higher than that for Species I (about 0.1 µmol/h mg protein) and Species III (about 0.2 µmol/h mg protein). Therefore, choices (A) and (C) are incorrect. Differences exist, so choice (D) is incorrect. (LO 2.12)

127. **(B)** The plasma osmolarity of Species II returned to near normal levels within 10 days. The osmolarity of Species I and Species III did trend back toward their initial values but not as rapidly as Species II did. Therefore, choices (A) and (C) are incorrect. Choice (D) is incorrect because it is possible to determine adaptation based on the figure. (LO 2.12)

128. **(B)** The fish kept in freshwater were the control groups. Without these groups, researchers would not have baselines against which to compare results. There is no evidence that the researchers did not want to kill all of the fish or that there was not enough saltwater, so choices (A) and (C) are not correct. The description of the experiment clearly shows a necessity for control groups, so choice (D) is not correct. (LO 2.21)

129. **(C)** In freshwater, fish absorb ions from the water around them. When placed into saltwater, this same mechanism causes the fish to uptake ions rapidly. Pumping these ions out by active transport returns the blood to normal salt levels. Freshwater fish should have tissues that are hypotonic to saltwater, not hypertonic. If their tissues were hypertonic, the tissues would swell, not shrivel. Therefore, choice (A) is incorrect. Adding solutes to blood by pumping sodium into the plasma would make it more difficult for cells to maintain homeostasis, not easier. Therefore, choice (B) is incorrect. Exposure to high salinity levels does not cause fish to preferentially lay eggs that are better adapted to living in saltwater. Selection against fish that are poorly adapted to high salinity might lead to this effect. However, simple exposure to an environment won't cause this change. Therefore, choice (D) is incorrect. (LO 2.17)

130. **(D)** From the graph, Form B protein has a relative mRNA level of about 2 in saltwater that falls to about 0.5 in freshwater. For Form A protein, the values of mRNA are about 5 in freshwater and about 1 in saltwater. Choice

(A) is incorrect because 1 (for saltwater) is not greater than 5 (for freshwater). Choice (B) is incorrect because 2 (for Form B in saltwater) is not greater than 5 (for Form A in freshwater). Choice (C) is incorrect because 1 (for Form A in saltwater) is not greater than 2 (for Form B in saltwater). (LO 2.24)

131. **(D)** From the graph, we can see that the mRNA for Form B protein has higher expression levels in saltwater and that the mRNA for Form A protein has higher expression levels in freshwater. In saltwater, it is necessary to pump ions out of the body to maintain homeostasis. In freshwater, however, it is necessary to scavenge ions from the water. Choice (A) has a correct description of the data from Figure III. However, the reasoning for why the pumps are needed is backward. Although Form B protein may have a different ATP binding site than does Form A protein, as suggested in choice (B), that cannot be determined from Figure III. Choice (C) would require adult fish to have different DNA than they did when they were young. Gene expression changes based on age, not the DNA, so choice (C) is incorrect. (LO 2.15)

132. **(C)** A lower leaf water potential indicates a stronger pull on a column of water. The longer the plant is at a lower leaf water potential, the more water it will use. Choice (A) is incorrect because Species B had a lower leaf water potential than Species A. Species A did have a higher leaf water potential for a longer period of time. However, that would lead to a *weaker* pull on the column of water and *less* water use. Therefore, choice (B) is wrong. Choice (D) is incorrect because there were differences in the leaf water potential. (LO 2.24)

133. **(A)** Between hours 5 and 8, the slope for Species B is steeper. However, it flattens out during hours 12 to 14 when Species A's slope is still falling. Choice (B) has this situation reversed. Between hours 5 and 12, Species B's water potential falls rapidly and then levels out; the average slope for hours 5 to 8 is steeper. Therefore, choice (C) is incorrect. Choice (D) is incorrect because there were differences in the leaf water potential. (LO 2.24)

134. **(A)** Leaf water potential is a measure of the suction applied by a leaf on the water column in the xylem. In order for water to flow from the roots into the leaf, the water potential in the leaf has to be lower than the water potential in the soil. The figure shows the leaf water potential decreasing as the day progresses toward noon and then increasing as the day continues. More light is available for photosynthesis as the day approaches noon. More photosynthesis leads to a higher need for gas exchange, so plants open their stomata more. This decreases the water potential in the leaf. Higher humidity *increases* the water potential of the air, so choice (B) is incorrect. The leaf water potential *decreases* as the soil dries out, so choice (C) is incorrect. The figure shows the leaf water potential decreasing as the day progresses toward noon when the light intensity is high. Additionally, plants generally open, not close, stomata in response to light. Therefore, choice (D) is incorrect. (LO 2.36)

135. **(A)** Species B develops lower leaf water potential earlier and that leaf water potential stays lower longer than in Species A. Opening stomata earlier is consistent with the lower leaf water potential. If Species B had a thicker cuticle, its leaf water potential should have been higher than that of Species A. Since that is not the case, choice (B) is wrong. Hairy leaves do slow air flow. However, this increases, not decreases, the leaf water potential, so choice (C) is wrong. Fewer stomata lead to a higher, not a lower, leaf water potential because they cause the plant to retain more water in the leaf. Therefore, choice (D) is incorrect. (LO 2.22)

136. **(B)** On day 309, the stressed NM6 had a water potential of about –1.1 MPa while the stressed NE308 had a water potential of about –1.0 MPa. On day 309, the watered NM6 had a water potential of –0.4 while the watered NE308 had a water potential of –0.3. The differences between the water potentials for the watered and stressed plants was very large. Choice (A) is incorrect because NE308 had a higher water potential than NM6 in most cases. Choice (C) is incorrect because there were clear differences between watered and stressed plants. Since choice (B) contains an accurate statement, choice (D) is incorrect. (LO 2.17)

137. **(C)** The water potential of NM6 stressed plants on day 336 was approximately –1.6 MPa. This is much lower than the water potential of NE308 stressed plants, which was approximately –1.1 MPa. Since there are differences in the water potentials for the clones, choice (A) is incorrect. Choice (B) is the opposite of what is correct. The results are the average of a number of samples. Only a single data point can be an outlier. It is also not a good idea to simply disregard outliers without a reason. Therefore, choice (D) is incorrect. (LO 2.15)

138. **(C)** At two times during the experiment, the water potential for NM6 was different than that for NE308. Choice (A) has this backward. Choice (B) is incorrect because experimental error was accounted for in the standard error of the mean. Since it is possible to determine the water potential from the figure, choice (D) is incorrect. (LO 2.15)

139. **(D)** On the *x*-axis, 75 kg is halfway between 50 and 100. The corresponding point on the *y*-axis is about 2,500 kcal/day. Likewise, 45 kg on the *x*-axis corresponds to about 1,700 kcal/day on the *y*-axis. Since neither of these data points is exactly the same, choice (A) cannot be correct. The Western diet population is likely to be more sedentary than a population that is not getting enough to eat. Therefore, choice (B) is the opposite of what would be expected. A Western diet has many more calories than a subsistence diet. This is the opposite of what is described in choice (C). (LO 2.4)

140. **(B)** The data show a direct, although nonlinear, relationship between average mass and TEE. Larger organisms have more cells to maintain and therefore have a larger total energy expenditure than smaller organisms. Obviously,

if this relationship described in choice (B) is true, then choice (A), which states that there is no relationship between TEE and average mass, must be incorrect. Choice (C) states the opposite of the correct conclusion and is thus incorrect. If you divide the TEE by the mass, the highest TEE per unit mass on the graph is at the far left of the graph. Humans are heavier than this, so choice (D) is incorrect. (LO 2.2)

141. **(B)** Dividing the TEE by the mass gives a value that is larger as the organism gets smaller. The metabolic rate is the slope of the graph, which decreases as the mass increases. Since there is in fact a relationship between the metabolic rate and the size of an organism, choice (A) is incorrect. Although a larger organism has a larger TEE, its energy expenditure per unit mass is lower than that of a smaller organism. Therefore, choice (C) is incorrect. Finally, choice (D) is incorrect because humans are found in the middle of the graph and have a metabolic rate per unit mass somewhere in the middle of all the species plotted. (LO 2.2)

142. **(D)** Overfishing, pollution, and habitat loss all lead to lower numbers of crabs. Therefore, any or all of these could have caused the decline in crab population from 1990 to 1999. (LO 2.23)

143. **(B)** If female crabs were harvested at the same rate as male crabs between 1990 and 1995, we would expect an even number of each sex. Prohibiting the harvest of females from 2005 to 2010 would lead to more females than males in the population during those years. Harvesting only male crabs each year would cause fewer total males during the winter. Since there are approximately equal numbers of males and females between 1990 and 1995, there is no evidence of selection against males during these years. Therefore, choice (A) is not correct. No information is given about pollution. Therefore, choice (C) is not a good choice. Choice (D) is not correct because selection against females would lead to the opposite pattern than the one seen. (LO 2.3)

144. **(D)** The limiting factor in the amount of offspring generated each year is the number of females because one male can fertilize many females. Choice (A) is incorrect because no information is presented that suggests that females have a higher natural death rate than males. Choice (B) is incorrect because no evidence is presented to suggest that there is a relationship between female population size and disease. There is also no information that males bite more, so choice (C) is incorrect. (LO 2.3)

145. **(B)** The slope of the curve is steepest at 15 years of age. Before and after this point, the slope becomes less steep, making choices (A), (C), and (D) incorrect. (LO 2.1)

146. **(C)** Harvesting 20-year-old trees captures the growth produced during the most active years. Harvesting after only 10 years gives 100 m^3/ha, making choice (A) a poor choice. Between 10 and 15 years, about 250 m^3/ha is

added. Harvesting at age 15, as choice (B) suggests, would prevent the addition of another 175 m^3/ha. After 20 years, the rate of growth slows to less than it was in the first 10 years. Waiting until 35 years will give less return than harvesting earlier. Therefore, choice (D) is incorrect. (LO 2.1)

147. **(B)** Since the students measured respiration over time and calculated the rates for peas with different amounts of exposure to water, it is most reasonable to think that they wanted to know how the rate of respiration changed with the duration of exposure to water. Choice (A) focuses on photosynthesis, which does not begin until the photosynthetic portions of the pea are produced (leaves primarily). Although the experiment does demonstrate that water exposure leads to germination, there would be no need for durations of exposure if this was the main question of the experiment. Therefore, choice (C) is incorrect. Since the students did not measure oxygen, choice (D) cannot be the correct answer. (LO 2.23)

148. **(D)** When seeds break dormancy and germinate, they rely on stored chemical energy to grow. Choice (A) suggests that germinating plants use light energy, which is true only after leaves have been produced. Breaking down water into hydrogen and oxygen requires a significant input of energy, so choice (B) cannot be correct. Molecular nitrogen is not a good energy source because it requires energy from plants and symbiotic bacteria for the nitrogen to become usable. Therefore, choice (C) is incorrect. (LO 2.2)

149. **(C)** The peas that weren't exposed to water provided the baseline for comparing the other treatments. In other words, these peas were the control. Choice (A) is incorrect because the students didn't measure the amount of water absorbed from the air. Choice (B) is incorrect because the students did not measure anything in units of energy (calories, for example). The students didn't measure weight, so choice (D) is incorrect. (LO 2.3)

150. **(D)** The change in volume in the glass beads was due to changes in temperature and pressure, which would have been the same for the other respirometers. This change was subtracted from the volume in the other respirometers to account for these sources of error. Glass beads neither respire nor remove carbon dioxide from the air. Therefore, choices (A) and (B) are incorrect. The information presented states that the data from the glass beads were subtracted from the other respirometers' data. Therefore, they were really important and necessary and thus choice (C) is wrong. (LO 2.23)

151. **(C)** The slopes of the lines in Figure I are in units of μL/min. This unit is plotted on the *y*-axis in Figure II. Each line in Figure I represents seeds exposed to water for differing numbers of days. Days after exposure to water is displayed on the *x*-axis in Figure II. There is clearly a relationship between the two figures, so choice (A) is wrong. Choice (B) has this relationship backward. Photosynthesis was not measured in the experiment, so choice (D) is not correct. (LO 2.23)

152. **(B)** Standard error is a measure of the variability in the data. Since the variability in the data is greater than the difference between the two means, there is no evidence that the means are actually different. Maltose has an average of 11 bubbles per minute with a standard error of 2.1. We can be reasonably confident that the true mean for maltose is probably at least as high as its average minus its standard deviation (11 – 2.1), which is 8.9. Sucrose has an average of 9 bubbles per minute with a standard error of 2.4. We can be reasonably confident that the true mean for sucrose is at least as low as its average plus its standard deviation (9 + 2.4), which is 11.4. If you were to plot these values on a graph, using the standard errors as error bars, the error bar for sucrose would be above the average for maltose, and the error bar for maltose would be below the average for sucrose. Therefore, these averages are not significantly different. Although students did measure more bubbles on average in the flasks containing maltose, the variability in the data makes these differences not meaningful. Therefore, choice (A) is incorrect. Choice (C) shows a misunderstanding of the standard error. Although there is no evidence that a significant difference exists, this doesn't support the idea that there isn't enough information to compare the results. Therefore, choice (D) is not correct. (LO 2.9)

153. **(A)** Since the yeast could not metabolize the lactose, it is reasonable to believe that the yeast lacks the enzyme to break down lactose. The molecular formula for lactose is given. It contains 12 carbon atoms, so choice (B) is wrong. Since all of the solutions were the same molarity and the yeast survived in two of the other solutions, it is not reasonable to think that there was any osmotic problem in the lactose solution, as choice (C) incorrectly suggests. The students did not test the viability of the yeast in the lactose flask. Therefore, there is no evidence of toxicity, meaning that choice (D) is wrong. (LO 2.2)

154. **(D)** Using equal molar concentrations of the compounds should standardize the number of carbon atoms and the amount of carbon dioxide possible, the number of bubbles produced per minute, and the osmotic conditions. (LO 2.3)

155. **(A)** The rate of respiration depends on the availability of the reactants. Mixing the solution with air dissolves oxygen in the water and increases the rate of respiration. Since the question said that the yeast were added after the mixture was shaken, choice (B) is incorrect. Shaking could have caused more nucleation sites as yeast that were settled were resuspended. However, the shaking occurred before the yeast were added. Choice (C) is therefore incorrect. Choice (D) is wrong because there is no connection between the production of amino acids and the generation of carbon dioxide. Additionally, atmospheric nitrogen is not a form that can be readily used by plants. It must first be fixed by bacteria, such as *Rhizobium*. (LO 2.24)

156. **(B)** This bar graph shows the independent variable (the types of sugar) on the *x*-axis and the dependent variable (fermentation rate) on the *y*-axis. It uses the standard errors as error bars. Pie charts are used to indicate fractions of a whole, and these data do not fit that description. Therefore, choice (A) is not correct. Choice (C) shows the standard errors plotted as separate bars. This format doesn't clearly indicate how the means vary. Since the data are categorical and not continuous numerical data, a line graph, as shown in choice (D), is not appropriate. (LO 2.9)

157. **(B)** Chemicals fit into active sites because of their charge and shape. Chlorinating sucrose changes both the charge and shape of the molecule. As a result, sucralose does not fit into and bind to the active site of sucrase and therefore it can't be broken down. Choice (A) is the opposite of what is true. Choice (C) confuses active sites and allosteric sites. Allosteric sites are regulatory sites that change the shape of active sites. Sucralose appears to be approximately the same size as sucrose in the diagram, so choice (D) is not correct. (LO 2.5)

158. **(B)** ATP is constantly being produced by attaching phosphate to ADP using energy from the breakdown of other compounds, such as glucose, to drive the endergonic reaction. ATP is then used to cause muscle contraction. One common misconception about ATP is that it contains a lot of energy. Glucose and fat are much more energy dense than ATP. ATP simply contains small but useful amounts of energy. Choice (A) suggests that there was a math error. The equation below shows the correct math.

$$10{,}309 \text{ kcal} \times \frac{1 \text{ mol ATP}}{7.3 \text{ kcal}} \times 507 \text{ g/mol} \times \frac{1 \text{ kg}}{1{,}000 \text{ g}} \times 2.2 \text{ lb/kg} \approx 1{,}575 \text{ lb}$$

Therefore, choice (A) is incorrect. Although the food the bicyclist ate may have contained small amounts of ATP, there is no physical way he could have eaten that much ATP even if consumed ATP were able to be transported to cells. Therefore, choice (C) is incorrect. An O_2 molecule weighs *less* than a CO_2 molecule because CO_2 has a carbon atom in addition to the 2 oxygen atoms. Therefore, choice (D) is incorrect. (LO 2.1)

159. **(B)** If the levels of the carrier protein (SHBG) are low, fewer hormones would be transported to cells. In order for choice (A) to be true, fewer carrier proteins would somehow need to carry more hormones per molecule. This would not happen. For choice (C) to be correct, SHBG would need to carry even more hormones per molecule than it would to make choice (A) correct. Therefore, choice (C) is incorrect as well. Low amounts of hormones might eventually lead to an upregulation of SHBG. However, this occurrence certainly wouldn't be an immediate effect after discontinuing steroid use, making choice (D) wrong. (LO 2.15)

160. **(C)** The carrier protein SHBG must have an internal region with the correct shape to carry sex hormones. That internal region must also have the same polarity (nonpolar) as the molecules it carries. The outer portion must be hydrophilic so that SHBG can dissolve in the water found in blood. Choice (A) has this backward. To be carried, the hormone needs to be inside the carrier protein, not bound to the outside. Therefore, choice (B) is incorrect. Since polar molecules do not dissolve in nonpolar portions of phospholipids, choice (D) is wrong. (LO 2.4)

161. **(A)** The graph for the forward reaction depicts an endergonic reaction because the final energy is higher than the starting energy. Choices (B) and (D) are incorrect because the reaction is not exergonic. Hydrolysis of ATP releases free energy and is therefore an exergonic reaction. Therefore, choice (C) is incorrect. (LO 2.3)

162. **(B)** Reactions that release free energy occur more readily than those that absorb free energy. Therefore, the backward reaction would happen more quickly than the forward reaction. Choice (A) is the opposite of this. Forward and backward reactions would happen at the same rate if there was no energy change from product to reactant, so choice (C) is incorrect. Enzymes only lower the activation energy. They don't alter the overall free energy absorbed or released. Therefore, the forward and backward reactions are sped up with a catalyst, making choice (D) incorrect. (LO 2.3)

163. **(C)** The products do have more energy than the reactants. Therefore, the change in free energy (ΔG) is positive. Entropy is only one portion of the equation, so it alone cannot make the free energy change equal zero. Therefore, choice (A) is incorrect. Since the products have more energy than the reactants, the change in free energy is positive, not negative. Therefore, choice (B) is incorrect. Since the ΔG can be determined from looking at the graph, choice (D) is incorrect. (LO 2.3)

164. **(A)** If the reaction is endothermic, heat energy is absorbed (ΔH is positive). This tends to make ΔG positive unless $T\Delta S$ is larger than ΔH. If $T\Delta S$ is larger than ΔH, then there would be enough entropy to make ΔG negative. This is true because we would be subtracting a larger number from a smaller number, which would make ΔG negative. The graph for the backward reaction shows an exergonic reaction. That reaction could be either exothermic or endothermic depending on the size of $T\Delta S$. Therefore, choices (B) and (C) are not correct. Since choice (A) is true, choice (D) must be incorrect. (LO 2.3)

165. **(B)** Chemical reactions convert one form of energy into another. If a process (such as combining two elements to create a molecule) decreases entropy by a certain amount, then the reverse process must increase entropy by the same amount (hence the difference in sign). Choice (A) is incorrect because if one process creates disorder, the reverse process cannot also create disorder. Although entropy is often described as the amount of randomness in a

system, the change in entropy depends on the specific chemicals and how they change, which is not random, so choice (C) is incorrect. Choice (D) is incorrect because the amount of entropy of a particular reaction does not change with concentration. (LO 2.5)

166. **(B)** Line *A* is the activation energy for the uncatalyzed forward reaction. This is the energy that must be added to the reaction in order to make the reaction proceed. Choice (A) is incorrect because the activation energy for the catalyzed forward reaction is indicated by line *B* on the graph. Choice (C) is incorrect because the total energy change for the forward reaction is represented by line *D*. Choice (D) is incorrect because the activation energy for the uncatalyzed backward reaction is indicated by line *C*. (LO 2.3)

167. **(A)** Line *B* is the activation energy for the catalyzed forward reaction. Line *A* is the activation energy for the uncatalyzed forward reaction. Therefore, choice (B) is incorrect. Choice (C) is incorrect because the total energy change for the forward reaction is represented by line *D*. Choice (D) is incorrect because the activation energy for the uncatalyzed backward reaction is indicated by line *C*. (LO 2.3)

168. **(D)** To determine the activation energy for the uncatalyzed backward reaction, read the diagram from right to left as opposed to from left to right. The uncatalyzed reaction is the solid line with a higher peak. Line *C* indicates this amount of energy. Choice (A), the activation energy for the catalyzed forward reaction, is indicated by line *B*. Choice (B), the activation energy for the uncatalyzed forward reaction, is indicated by line *A*. Choice (C), the total energy change for the forward reaction, is indicated by line *D*. (LO 2.3)

169. **(C)** Line *D* shows the difference between the reactants and the products. This is the total energy change for both the forward and backward reactions. Choice (A), the activation energy for the catalyzed backward reaction, is indicated by line *E*. Choice (B), the activation energy for the uncatalyzed forward reaction, is indicated by line *A*. Choice (D), the activation energy for the uncatalyzed backward reaction, is indicated by line *C*. (LO 2.3)

170. **(A)** To examine the backward reaction, read the graph from right to left. The dotted line is lower and therefore represents the catalyzed reaction. The activation energy is the amount of energy necessary to get the reaction to begin. Line *E* shows the energy added from the beginning of the backward reaction in order to make the reaction proceed. The activation energy for the uncatalyzed forward reaction is indicated by line *A*, so choice (B) is incorrect. The total energy change for the forward reaction is indicated by line *D*, so choice (C) is incorrect. The activation energy for the uncatalyzed backward reaction is indicated by line *C*, so choice (D) is incorrect. (LO 2.3)

171. **(B)** The experimental results demonstrate that ABA prevents wilting. Therefore, it must be doing something to allow the plant to conserve water. There is no mention of wasps, caterpillars, fruit ripening, or cell division in the question. Therefore, choices (A), (C), and (D) are not correct. (LO 2.21)

172. **(A)** Tumors are characterized by uncontrolled cell growth and division. Auxin causes cell growth, and cytokinin causes cell division. Opine synthesis provides a food source for the bacterium and does not contribute directly to the growth of the tumor. Therefore, choices (B) and (C) are wrong. The origin of replication is a sequence that allows the plasmid to be replicated in the bacterium, so choice (D) is incorrect. (LO 2.28)

173. **(C)** By providing itself with an exclusive energy source, the bacterium prevents competition from other organisms. Although building a compound to export it and then break it down does cost energy, the benefit of an exclusive food source outweighs the cost. Since this is not entirely energetically wasteful and since there is a benefit to this mechanism, choices (A) and (D) are incorrect. Vascular tissue proliferates near many animal tumors. However, no evidence is presented in the question that suggests that this is true of plant tumors. Therefore, choice (B) is incorrect. (LO 2.27)

174. **(C)** The bacterium would still be able to transfer DNA into the host. Since the DNA that was transferred would now have a gene that encodes a GFP, and assuming there is a promotor to drive protein synthesis, GFP should be produced and the cell should glow. Without antibiotic selection or hormones to amplify the number of cells, the number of transformed, glowing cells should be low. No tumors would form. Since no tumors would form, choice (A) is incorrect. Auxin, cytokinin, and opine synthesis do not produce proteins involved in transferring DNA, so choice (B) is incorrect. The plasmid would still replicate as long as the origin of replication is present. Therefore, choice (D) is incorrect. (LO 2.35)

175. **(D)** Deleting the origin of replication (ori) would make it impossible for the plasmid to replicate. In order for replication to begin, the two DNA strands must be separated to form a replication bubble. The enzymes responsible for separating the DNA strands recognize the DNA sequence at the ori and specifically bind to it (they won't bind anywhere else). Once the strands are separated, they can be copied. If the ori is deleted, these enzymes would not be able to bind to the DNA strand, and replication would not happen. Choice (A) is incorrect because there would be no plasmid to transfer to the plant cells, so there would be no tumors or glowing. Choice (B) is incorrect because the ori does not produce a protein involved in transferring DNA. Choice (C) is incorrect because, without the presence of the origin of replication, the plasmid would not replicate. As a result, no DNA could be transferred to plant cells. Without the DNA from the bacterium, the plant cells could not be transformed and would not contain the GFP gene. Therefore, the cells would not glow. (LO 2.36)

176. **(B)** The necessary components would include ori, virulence, transfer, and the border sequences. Auxin, cytokinin, and the opine genes cause processes that are not necessary to transfer DNA from the bacterium into the host cell. Therefore, choices (A), (C), and (D) are incorrect. (LO 2.32)

177. **(A)** A selectable marker must be introduced into the plant cell so that cells that don't take up the transfer DNA can be killed, leaving the ones that did take up the transfer DNA to survive. Choice (B) protects bacteria against kanamycin. At this point in the process, it is important to leave plant cells with the transferred DNA alive and kill those plant cells without it. Choice (C) is incorrect because the resistance gene needs to be in the cell that is susceptible to the toxin. Choice (D) is a clever way to disable the antibiotic if transformation happens, but this certainly doesn't match the function needed in this case. (LO 2.36)

178. **(A)** This positive feedback loop is perpetuated as long as the fetus pushes against the stretch receptors in the cervix. If this doesn't happen, nothing in the diagram suggests that further oxytocin and prostaglandin will be secreted. Without stimulation from the fetus pushing on the cervix, neither prostaglandin levels nor oxytocin levels rise, so choices (B) and (C) are incorrect. Choice (D) relies on a misunderstanding of the diagram. The arrows indicate stimulation of the placenta and uterine muscles, not their ability to break down oxytocin and prostaglandin. (LO 2.19)

179. **(D)** Intravenous oxytocin is used to induce labor. The oxytocin stimulates both the production of prostaglandins and muscle contractions. Prostaglandins stimulate more muscle contractions. Increased muscle contractions lead to more pressure on the cervix and more oxytocin production by the body. Choice (A) is incorrect because stretch receptors detect stretch. They do not cause cervical dilation. Although oxytocin does stimulate milk production, the fetus does not need milk until after being delivered. Therefore, choice (B) is not the correct response. Childbirth uses a positive feedback cycle that leads to an amplification effect. Since this is not a steady-state phenomenon, choice (C) is incorrect. (LO 2.20)

180. **(D)** No positive feedback will occur if the fetus is not pressing against the cervix because the stretch receptors will not stimulate the production of additional oxytocin. Less oxytocin will not lead to more prostaglandin, so choice (A) is incorrect. The diagram does not mention sensory mechanisms in the uterus or uneven contractions, so choice (B) is not correct. No mechanism is shown that supports the idea in choice (C) that contractions would get stronger regardless of positive feedback from the stretch receptors. (LO 2.19)

181. **(B)** Released prostaglandin stimulates muscle contractions, as shown in the diagram. An amniotomy is done to stimulate the positive feedback cycle that leads to childbirth. Prostaglandin causes positive feedback that produces more prostaglandin, not less. Therefore, choice (A) is not correct.

Prostaglandin stimulates muscle contractions, which push the fetus, which activates stretch receptors in the cervix. This leads to oxytocin production by the pituitary. Prostaglandin stimulates oxytocin production indirectly through this chain of events. It does not lead to less oxytocin being produced by the pituitary, so choice (C) is wrong. Choice (D) is incorrect because the diagram clearly shows what would happen if prostaglandin levels increased. (LO 2.19)

182. **(A)** Simple sugars, like sucrose, fructose, and glucose, are readily available for entry into glycolysis. These compounds are not good long-term energy storage molecules. If storage was needed, the sugars would be converted into starch or fats, which are much more energy dense. Therefore, choice (B) is not correct. There is no nitrogen in these sugars, so choice (C) is wrong. Choice (D) is wrong because dilute solutions, such as nectar, contain plenty of water. (LO 2.8)

183. **(B)** Amino acids are necessary for making proteins, and pollen does contain amino acids. Choice (A) is incorrect because nitrogen is not necessary to build carbohydrates. Since nucleic acids do not contain sulfur, sulfur would not be needed to build nucleic acids. Therefore, choice (C) is incorrect. Since choices (A) and (C) are both false, choice (D) is not the correct answer. (LO 2.8)

184. **(B)** The bicyclist used respiration to generate the 935 kcal from food. This respiration released carbon dioxide, so his claim is technically incorrect. However, it is true that a car driven by an internal combustion engine would release much more carbon dioxide by burning fuel than the man did by respiring while bicycling. Choice (A) is incorrect because riding a bike uses more energy, and therefore a person releases more carbon dioxide while riding a bike than when not riding a bike. Since the bicyclist used 935 kcal to produce the ATP to power his muscles, he released carbon dioxide through respiration. If the bicyclist had claimed that bicycling was a zero *net* emission activity, he would have been correct because the food he ate did first absorb carbon dioxide from the atmosphere. Since that claim was not made in choice (C), choice (C) is incorrect. Bicycling certainly produces less carbon dioxide than driving a car does, so choice (D) is incorrect. (LO 2.2)

185. **(B)** If sex were determined entirely by an X and Y sex chromosome system, there should be a 1:1 ratio of males to females in both habitats. Since this did not occur, choice (B) is not consistent with the data presented. Predation on males in the suburban environment would lead to fewer males in that environment. This is reflected in the table, so choice (A) is consistent with the data. Since the question asked for the hypothesis that is not consistent with the data presented, choice (A) is not the correct answer. Chemicals, such as those used to treat suburban lawns, could affect the sex ratios. Since choice (C) contains a hypothesis that is consistent with the data presented, choice (C) is not the correct answer. The high mortality of males due to competition

could also skew the sex ratio in the suburban population. Since choice (D) contains a hypothesis that is consistent with the data presented, choice (D) is not the correct answer. (LO 2.22)

186. **(A)** Raising frogs in water from each environment, but under controlled conditions in the lab, removes all but one variable from the experiment. Observing frogs in nature does not give any information on the cause of the sex shift, so choice (B) is incorrect. Adding estrogen to forested lakes might change the sex ratio, but it doesn't give any information about the cause of the skewed sex ratio that already exists in the suburban lakes. Therefore, choice (C) is wrong. Converting forested lakes to suburban environments is likely to lead to the skewed sex ratio observed in the suburban lakes. However, this conversion is not likely to provide any information about the cause of the skewed sex ratio that already exists in the suburban lakes. Therefore, choice (D) is incorrect. (LO 2.21)

187. **(B)** If the sex ratio of frogs collected from the stomachs of predators skewed toward males in suburban environments but not in forested lakes, this would support the hypothesis that there is higher predation pressure on males in suburban environments. Observations of frogs' daily cycles of activity would not directly test the amount of predation in each environment, so choice (A) is not correct. Neither transferring predators nor transferring frogs would test the levels of males being consumed. Therefore, choices (C) and (D) are not correct. (LO 2.23)

188. **(D)** Approximately 10% of the energy at one trophic level is transferred to the next, and 200 J is 10% of 2,000 J. Choice (A) underestimates the amount of photosynthetic plankton. There would need to be 200,000 J of energy in photosynthetic plankton to support 2,000 J of energy in small fish. Choice (B) underestimates the amount of energy in tuna fish. Choice (C) underestimates the amount of energy in animal plankton. There would have to be 20,000 J of energy in animal plankton to support 2,000 J of energy in small fish. (LO 2.3)

189. **(A)** Feeding at the lowest possible trophic level maximizes the amount of energy available to people. At each higher trophic level, energy is lost due to the organisms' requirements to maintain cells, grow, and reproduce. If there were 200,000 J of energy in photosynthetic plankton, it could support 20,000 J of either people or animal plankton. If the people ate the animal plankton, that trophic level could support 2,000 J of either people or small fish. If the people ate the small fish, that trophic level could support 200 J of either people or tuna fish. If the people ate the tuna fish, that trophic level could support 20 J of people. As a result, choices (B), (C), and (D) are not correct. (LO 2.3)

190. **(D)** Increasing the number of photosynthetic plankton would lead to an increase in energy available at every successive trophic level. Removing mercury, as choice (A) suggests, would make the tuna healthier, but that action would not affect the energy available. Harvesting any organisms from the lower trophic levels would make less energy available to support tuna fish. Therefore, choices (B) and (C) would lead to a decrease in the number of tuna fish that could be harvested. (LO 2.4)

191. **(B)** Pollutants, such as mercury, are biologically magnified at higher trophic levels because each higher trophic level accumulates mercury that was absorbed at each lower trophic level. Tuna fish, therefore, have accumulated mercury that the small fish accumulated from the animal plankton that was accumulated from the photosynthetic plankton. The tuna fish have the highest concentrations of mercury, so choice (A) is incorrect. Animal plankton have accumulated mercury from only a single trophic level below them. Therefore, animal plankton do not have the highest concentrations of mercury, making choice (C) incorrect. Photosynthetic plankton have the lowest concentrations of mercury since they accumulate mercury only from the ocean and not from any trophic level below them. Therefore, choice (D) is incorrect. (LO 2.9)

192. **(D)** Since hydrogen ions accumulate in the thylakoid space, this space becomes more acidic and therefore turns red. The stroma, on the other hand, loses hydrogen ions. Therefore, the pH rises (becomes more basic), and the stroma turns blue. Choice (A) is the exact opposite of the correct answer. Choice (B) ignores the fact that hydrogen ions are removed from the stroma. Choice (C) suggests that both the stroma and thylakoid space have neutral pH, which does not happen because of the active pumping of ions. (LO 2.5)

193. **(C)** Since there is no hydrogen ion pumping with the lights off, there should be no difference in pH across the membrane. In that case, both the stroma and the thylakoid space would appear yellow. Choices (A), (B), and (D) are all incorrect because the different colors indicate different pH in different compartments. Without any hydrogen ion pumping, there would be no pH difference. (LO 2.5)

194. **(D)** The mitochondrion needs pyruvate and oxygen to run the electron transport chain (ETC), which is located on the inner mitochondrial membrane. Protons are moved from the matrix, which is labeled B in the diagram, to the intermembrane space, which is labeled D in the diagram. This causes low pH (acidic) in the intermembrane space and results in the dye turning red. The matrix turns blue because removing protons causes high pH (basic). Choice (A) has this relationship backward. Choice (B) shows the inner membrane changing pH, which shouldn't happen under any circumstances. Choice (C) would happen if the electron transport chain stopped working for some reason. (LO 2.5)

195. **(C)** Without oxygen, the electron transport chain (ETC) would have no place for the electrons to go. Therefore, the electrons would stop flowing through the ETC, and there would not be a pH differential. Choices (A), (B), and (D) are all incorrect because the different colors indicate different pH in different compartments. Without any electron flow, there would be no pH difference. (LO 2.5)

196. **(D)** Running the reactions of the Krebs cycle backward would take inorganic carbon dioxide and attach it to other compounds to make organic compounds. Autotrophs are organisms that make organic compounds from inorganic compounds, so choice (D) is correct. Heterotrophs can't make organic compounds from inorganic compounds, so choice (A) is incorrect. Methane does not appear in the diagram of the Krebs cycle, so choice (B) is incorrect. Nitrate and nitrogen are not shown in either the diagram or in the information provided, so choice (C) is wrong. (LO 2.2)

197. **(A)** To run the Krebs cycle backward, one would need all of the items produced by running the Krebs cycle forward. Since ATP, NADH, and $FADH_2$ are produced by the forward cycle, they would be needed to power the backward cycle. Choice (B) is incorrect for two reasons. First, sunlight does not contain electrons. Second, acetyl CoA would be produced, not used. A source of energy would be needed. However, the cycle would harvest acetyl CoA, not need it. Therefore, choice (C) is incorrect. The chemicals listed in choice (D) are needed to run the cycle forward, not backward. (LO 2.4)

198. **(A)** Fixation of carbon dioxide, the reaction catalyzed by rubisco, is an endergonic reaction. Therefore, the back reaction is exergonic. Without an input of energy, the Calvin cycle has a natural tendency to run backward. In the light, there is an abundant source of energy in the form of ATP. This ATP is made in the light-dependent reactions and is used in the Calvin cycle to drive the reaction that makes GP from 3-PGA, as shown in the diagram. Since 3-PGA is constantly being turned into GP, the reaction that fixes carbon dioxide can proceed. If rubisco were mutated such that rubisco was not turned off in the dark, the Calvin cycle would run backward. Rubisco does not catalyze a reaction that generates carbon and oxygen, so choice (B) is not correct. Excess 3-PGA would drive the cycle backward and wouldn't cause the stomata to close. Therefore, choice (C) is incorrect. If the Calvin cycle were run backward, it would release carbon dioxide rather than fix it. The ability to turn rubisco off in the dark is vital to prevent the Calvin cycle from running backward, so choice (D) is wrong. (LO 2.4)

199. **(A)** The reaction catalyzed by rubisco combines one carbon from carbon dioxide with a 5-carbon compound to produce two molecules containing 3-carbon atoms (for a total of 6 carbons). If rubisco uses an oxygen rather than a carbon dioxide to react with the RuBP, there would only be 5 carbons present. Therefore, the reaction would produce a 3-carbon compound and a

compound that is missing a carbon (a 2-carbon compound). ATP production requires an input of energy. The reaction of rubisco and oxygen would not provide energy. Therefore, choice (B) is incorrect. If the 2-carbon compound were turned into carbon dioxide, it might produce two molecules of carbon dioxide. However, this is not the result of the reaction of oxygen with RuBP that rubisco catalyzes, so choice (C) is incorrect. Splitting a 5-carbon compound without adding another carbon would not yield a total of 6 carbons, so choice (D) is incorrect. (LO 2.4)

200. **(D)** The diagram shows $FADH_2$ donating its electrons to an electron carrier after the first hydrogen pump. In contrast, NADH uses that first hydrogen pump. Choices (A) and (B) are incorrect because NADH provides *more* energy than does $FADH_2$. Both NADH and $FADH_2$ donate electrons to the electron transport chain. $FADH_2$ does not donate electrons to hydrogen ions. Therefore, choice (C) is incorrect. (LO 2.5)

201. **(C)** The mitochondrion uses hydrogen ions' tendency to flow down their concentration gradient to drive the ATP synthase to make ATP. These hydrogen ions flow from the intermembrane space into the matrix because there are more hydrogen ions in the intermembrane space and fewer hydrogen ions in the matrix. Recall that the more hydrogen ions there are, the lower the pH and the more acidic a solution is. The fewer hydrogen ions there are, the higher the pH and the more basic a solution is. Anything that lowers the pH of the intermembrane space and increases the pH of the matrix will allow the ATP synthase to make more ATP. If you examine the diagram and find the reaction that makes water, you will see that it takes two hydrogen ions and half of an O_2 to make one water molecule. By consuming hydrogen ions, this reaction increases the pH. Normally, the mitochondrion performs this reaction in the matrix, where it increases the pH and helps the ATP synthase to make ATP. If this reaction was performed in the intermembrane space, however, it would raise the pH in the intermembrane space. If decreasing the pH in the intermembrane space allows the ATP synthase to make more ATP (as discussed above), then raising the pH in the intermembrane would lead to less ATP production. Choices (A) and (B) are incorrect because the reaction does not produce hydrogen ions; it consumes them. If the reaction did produce hydrogen ions, it would lead to a lower pH, not a higher pH as choice (A) suggests. Additionally, a higher pH in the intermembrane space would lead to less ATP made. Choice (D) is incorrect because consuming hydrogen ions to make water in the intermembrane space would work against the mechanism that lowers the pH in the intermembrane space and would lead to less ATP production. (LO 2.5)

202. **(B)** If there was no final electron acceptor (oxygen), electrons would stop moving through the electron transport chain because there would be no place for the electrons to go. The organism might switch to using fermentation, or the organism might die. However, these are not the most direct consequences. Therefore, choices (A) and (D) are incorrect. The

electron transport chain would stop, not speed up, so choice (C) is incorrect. (LO 2.5)

203. **(A)** The diagram shows that adding ATP is necessary to release the myosin head from actin. Without ATP, the muscles would be contracted and would not release. If the myosin was stuck to actin, there would be no relaxation. Therefore, choice (B) is incorrect. Choice (C) is incorrect because there would be no problem binding to the actin fiber; there would only be a problem releasing from it. Oxygen cannot compensate for a lack of ATP, so choice (D) is incorrect. (LO 2.5)

204. **(D)** The *x*-axis shows the partial pressure of oxygen, which is a measure of oxygen concentration. The *y*-axis shows the percentage of saturation of binding sites, which is a measure of binding affinity. Choice (A) is incorrect because there is no mention of energy use. No amino acid sequences are presented, so choice (B) is wrong. Choice (C) is incorrect because it is vague and because the graph doesn't include any information on fetal myoglobin. (LO 2.17)

205. **(B)** Oxygen will move most rapidly when there is the greatest difference in percentage saturation. At 10 mmHg, adult hemoglobin is less than 5% saturated, while myoglobin is approximately 85% saturated. At 1 mmHg, the two molecules have almost the same percentage saturation (near 0%), so choice (A) is incorrect. At 40 mmHg, adult hemoglobin is around 60% saturated while myoglobin is 100% saturated. However, the difference between these saturation levels is less than at 10 mmHg. Therefore, choice (C) is incorrect. The curves are even closer at 80 mmHg, so choice (D) is also wrong. (LO 2.17)

206. **(C)** In order for oxygen to pass efficiently from maternal hemoglobin to fetal hemoglobin, fetal hemoglobin must bind that oxygen more tightly. Fetal hemoglobin is a protein made by the fetus, not a stage of development of hemoglobin. Therefore, choice (A) is incorrect. The fetus does require a large amount of oxygen to generate ATP for growth and development. However, this has little to do with the binding affinity of the different hemoglobin molecules. Therefore, choice (B) is incorrect. Proteins like hemoglobin are constantly recycled by the body. No information is presented on how long adult and fetal hemoglobin are retained before recycling, so choice (D) is not correct. (LO 2.31)

207. **(D)** Oxygen needs to be transferred to muscle cells rapidly. Therefore, myoglobin needs a high affinity for oxygen. Choice (A) is incorrect because myoglobin facilitates the transfer of oxygen from hemoglobin to myoglobin. Oxygen travels through the bloodstream from the lungs to the muscle cells. Oxygen does not bypass the hemoglobin, so choice (B) is wrong. Oxygen dissolves in blood before it can diffuse into a red blood cell and bind to hemoglobin, so choice (C) is incorrect. (LO 2.37)

208. **(D)** This shift in binding affinity allows the efficient uptake of oxygen in the lungs. Oxygen uptake does not occur in the heart, so choice (A) is incorrect. Choice (B) has the cause and the effect backward. The change in pH affects the binding affinity of hemoglobin, not the other way around. Buffers release and accept protons. No information is provided about hemoglobin acting as a buffer, so choice (C) is incorrect. (LO 2.33)

209. **(C)** As blood flows through capillaries in actively respiring tissue, such as muscle, the carbon dioxide that diffuses out of the cells into the blood decreases the pH. At this lower pH, hemoglobin's affinity for oxygen decreases, and oxygen is released from the hemoglobin. Lactic acid is produced by glycolysis and is not related to carbon dioxide levels, so choice (A) is incorrect. A decrease in pH would decrease the amount of oxygen bound to hemoglobin and therefore in the red blood cells. Therefore, choice (B) is wrong. No information is presented about the effect of pH on myoglobin, so choice (D) is incorrect. (LO 2.33)

210. **(A)** Figure I shows that as the percent body fat (% BF) increases, so does the body mass index (BMI). Since BMI and % BF are clearly related, choice (B) is incorrect. There is a correlation between BMI and % BF. However, it is not possible to calculate the % BF directly from the BMI because there is significant variability shown in Figure I. Therefore, choice (C) is incorrect. No information is provided about which of the two is a better indicator of physical health, so choice (D) is wrong. (LO 2.24)

211. **(B)** A BMI of 30 is in the obese category, while a % BF of 12 is essential fat for a woman. Choice (A) could result from a BMI of 12 and a % BF of 33. Choice (C) is incorrect because the woman's % BF is not high enough to be classified as obese. Choice (D) is incorrect because her % BF is lower than the acceptable range. (LO 2.24)

212. **(C)** BMI is based solely on height and weight. It does not differentiate between the source of the weight. As a result, an athlete with a lot of muscle mass may weigh the same as a person with an equal mass of adipose tissue. Since both of the measures were taken after the race, choice (A) is wrong. Choice (B) is incorrect because % BF *does* take into account the weight of adipose tissue. Choice (D) is incorrect because no error in measurement is necessary to explain the data. (LO 2.24)

213. **(C)** The data presented suggest that the ability of the mealworms to use Styrofoam as a food source depends on intestinal flora. If bacteria isolated from the intestinal tract of mealworms can grow on polystyrene agar on a Petri plate, we can then conclude that the bacteria can metabolize polystyrene. Coupled with the fact that antibiotics that kill bacteria make mealworms unable to use Styrofoam as an energy source, we can conclude that the bacteria make it possible for the mealworm to use polystyrene as an energy source. Choice (A) is incorrect because that experiment would not

provide any information about the cause of the mealworm's ability to use polystyrene as an energy source. If the fruit flies gained the ability to use polystyrene as an energy source, we would know that something in the fecal matter caused this ability, but we still would not know what the causal agent was. If the fruit flies did not gain the ability to use polystyrene as an energy source, we might conclude that the fruit fly intestine simply couldn't harbor the bacteria. Since the gene for polystyrene degradation likely resides in the intestinal flora, sequencing the genome of the mealworm would not provide any useful information. Therefore, choice (B) is not correct. Choice (D) might show that mealworms prefer polystyrene or some other food source, but this experiment would not give any information about why mealworms can use polystyrene. (LO 2.24)

214. **(B)** All new genes have their origins in mutation. If the mutation already existed in another bacterium, it could be transferred, as choice (A) suggests. However, this would not be the *original source* of the mutation, so choice (A) is incorrect. Bacteria do not have nuclei and therefore cannot perform meiosis. Therefore, choice (C) is incorrect. For choice (D) to be correct, the gene would already have to exist and then be transferred horizontally from plants, which is a highly unlikely scenario at best. (LO 2.27)

215. **(C)** In order for the inside of the neuron to be negatively charged relative to the outside, more positive ions need to be pumped out than in. In this choice, we see 3 Na^+ being pumped out of the cell for every 2 K^+ pumped in. Choices (A), (B), and (D) all create a positive charge in the cytoplasm. (LO 2.9)

216. **(C)** Adding CO_2 to the blood pushes the equilibrium to the right, which releases hydrogen ions and lowers the blood pH. This scenario leads to an increased breathing rate. This causes CO_2 to escape through the lungs, raising the pH. This is a negative feedback loop. Choice (A) mentions *decreased* ventilation in response to low pH. This scenario would not allow CO_2 to escape and would not cause the blood pH to rise. Secretion of carbonate would further shift the equilibrium to the right and lower pH, not raise it. Therefore, choice (B) is incorrect. Storing carbonate in the bone, as choice (D) suggests, would have the same effect as secreting carbonate in the urine. (LO 2.15)

217. **(A)** The sweet potato is isotonic at about 0.8 M, while the russet potato is isotonic at about 0.3 M. The sweet potato has a higher solute concentration and therefore a lower solute water potential. This same reasoning explains why choices (B) and (D) are incorrect. Choice (C) is wrong because the solute water potentials can be easily determined using the data from the graph. (LO 2.9)

218. **(D)** Since the sweet potato has a higher solute concentration, as established in Question 217, it is hypertonic to the russet potato. Water will flow into the sweet potato from the russet potato, changing the mass of each as described in choice (D). Eukaryotic cells do not exchange genetic information as described in choice (A). According to the information in the graph, when the potato cubes were placed in 0 M sucrose solution, water flowed into the cells. Solutes did not diffuse out of the cells. Therefore, it is unlikely that solutes would flow from one type of potato to another, making choice (B) incorrect. Choice (C) would be correct if the cells were isotonic to one another, which they aren't. (LO 2.9)

219. **(A)** Photosynthesis requires light. If the sample was kept in total darkness, no photosynthesis took place. Choice (B) is incorrect because respiration must have taken place for the dissolved oxygen level to fall. Choices (C) and (D) are incorrect because no photosynthesis took place in the dark. (LO 2.10)

220. **(C)** Some oxygen must have been produced by photosynthesis or the line would be the same as the line for 0% light. This also means that choice (A) is incorrect. Since the dissolved oxygen levels fell, more oxygen was being consumed than produced. Since oxygen was being consumed, some respiration must have taken place. This makes choice (B) incorrect. If choice (D) was correct, the slope of the line should be positive, which it is not. (LO 2.12)

221. **(D)** Since oxygen levels rose, photosynthesis must have generated more oxygen than what was consumed by respiration. Choice (A) is incorrect because photosynthesis clearly took place since the 66% line does not match the 0% line. Since oxygen was being consumed, some respiration must have taken place. This makes choice (B) incorrect. Choice (C) is incorrect because there was a net increase in oxygen. This indicates that there was more photosynthesis generating oxygen than there was respiration consuming oxygen. (LO 2.12)

222. **(D)** *Mycobacterium tuberculosis* is an intracellular parasite. Therefore, it must enter the phagocyte. Avoiding recognition as non-self would prevent the bacterium from being ingested. The strategies in choices (A), (B), and (C) all occur after ingestion. (LO 2.22)

223. **(D)** If the FECB were involved in oxygen uptake for fluke muscles, the blood should flow to the fluke muscles after oxygen exchange. Choice (A) is incorrect because a thick layer of mucus would prevent gas exchange. Sending oxygenated blood from one organ that absorbs oxygen to another organ that performs the same function is pointless. Therefore choice (B), which involves blood flowing from the FECB to the lungs, and choice (C), which involves blood flowing from the lungs to the FECB, are incorrect because both organs absorb oxygen. (LO 2.33)

224. **(A)** If the function of the FECB is to dissipate heat, more blood should flow through the FECB and the temperature of the blood should decrease during fast swimming. Choices (B) and (C) might provide interesting information. However, they do not test the hypothesis about heat dissipation under different swimming speeds. Choice (D) doesn't measure blood temperature, so it won't address heat dissipation. (LO 2.35)

225. **(B)** Consuming caffeine leads to less ADH and fewer aquaporins in the collecting duct. This leads to less water reabsorption and more dilute urine. Choice (A) is incorrect because the urine is more dilute. Additionally, water is not pumped; it is passively transported. Aquaporins specifically transport water. Salt should not be able to move through aquaporins, so choice (C) is incorrect. The passage does not mention salt leaking into the collecting duct, so choice (D) is incorrect. (LO 2.28)

226. **(A)** The dehydrated person needs to reabsorb water. Inserting aquaporins into the collecting duct accomplishes this. Having few aquaporins in the collecting duct prevents the reabsorption of water, so choice (B) is not correct. Adding sodium pumps to the collecting duct reabsorbs sodium, making the urine more dilute. This is the opposite of what is needed, so choice (C) is incorrect. Choice (D) is incorrect because secreting urea decreases the collecting duct's ability to reabsorb water. Some urea is typically reabsorbed at the bottom of the collecting duct, not secreted. (LO 2.28)

227. **(B)** A long Loop of Henle creates a very concentrated interstitial fluid in the medulla of the kidney. This allows urine to become extremely concentrated as it is descending in the collecting duct. A short Loop of Henle leads to less concentrated urine, so choice (A) is incorrect. The collecting duct is capable of allowing reabsorption of water only up to the concentration of the medulla, regardless of the number of aquaporins, so choice (C) is incorrect. Choice (D) is incorrect because the ascending limb of the Loop of Henle needs to be impermeable to water or it would not be able to create the concentration gradient for the kidney to work. (LO 2.26)

228. **(B)** The diagram that follows compares the movement of toxins from the blood into dialysis fluid in concurrent and countercurrent flow. The numbers represent the amount of toxin in each compartment. In both the concurrent and countercurrent flow shown, blood flowing into the apparatus has 100 units of toxin in each case. In the diagram of concurrent flow, blood and dialysis fluid are both flowing from left to right. In the diagram of countercurrent flow, blood is shown moving from right to left while dialysis fluid flows in the opposite direction from left to right. Countercurrent flow can eliminate nearly all of the toxin because, as blood flows from right to left in the diagram of countercurrent flow on page 330, it encounters progressively cleaner dialysis fluid, which is flowing from left to right.

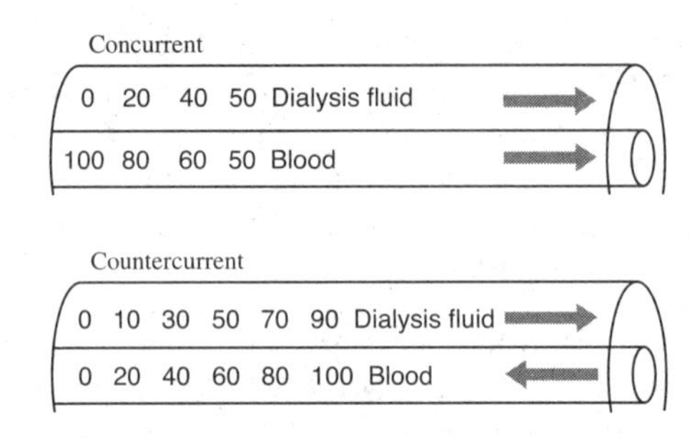

By contrast, concurrent flow can eliminate only half of the toxins. As toxin-laden blood (100 units of toxin) enters the apparatus, it immediately encounters the cleanest dialysis fluid (0 units of toxin). As the two fluids move with each other along the apparatus, toxins diffuse from the blood into the dialysis fluid. Once the toxin reaches equal concentration in both tubes (50 units in the diagram), there is no net diffusion of toxin out of the blood. Therefore, choice (A) is incorrect. Choice (C) is incorrect because the amount of diffusion time depends on the length of the tube and on the flow rate, which are assumed to be the same in this example. Since it has already been established that countercurrent flow is more efficient than concurrent flow, choice (D) is incorrect. (LO 2.12)

229. **(D)** The amount of surface area per unit volume for exchange is dependent on the diameter of the inner tube, which is the same for Set 1 and Set 2. Since wider tubes, such as in Set 3, have lower ratios of surface area to unit volume, choice (C) is incorrect. Since Set 1 and Set 2 have the same amount of surface area per unit volume, both choices (A) and (B) are incorrect. (LO 2.7)

230. **(C)** The largest inner tube has more total surface area because if its diameter is the largest, so is its circumference. In terms of the rate of exchange, total surface area is less important than the surface area per volume. Set 1 and Set 2 have less surface area than Set 3, so choices (A), (B), and (D) are incorrect. (LO 2.7)

231. **(C)** Since it has the lowest ratio of surface area to volume, Set 3 is the least efficient at diffusion. Set 1 and Set 2 should be approximately equivalent since there is the same ratio of surface area to volume. (Note that Set 1 may be marginally better since there is more dialysis volume into which toxins can dissolve.) However, both Set 1 and Set 2 should be *more* efficient than Set 3. Therefore, choices (A), (B), and (D) are incorrect. (LO 2.7)

232. **(C)** If the volume is the same, this becomes a question about surface area alone. A highly convoluted shape like III maximizes surface area. A sphere minimizes surface area. The surface area of a cube will be intermediate. Therefore, choices (A), (B), and (D) are incorrect. (LO 2.7)

233. **(A)** The sphere should have the lowest ratio of surface area to volume. The surface area of a cube will be intermediate. Therefore, choice (B) is incorrect.

Choice (C) is incorrect because a highly convoluted shape like III maximizes surface area. Since they all produce different ratios of surface area to volume, choice (D) is incorrect. (LO 2.7)

234. **(A)** Since the line increases from left to right, the reaction rate increases in response to increases in the enzyme concentration. Choice (B) is the opposite of the correct answer. The reaction rate does depend on the enzyme concentration, so choice (C) is incorrect. Choice (D) is incorrect because choice (A) is a good description of the relationship shown. (LO 2.24)

235. **(A)** Since the line increases from left to right, the reaction rate increases in response to increases in the substrate concentration. Choice (B) is the opposite of the correct answer. The reaction rate does depend on the substrate concentration, so choice (C) is incorrect. Choice (D) is incorrect because choice (A) is a good description of the relationship shown. (LO 2.24)

236. **(C)** With more enzymes present, more active sites are available where reactions can take place. A higher concentration of the enzyme has no effect on the number of substrate molecules, so choice (A) is incorrect. Higher concentrations of the enzyme do not make active sites work faster, nor do they open more active sites on an enzyme molecule. Therefore, choices (B) and (D) are incorrect. (LO 2.17)

237. **(D)** Increasing the concentration of the substrate increases the number of collisions between enzymes and substrates, making reactions more likely to occur. Increasing the substrate concentration does not make more active sites on enzymes available, nor does it make the enzymes more active. Therefore, choices (A) and (B) are incorrect. Substrate concentration alone does not increase reaction temperature. If the reaction is exothermic, increasing the concentration of the substrate might increase the temperature. However, no information is given about this. In addition, the reaction might easily be endothermic and have the opposite effect. Therefore, choice (C) is incorrect. (LO 2.17)

238. **(B)** The highest rate of reaction (0.81/s) occurs when the pH is 6. This reaction rate is higher than at pH 3 (0.19/s), pH 7 (0.31/s), or pH 9.5 (0.01/s). Therefore, choices (A), (C), and (D) are all incorrect. (LO 2.17)

239. **(B)** Interactions among amino acid side chains are affected by changes in the concentration of hydrogen ions. As a result, the shape of the protein changes. The phosphate ion (PO_4^{3-}) is not related to pH, which is the $-\log[H^+]$. Therefore, choice (A) is incorrect. Macrophage digestion is not dependent on pH, so choice (C) is incorrect. The speed of molecular motion is dependent on temperature, not pH. Therefore, choice (D) is incorrect. (LO 2.17)

240. **(C)** The highest rate of reaction (0.81/s) occurs when the temperature is 40°C. This reaction rate is higher than at 5°C (0.11/s), 20°C (0.31/s), or

70°C (0.05/s). Therefore, choices (A), (B), and (D) are all incorrect. (LO 2.17)

241. **(A)** At higher temperatures, increased molecular motion can cause a protein to lose its shape (denature). As a result, the shape of the active site changes so that it does not work as well. If more collisions occurred, the reaction rate should increase, not decrease. Therefore, choice (B) is incorrect. Primary structure depends on covalent peptide bonds between amino acids. These peptide bonds do not break at the temperatures shown in the figure. Therefore, choice (C) is incorrect. Peroxide does degrade spontaneously at higher temperatures. However, this should lead to an apparent increase in the reaction rate, not a decrease. Therefore, choice (D) is incorrect. (LO 2.24)

242. **(B)** Temperature is a measure of average kinetic energy. As the temperature falls, there is less molecular motion. As a result, fewer collisions of substrates with enzyme active sites occur. Choice (A) is incorrect because decreased temperature is unlikely to denature enzymes. It is standard practice to freeze enzymes to preserve them. This would not work if enzymes were denatured at low temperatures. Primary structure depends on covalent peptide bonds between amino acids. These peptide bonds do not break at the temperatures shown in the figure. Therefore, choice (C) is incorrect. Peroxide degrades spontaneously at higher temperature, not at lower temperatures. Even if this did occur at lower temperatures, though, this would increase the apparent reaction rate, not decrease it. Therefore, choice (D) is incorrect. (LO 2.24)

243. **(B)** Photosynthesis releases oxygen as a result of the splitting of water by photosystem II. This gas is released into the spongy mesophyll, where it increases the buoyancy of the leaf. Choice (A) is incorrect because photosynthesis consumes carbon dioxide. It doesn't produce carbon dioxide. Unless the pressure changes significantly, any nitrogen in the solution should stay in the solution. Therefore, choice (C) is incorrect. Although water is broken into oxygen and hydrogen in photosynthesis, there is no hydrogen gas generated in this experiment. The electrons needed to make the bond between the two hydrogens are harvested for the electron transport chain. Therefore, choice (D) is incorrect. (LO 2.8)

244. **(A)** Shade-grown leaves had a higher rate of photosynthesis than sun-grown leaves until 60% of maximum light. At that point, sun-grown leaves had a higher rate of photosynthesis. Choice (B) is the opposite of what the data show. Choices (C) and (D) are both wrong because neither type of leaf performed photosynthesis fastest in all light conditions. (LO 2.21)

245. **(C)** Each leaf produces a certain amount of light-absorbing photosystems. Once all of these reaction centers are full, additional light cannot excite more electrons. Choice (A) is incorrect because the plateau always begins before the maximum amount of light is encountered. Since there is no information about damage to the electron transport chain, choice (B) is incorrect. Photosynthesis

does not speed up because electrons move faster. It speeds up because more electrons are moving. Therefore, choice (D) is incorrect. (LO 2.15)

246. **(C)** Energy must be used to move molecules up their concentration gradient as shown in process C. Processes A and B show molecules moving down their concentration gradients. Thus, no input of energy is required, and choices (A) and (B) are both incorrect. Since process C does require the input of energy to move molecules up their concentration gradient, choice (D) is incorrect. (LO 2.12)

247. **(A)** Nonpolar substances are not excluded by the nonpolar fatty acid portion of the membrane and can therefore pass through the membrane without a channel protein. Choice (B) is incorrect because fatty acid tails are nonpolar. Any substance can be moved against its concentration gradient, so choice (C) is incorrect. Since the molecules in process A can be nonpolar, choice (D) is incorrect. (LO 2.10)

248. **(C)** The diagram shows hydrogen ions flowing through the ATP synthase, down their concentration gradient. Although sunlight and the breakdown of glucose into carbon dioxide might provide energy to create the hydrogen ion gradient, they are not shown in the diagram and are not the immediate source of energy. Therefore, choices (A) and (B) are incorrect. In a mitochondrion, the membrane folds around in three dimensions. Thus, it would be equally correct to draw the diagram in the opposite orientation so that the hydrogen ions appear to be flowing up. Gravity has little to no effect at this scale, so choice (D) is incorrect. (LO 2.12)

249. **(B)** The concentration gradient would exist in the opposite direction needed. Thus, the ATP synthase would run in the reverse direction. This would cause the breakdown of ATP into ADP. Choice (A) is incorrect because the hydrogen ions would still move from high concentrations to low concentrations down their gradient. The knob portion of the ATP synthase that sticks out from the membrane is where ATP is formed or broken down. If this portion were in the region with a high concentration of hydrogen ions, it would spin backward, breaking ATP down into ADP. The hydrogen ions would move down their concentration gradient, not against it. By returning through the channel provided, hydrogen ions would spin the rotor in the synthase. As a consequence, a reaction would happen. Therefore, choice (C) is incorrect. Choice (D) is incorrect because choice (B) is the correct choice. (LO 2.12)

250. **(A)** With a high enough concentration of ATP, breakdown of ATP would be favored. This would cause the pumping of hydrogen ions against their concentration gradient. Choice (B) is incorrect because the hydrogen ions would flow *up* their concentration gradient. Choice (C) is incorrect because hydrogen ions cannot pass through the enzyme without spinning the rotor and causing a reaction. Choice (D) is incorrect because choice (A) is the correct choice. (LO 2.12)

251. **(C)** If there was no knob portion, there would be no reaction involving ATP. The enzyme would simply act as a channel through which hydrogen ions could pass. Therefore, choices (A) and (B) are incorrect. Choice (D) is incorrect because choice (C) is the correct choice. (LO 2.10)

252. **(A)** Diagram I shows an equal flow of water into and out of the cell. Therefore, that cell is isotonic to the solution; both have the same concentration of solutes. Choice (B), Diagram II, shows more water flowing into the cell than out of the cell. Thus, that cell is hypertonic to the environment. Choice (C), Diagram III, shows more water flowing out of the cell than into the cell. Thus, that cell is hypotonic to the environment. Choice (D) is incorrect because it is possible to tell the relative solute concentration from these diagrams. (LO 2.12)

253. **(B)** Diagram II shows more water flowing into the cell than out of the cell. Thus, that cell is hypertonic to the environment. Choice (A) is incorrect because Diagram I shows an equal flow of water into and out of the cell. Therefore, that cell is isotonic to the solution; both have the same concentration of solutes. Choice (C), Diagram III, shows more water flowing out of the cell than into the cell. Thus, that cell is hypotonic to the environment. Choice (D) is incorrect because the cell in Diagram II is hypertonic. (LO 2.12)

254. **(A)** Red blood cells are typically isotonic to the plasma. If the cells were hypertonic, choice (B), they would swell and burst. If the cells were hypotonic, choice (C), the cells would shrivel. Neither condition would be healthy. Choice (D) is incorrect because Diagram I shows a normal cell. (LO 2.12)

255. **(A)** Since the cells are shriveled, net water flow is out of the cells. Therefore, the saline is likely hypertonic to the cells. If the saline was hypotonic, the cells would swell. Therefore, choice (B) is incorrect. If the saline was isotonic, the cells would appear normal. Therefore, choice (C) is wrong. *Hydrophobia* is the medical term for rabies. Hydrophobia cannot be determined from the description provided. Therefore, choice (D) is incorrect. (LO 2.15)

256. **(B)** When placed into distilled water, the cell will be hypertonic to the solution since the plasma has solutes in it. Therefore, net water flow will tend to be into the cell. Diagram I, choice (A), depicts an isotonic cell. For a cell to be isotonic to distilled water, it would have to contain no solutes, which is impossible. Diagram III, choice (C), depicts a cell that is hypotonic to the environment and therefore plasmolyzes. This is impossible for a cell in distilled water since the plasma would have to have less than zero solute in it. Choice (D) is incorrect because Diagram II shows a cell that could be in distilled water. (LO 2.12)

257. **(B)** Turgor pressure is a result of the pressure of the membrane against the cell wall because the cell has more solute than the environment. In Diagrams I and

III, the turgor pressure is zero because there is no pressure against the cell wall. Therefore, choices (A) and (C) are incorrect. Choice (D) is incorrect because there are differences in the turgor pressure among the images. (LO 2.12)

258. **(B)** Healthy plant cells are hypertonic to their environment and have positive turgor pressure. Diagram I represents a cell at the wilting point, and Diagram III is a plasmolyzed cell. Therefore, choices (A) and (C) both show unhealthy cells. Since Diagrams I and III are from unhealthy cells, choice (D) is incorrect. (LO 2.12)

259. **(A)** When there is no net flow of water into or out of the cell, the cell is isotonic and the turgor pressure is zero. Diagram II shows a cell with turgor pressure, so choice (B) is incorrect. Diagram III shows a cell that is past the point where the turgor pressure just reached zero. Therefore, choice (C) is incorrect. Neither Diagram II nor Diagram III shows a plant at the wilting point since Diagram II has turgor pressure while Diagram III is well past the wilting point. Therefore, choice (D) is incorrect. (LO 2.12)

260. **(B)** Solute water potential is lower for a hypertonic solution. Since water is flowing into the cell in Diagram II, it has a lower solute water potential than the environment. The cell in Diagram I has the same water potential as the environment, while the cell in Diagram III has a higher solute water potential than the environment. Therefore, choices (A) and (C) are incorrect. Since all of the cells do not have the same solute water potential as the solutions they are in, choice (D) is wrong. (LO 2.12)

261. **(D)** If salt was added to the solution surrounding the artificial cell described, water would flow through the aquaporins out of the cell. As a result, the diameter of the cell would decrease. There is no reason to believe that the proteins would dissociate from the membrane, so choice (A) is incorrect. Since the diameter of the cell should decrease, the diameter wouldn't stay the same, so choice (B) is incorrect. The diameter of the cell should decrease, not increase, so choice (C) is incorrect. (LO 2.12)

262. **(C)** Sun-grown leaves have an average of about 32 stomata/view. Shade-grown leaves have an average of about 17 stomata/view. The data do not measure cuticle thickness, so choice (A) is not correct. Since the sun-grown leaves have more average stomata/view than shade-grown leaves, choice (B) is incorrect. Since the error bars don't overlap, we can't say whether or not statistically significant differences exist. Therefore, choice (D) is incorrect. (LO 2.24)

263. **(C)** Since sun-grown leaves are exposed to more light, they perform more photosynthesis and generate more oxygen. Therefore, sun-grown leaves have a greater need for more stomata to perform more gas exchange. Choices (A) and (B) are incorrect because shade-grown leaves neither perform more photosynthesis nor use more water. Choice (D) is incorrect because sun-grown leaves do not have fewer stomata. (LO 2.24)

264. **(B)** Differential gene expression allows cells to have the same DNA but produce different types of leaves. Since the leaves are from the same plant, the leaves all have the same DNA. Therefore, choice (A) is incorrect. Although grafting can occur, it does not explain the developmental difference in sun-grown leaves and in shade-grown leaves. Therefore, choice (C) is incorrect. Genetically identical cells can produce different structures like roots, leaves, and flowers. Thus, increasing or decreasing stomatal density should not be that difficult. Therefore, choice (D) is incorrect. (LO 2.40)

265. **(A)** The fan may have caused evaporation from the surface of the flask whether or not a branch was in the flask. Including a control would establish the extent of that evaporation. The control should have been a 50 mL flask filled with water, covered with Parafilm, and without a branch. Choice (B) would replace controlled air movement in the same space with uncontrolled air movement. It would also introduce different light and temperature conditions. Most of the mass change in the experiment would be due to water loss, so mass is an appropriate measure to use. Therefore, choice (C) is incorrect. Adding a second variable would not improve the experiment. It would simply complicate things. Therefore, choice (D) is incorrect. (LO 2.35)

266. **(C)** There are many differences between a refrigerator and a window, including air flow and light conditions. A better experiment would use two identical refrigerators with identical lights but set at different temperatures. Choice (A) is incorrect because two temperatures should be enough to see a difference if one exists. Although more temperatures would improve the experience, they are not as important as consistent light conditions. A larger sample size is always useful. However, using more samples of a poor design only yields results that are impossible to interpret. Therefore, choice (B) is incorrect. Choice (D) is incorrect because all of the criticisms are valid, with some being more important than others. (LO 2.35)

267. **(B)** Since the error bars overlap, the data do not show statistically significant differences. Choices (A) and (C) are incorrect because the data do not show significant differences. Since choice (B) is a reasonable assertion, choice (D) is incorrect. (LO 2.32)

268. **(A)** This diagram shows hydrogen ions moving down their concentration gradient while lactose is moved up its gradient. Choice (B) shows both hydrogen ions and lactose moving up their gradients and is therefore incorrect. Choice (C) shows hydrogen moving up its concentration gradient while lactose is moving down its concentration gradient. This is the opposite of what is described in the prompt and is therefore incorrect. Since the question states that the hydrogen ion gradient is set up by pumping hydrogen ions out of the cell, the extracellular fluid should have a higher concentration of hydrogen ions. Therefore, choice (D) is incorrect. Additionally, choice (D) shows lactose going down its concentration gradient, which is incorrect. (LO 2.11)

Grid-In Questions (pages 127–130)

When working through the grid-in, long free-response, and short free-response questions that involve math, remember that your answer may vary slightly from the answer provided due to rounding or due to reading a graph differently. See page ix of the Introduction of this book for more information regarding how to avoid rounding errors. As a general rule, if you are within 5% of the actual answer, then your answer will be marked as correct.

269. **–19.5** To answer this question, use the formula for the solute potential of a solution, which is $\Psi_S = -iCRT$. See the Reference Tables at the end of this book for a breakdown of this formula. The isotonic point for the sweet potato is 0.8 M. This is the value C in the equation for solute water potential. The van 't Hoff factor (i) is 1 for all molecular compounds like sucrose. The universal gas constant, R, is given on the Reference Tables as 0.0831 liter bars/mole K). The temperature in Kelvin is 273 + 20°C = 293 K.

$$\Psi_S = -1 \times 0.8 \times 0.0831 \times 293 = -19.5$$

(LO 2.12)

270. **–7.3** To answer this question, use the formula for solute potential of a solution, which is $\Psi_S = -iCRT$. The isotonic point for the russet potato is 0.3 M. This is the value C in the equation for solute water potential. The van 't Hoff factor (i) is 1 for all molecular compounds like sucrose. The universal gas constant, R, is given on the Reference Tables as 0.0831 liter bars/mole K). The temperature in Kelvin is 273 + 20°C = 293 K.

$$\Psi_S = -1 \times 0.3 \times 0.0831 \times 293 = -7.3$$

(LO 2.12)

271. **7.2** To answer this question, use the formula for water potential, which is $\Psi = \Psi_P + \Psi_S$. See the Reference Tables at the end of this book for a breakdown of this formula. A beaker of distilled water that is open to the atmosphere has a water potential of zero. When placed into the distilled water, the cell will come into equilibrium with the outside environment. Therefore, the total water potential of the cell (Ψ) will be 0. Since the plant cell has a solute water potential (Ψ_S) of –7.2 bars, the pressure potential (Ψ_P) will have to balance this and is therefore +7.2 bars. Rearrange the formula for water potential as follows. (Note that you will be subtracting a negative number, which will make the result positive.)

$$\Psi_P = \Psi - \Psi_S = (0) - (-7.2) = 7.2$$

(LO 2.12)

272. **3.9** To answer this question, use the formula for water potential, which is $\Psi = \Psi_P + \Psi_S$. The solution inside the beaker has a solute water potential (Ψ) of –3.3 bars. When placed into water, the cell will come into equilibrium with this value. Since the inside of the cell has a solute potential (Ψ_S) of –7.2 bars, adding 3.9 bars of pressure will cause the inside and outside values to both equal –3.3 bars. Rearrange the formula for water potential as follows:

$$\Psi_P = \Psi - \Psi_S = (-3.3) - (-7.2) = 3.9$$

(LO 2.12)

273. **0.57** Any answer between 0.53 and 0.60 would be acceptable. The most serious drought stress for NM6 was about –14 bars. Since the solute concentration is needed, rearrange the formula for the solute potential of the solution ($\Psi_S = -iCRT$):

$$C = \frac{\Psi_S}{-iRT} = \frac{-14 \text{ bars}}{(-1)(0.0831 \text{ liter bars/mole K})(293 \text{ K})} = \frac{-14}{-24.3483} = 0.57$$

(LO 2.12)

274. **0.49** Any answer between 0.45 and 0.50 would be acceptable. The most serious drought stress for NE308 was about –12 bars. Since the solute concentration is needed, rearrange the formula for the solute potential of the solution ($\Psi_S = -iCRT$):

$$C = \frac{\Psi_S}{-iRT} = \frac{-12 \text{ bars}}{(-1)(0.0831 \text{ liter bars/mole K})(293 \text{ K})} = \frac{-12}{-24.3483} = 0.49$$

(LO 2.12)

275. **4.0** The initial level of oxygen was 6 mg/mL. Over two days, this fell to 2 mg/mL (for when it was dark and there was 0% light intensity). To calculate the BOD, simply subtract the final value from the initial value:

$$6.0 \text{ mg/mL} - 2.0 \text{ mg/mL} = 4.0 \text{ mg/mL}$$

(LO 2.24)

276. **6.0** Net primary productivity is the amount of rise in oxygen above the starting amount. To calculate this, simply take the final oxygen value (12 mg/mL) and subtract the initial oxygen value (6 mg/mL):

$$12.0 \text{ mg/mL} - 6.0 \text{ mg/mL} = 6.0 \text{ mg/mL}$$

(LO 2.24)

277. **10.0** Gross primary productivity is the total amount of oxygen generated. Some of this oxygen was consumed by respiration. The oxygen consumed is called the biological oxygen demand (BOD), which can be measured by determining how much oxygen is consumed when there is no light for photosynthesis. To determine the BOD, you would put a sample of water in

a completely dark bottle and measure how much oxygen is consumed over a period of time. Since the DO fell from 6 mg/mL to 2 mg/mL, the BOD is 4 mg/mL of oxygen. The net primary productivity is how much the oxygen rose. In this case, the oxygen level rose from 6 mg/mL to 12 mg/mL. Gross primary productivity adds these two values (the BOD and the net primary productivity) together (4 + 6), which is 10. An easier way to calculate the gross primary productivity is to subtract the final dark value (2 mg/mL) from the final light value (12 mg/mL):

$$12.0 \text{ mg/mL} - 2.0 \text{ mg/mL} = 10.0 \text{ mg/mL}$$

(LO 2.24)

278. **1.54** To answer this question, you need to use the formula for temperature coefficient (Q_{10}). You will find this formula on the Reference Tables at the end of this book. That formula is:

$$Q_{10} = \left(\frac{k_2}{k_1}\right)^{\frac{10}{T_2 - T_1}}$$

Note that when $T_2 - T_1 = 10$, as it does in this instance, the exponent becomes 1. The equation thus becomes much simpler as it is simply then the ratio of the metabolic rate at the higher temperature to the metabolic rate at the lower temperature. For this particular question, $\frac{20}{13} = 1.54$. (LO 2.9)

279. **0.62** To answer this question, you need to use the formula for temperature coefficient (Q_{10}):

$$Q_{10} = \left(\frac{k_2}{k_1}\right)^{\frac{10}{T_2 - T_1}}$$

Note that when $T_2 - T_1 = 10$, as it does in this instance, the exponent becomes 1. The equation thus becomes much simpler as it is simply then the ratio of the metabolic rate at the higher temperature to the metabolic rate at the lower temperature. For this particular question, $\frac{42}{68} = 0.62$. (LO 2.9)

280. **–2.13** Since this question asks for the rate of change in the oxygen consumed by the chipmunks as temperature rises, and because the oxygen consumed decreases as the temperature increases for the chipmunks over the given range, a negative answer is appropriate. Take the respiration rate at 10°C (68 mL/hr) and subtract it from the respiration rate at 25°C (36 mL/hr):

$$36 - 68 = -32$$

Finally, divide –32 by the difference in the temperatures:

$$\frac{-32}{15} = -2.13$$

(LO 2.9)

281. **0.73** Since the respiration rate for the geckos increases as the temperature rises, a positive answer is appropriate. Take the respiration rate at 10°C (5 mL/hr) and subtract it from the respiration rate at 25°C (16 mL/hr):

$$16 - 5 = 11$$

Finally, divide 11 by the difference in the temperatures:

$$\frac{11}{15} = 0.73$$

(LO 2.9)

282. **–408.6** The question begins by talking about the breakdown of glucose into carbon dioxide. The first chemical formula shows this reaction and that it releases 686 kcal/mol (the negative sign indicates that the energy is released). The second chemical formula shows the breakdown of ATP. The question asks about the formation of ATP, which is the reverse reaction. Since energy is conserved, if the forward reaction releases 7.3 kcal/mol, then the backward reaction absorbs the same amount of energy (which would give it a positive sign). Since there are 38 moles of ATP generated and each mole of ATP absorbs 7.3 kcal, we multiply $38 \times 7.3 = 277.4$ kcal. Adding this value to the –686 kcal released by the oxidation of glucose reveals that –408.6 kcal are not captured in the form of ATP. (LO 2.9)

Long Free-Response Questions (pages 131–132)

When working through the grid-in, long free-response, and short free-response questions that involve math, remember that your answer may vary slightly from the answer provided due to rounding or due to reading a graph differently. See page ix of the Introduction of this book for more information regarding how to avoid rounding errors. As a general rule, if you are within 5% of the actual answer, then your answer will be marked as correct.

283. This question asks you to analyze an experiment conducted on chipmunks and geckos. The rubric that follows explains how to grade your response. Note that this question is worth 10 points total. (LO 2.24, LO 2.25, LO 2.32)

(a) Graph the results on the axes provided. (1 point for each of the following—3 points maximum)

- Correct orientation (dependent variable on the *x*-axis, independent variable on the *y*-axis)
- Correct labels, including units, title, and legend
- Correctly plotted data points

Your graph should look something like the following graph:

Respiration as a Function of Temperature in Chipmunks and Geckos

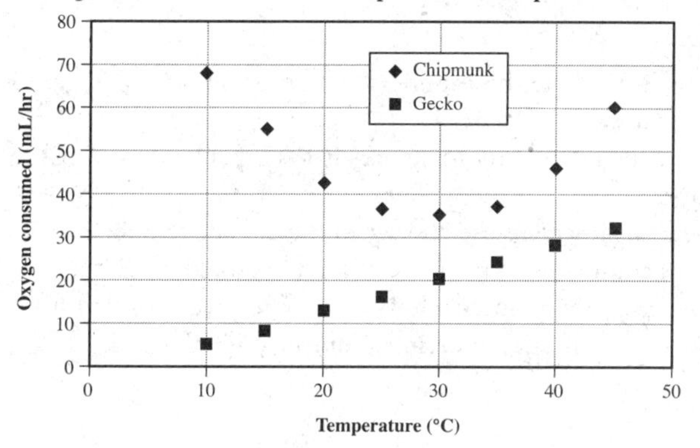

(b) For EACH animal species, describe the relationship between temperature and respiration, and explain the physiological basis for the shape of each curve. (2 points for each description and 2 points for each explanation—4 points maximum)

<u>Description of the Relationship Between Temperature and Respiration for Both Animal Species</u>

- Chipmunk: The respiration rate decreases until a moderate temperature is reached and then the respiration rate increases.
- Gecko: As the temperature increases, the respiration rate increases.

<u>Explanation of the Physiological Basis for the Shape of Each Curve for Both Animal Species</u>

- Chipmunks are thermoregulators. (Note that "endotherms" and "warm-blooded" would also be accepted in place of "thermoregulators.") Their body temperature stays relatively constant. As a result, they expend more energy when the outside temperature is cold in order to generate internal heat. When the outside temperature is hot, they use energy to cool themselves.
- Geckos are thermoconformers. (Note that "ectotherms," "poikilotherms," and "cold-blooded" would also all be acceptable in place of "thermoconformers.") Their body temperature is dependent on the environment. Since enzymatic reactions are temperature dependent, the reactions involved in respiration increase in rate as kinetic energy (temperature) increases.

(c) For ONE of the two species, identify ONE cost and ONE benefit for the type of metabolism used, and describe the conditions under which the benefit outweighs the cost for the organism chosen. (3 points maximum)

For this answer, be sure to choose *either* chipmunks *or* geckos. If you write about both, only the first species that you write about counts for your score.

Chipmunks

- One cost is that the chipmunks expend more energy than geckos at all temperatures.
- One benefit is that chipmunks are always capable of moving fast to evade predators.
- The benefit outweighs the cost when the outside temperature is cold and energy is abundant (in the form of food or stored chemical energy, such as fat); the chipmunk's speed allows it to avoid predators that it might not be able to avoid if it were a thermoconformer, like the gecko.

Geckos

- One cost is that geckos are slow at cooler temperatures and are thus easier prey when the temperature is cold.
- One benefit is that geckos require less energy and therefore less food than thermoregulators, like chipmunks.
- The benefit outweighs the cost when the environment is warm and food is scarce; using less energy allows thermoconformers like geckos to use less food and survive longer.

284. This question centers around two different strains of *Arabidopsis* plants. The question asks you to construct a graph, interpret the graph, and design a controlled experiment. The rubric that follows explains how to grade your response. Note that this question is worth 10 points total. (LO 2.18, LO 2.23, LO 2.32)

(a) Use the grid to construct a graph of the data. (1 point for each of the following—3 points maximum)

- Correct orientation (dependent variable on the *x*-axis, independent variable on the *y*-axis)
- Correct labels, including units, title, and legend
- Correctly plotted data points

Your graph should look something like the following graph:

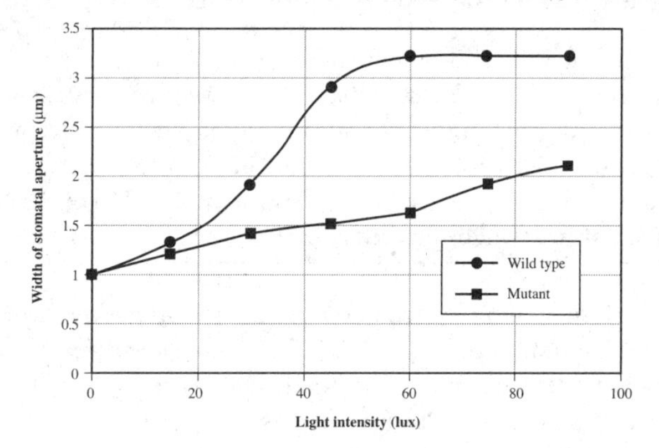

(b) Describe and explain the physiological reason for the shape of the curve for the wild-type plant between 0 and 60 lux and between 60 and 90 lux (1 point for each of the following—4 points maximum)

- For the wild-type plant, the stomata open more as the light increases from 0 to 60 lux.
- More light causes more photosynthesis and an increased need for gas exchange, which occurs through the stomata.
- From 60 lux to 90 lux, the stomata do not open wider with increasing light intensity.
- Photosystems are saturated and cannot absorb more photons of light above 60 lux, so there is no increased need for gas exchange OR the stomata have a physical limit as to how wide they can open.

(c) Describe a controlled experiment that would test the idea that the two different plant lines would respond differently to drought stress. (3 points maximum)

It is important to be specific about your experiment. A well-designed experiment will contain several of the following features, each of which might earn 1 point:

- State a hypothesis.
- Specify a control group.
- Use a large sample size (at least 3 in each treatment).
- Specify an independent variable.
- Specify a dependent variable.
- Describe how to control variables that might interfere with the results of the experiment.
- Repeat the experiment.
- Use mathematical procedures, such as calculating the average of several measures, calculating the rate of a process, or performing an appropriate statistical method.

Short Free-Response Questions (pages 133–134)

> When working through the grid-in, long free-response, and short free-response questions that involve math, remember that your answer may vary slightly from the answer provided due to rounding or due to reading a graph differently. See page ix of the Introduction of this book for more information regarding how to avoid rounding errors. As a general rule, if you are within 5% of the actual answer, then your answer will be marked as correct.

285. Using the information presented in the graph, determine whether there were statistically significant differences among the different treatments shown. Justify your answer. Determine whether the mealworms were able to use plastics as an energy source. Justify your answer. (LO 2.2)

A maximum of 4 points can be awarded for correctly answering this question.

(a) There were no statistically significant differences among the different treatments because the error bars all overlapped. (1 point)

(b) The mealworms were able to use plastics as an energy source. (1 point)

The remaining two points for this question can be earned as follows:

- One point is awarded for correctly comparing the mealworms that were fed plastic to the mealworms that were fed only water (the negative control). Either of the following would be acceptable statements:
  - Mealworms that were fed plastic lived whereas those that were fed only water died.
  - Mealworms that were fed plastic gained mass whereas those that were fed only water did not.
- One point is awarded for correctly comparing the mealworms that were fed plastic to the mealworms that were fed bran (the positive control). Both mealworms that were fed plastic and those that were fed bran gained mass.

286. Calculate the surface area:volume ratio for a 100-μm-long cell of each type and record your answers in the table provided. Assume that each cell type is approximated by a cylinder (see the Reference Tables at the end of this book). Use the value 3.14 for π. The surface area is given for both the plant root hair and the fungal cell. Compare the surface area:volume ratio of the two types of cells. Explain why this ratio is significant to the relationship between the two organisms. (LO 2.6)

A maximum of 3 points can be awarded for correctly answering this question.

(a)

	Diameter (μm)	Surface Area (μm^2)	Volume (μm^3)	Surface Area: Volume (μm^{-1})
Plant root hair	15	4,712	17,663	0.27
Fungal cell	5	1,570	1,963	0.80

To find the volume for both, use the formula for a cylinder:

$$V = \pi r^2 h$$

where V equals the volume, r is the radius (note that the radius is half of the diameter), and h is the height

For the plant root hair:

$$V = \pi r^2 h$$

$$V = 3.14 \times \left(\frac{15}{2}\right)^2 \times 100$$

$$V = 3.14 \times 56.25 \times 100$$

$$V = 17{,}662.5 \approx 17{,}663$$

For the fungal cell:

$$V = \pi r^2 h$$

$$V = 3.14 \times \left(\frac{5}{2}\right)^2 \times 100$$

$$V = 3.14 \times 6.25 \times 100$$

$$V = 1{,}962.5 \approx 1{,}963$$

To find the surface area:volume ratio for both, simply divide the surface area by the volume.

For the plant root hair:

$$\frac{4{,}712}{17{,}663} = 0.27$$

For the fungal cell:

$$\frac{1{,}570}{1{,}963} = 0.80$$

(1 point earned for correctly calculating the surface area:volume ratios as shown in the table above)

(b) The fungal cell has a higher surface area:volume ratio. (1 point)

Due to its higher surface area:volume ratio, the fungus is more efficient than the plant at absorbing minerals and water from the soil. (1 point)

287. Identify the primary functions of the chloroplast and the mitochondrion. Explain TWO reasons why these two organelles might be situated close together. The peroxisome contains the enzyme peroxidase that breaks down toxic hydrogen peroxide (H_2O_2) into water and oxygen. Propose a plausible explanation for locating the peroxisome near the chloroplast and the mitochondrion. (LO 2.5)

A maximum of 4 points can be awarded for correctly answering this question.

(a) The chloroplast performs photosynthesis, while the mitochondrion performs respiration. (1 point)

The chloroplast uses carbon dioxide, which is a waste product of the mitochondrion. This is one reason why these two organelles might be situated close together. (1 point)

The mitochondrion uses oxygen, which is a waste product of the chloroplast. This is another reason why these two organelles might be situated close together. (1 point)

(b) The peroxisome detoxifies hydrogen peroxide generated by the mitochondria and the chloroplasts as a result of oxygen metabolism. (1 point for correctly stating that either the mitochondria or the chloroplasts probably produce hydrogen peroxide)

288. Describe the cost of expressing UCP. UCP is present in the mitochondria of brown fat in human infants and in chipmunks during the winter. Identify one possible adaptive function of expressing this protein, and explain why that function would be beneficial. (LO 2.10)

A maximum of 3 points can be awarded for correctly answering this question.

(a) Proteins that travel through UCP do not make ATP and therefore consume stored energy. (1 point)

(b) UCP allows animals to generate heat by burning glucose without moving muscles. (1 point)

As a result of this adaptive function, a human baby or a chipmunk can keep warm in a cold environment. (1 point)

BIG IDEA 3—INFORMATION

Multiple-Choice Questions (pages 135–216)

289. **(B)** The first two columns in the F2 bracket both show close to 50% diapause. These were both crosses in which the F1 were backcrossed to a fly from Florida. If the F1 were backcrossed to a fly from Maine, 100% of them were able to enter diapause, so choice (A) is incorrect. The sex of the fly did not seem to matter since matings with both male and female flies produced either close to 50% diapause or 100% diapause depending on where the P were from. Therefore, choices (C) and (D) are incorrect. (LO 3.14)

290. **(D)** Since 100% of the F1 offspring were able to enter diapause, it can be inferred that the inability to enter diapause is recessive. This makes choices (A) and (C) incorrect since those answer choices say that the *ability* to enter diapause is recessive. The trait is autosomal because whether the male or the female has the recessive trait does not matter. Therefore, choice (B) is incorrect. (LO 3.14)

291. **(A)** The data are consistent with a dominant allele coding a functional copy of a protein for a step in the signal transduction pathway. Apparently, a single copy of the diapause plus allele is sufficient to turn on the diapause gene. Alleles are alternate forms of a gene. The diapause plus and the diapause minus alleles exist at the same gene locus. Therefore, you either get the diapause plus allele or the diapause minus allele. Since alleles at the same gene locus cannot segregate independently, choice (B) is incorrect. Choice (C) describes the diapause minus gene as being dominant, which is the opposite of what is actually occurring. There is no evidence of mate choice, so choice (D) is incorrect. (LO 3.14)

292. **(A)** The F2 results mentioned in the question show 100% dominant phenotypes. This is exactly what we would expect if the F1 were heterozygous and mated to a homozygous dominant genotype because half of the offspring would be homozygous dominant and the other half would be heterozygotes. We wouldn't see 100% homozygous dominant individuals, so choice (B) is not correct. Furthermore, we do not see a 9:3:3:1 ratio, as stated in choice (C), nor do we see a 3:1 ratio, as stated in choice (D). Therefore, both choice (C) and choice (D) are incorrect. (LO 3.14)

293. **(B)** The triplets shown in bold translate to Met-Pro-Leu-Stop:

5′-AC/**AUG/CCC/CUC/UAA**/UUCAC-3′

Choices (A) and (D) are RNA sequences that contain an AUG (Met) codon. However, the rest of each sequence is incorrect. Choice (C) includes T, which is not present in RNA. (LO 3.6)

294. **(A)** The triplets shown in bold translate to Met-Pro-Asn-Val:

5′-**AUG/CCC/AAU/GUU**/GA-3′

Choice (B) translates to Met-Pro-Leu-Stop. Choice (C) includes T, which is not present in RNA. Choice (D) translates to Met-Lys-Ala-Thr-Stop. (LO 3.6)

295. **(D)** The triplets shown in bold translate to Met-Lys-Ala-Thr-Stop:

5′-GUUC/**AUG/AAA/GCU/ACA/UAG**-3′

Choice (A) translates to Met-Pro-Asn-Val. Choice (B) translates to Met-Pro-Leu-Stop. Choice (C) includes T, which is not present in RNA. (LO 3.6)

296. **(C)** The DNA must be transcribed and then translated, as shown below. Note that the first nitrogen base and the last 5 nitrogen bases in the DNA strand do not code for the specific protein.

```
DNA 3′-T  TAC GTA AAA GAC ATC GGTGT-5′
mRNA 5′-A AUG CAU UUU CUG UAG CCACA-3′
Protein   Met- His- Phe- Leu- Stop
```

Choices (A), (B), and (D) do not code for the specific protein stated in the question. (LO 3.6)

297. **(A)** The DNA must be transcribed and then translated, as shown below. Note that the first 4 nitrogen bases in the DNA strand do not code for the specific protein.

```
DNA 3′-CCGA TAC CTT ACA AGT GGC ACT-5′
mRNA 5′-GGCU AUG GAA UGU UCA CCG UGA-3′
Protein      Met- Glu- Cys- Ser- Pro- Stop
```

Choices (B), (C), and (D) do not code for the specific protein stated in the question. (LO 3.6)

298. **(B)** The DNA must be transcribed and then translated, as shown below. Note that the first 4 nitrogen bases and the last 5 nitrogen bases in the DNA strand do not code for the specific protein.

```
DNA 3′-TTCC UAC UAA CGG AUA AUU GGAUA-5′
mRNA 5′-AAGG AUG AUU GCC UAU UAA CCUAU-3′
Protein      Met- Ile- Ala- Tyr- Stop
```

Choices (A), (C), and (D) do not code for the specific protein. (LO 3.6)

299. **(C)** The easiest way to look at DNA to see what it might code for, aside from using a computer, is to look at the DNA strand that is complementary to the strand that is read. This strand will be in the 5′ to 3′ orientation so it will read the same as mRNA except that it will have T instead of U. By using

this method, we see that the strand shown on the bottom of choice (C) reads 5′-G**ATG**ATG. . . . (Note that you will need to read this from right to left, which is the opposite way that you read text like this answer explanation.) This sequence translates as Met-Met. . . . It could also be read in a different frame as 5′-GA**TGA**TG. . . . This translates as STOP-STOP. Neither choice (A) nor choice (B) contain any Gs, so they could not encode a Met (specified by the codon AUG). The only Stop codon without a G is UAA. The top strand shown in choice (A) could code for STOP, and the bottom strand in choice (B) could code for STOP, but the question asks for a molecule that could code for both. Choice (D) seems to contain an ATG, but it is in the incorrect orientation and would be read as GTAGTA, which does not code for Met. (LO 3.5)

300. **(B)** The complementary sequence on the mRNA must be complimentary and antiparallel. G pairs with C, and A pairs with U. Choice (A) is complementary to 3′-UUC-5′. Choice (C) is complementary to 3′-UGG-5′. Choice (D) is complementary to 3′-ACC-5′. (LO 3.5)

301. **(B)** The mRNA that pairs with 3′-GAA-5′ is 5′-CUU-3′. This specifies Leu. Since each triplet specifies a single amino acid, choices (A), Gln, (C), Thr, and (D), Trp, are all incorrect. (LO 3.5)

302. **(A)** Reassigning the start codon AUG to a different amino acid would result in the loss of methionine. Choices (B), (C), and (D) are incorrect because they specify amino acids that are also specified by other codons. (LO 3.6)

303. **(B)** Evolution by natural selection predicts that there should be variation in any trait and that this variation should be more different in less closely related organisms. If life evolved independently, as suggested by choice (A), completely different genetic codes would be expected. Choice (C) is incorrect because evolutionary theory predicts that a situation like this would occur. A degenerate genetic code means that more than one codon exists for a particular amino acid. This is not relevant to the question, so choice (D) is incorrect. (LO 3.24)

304. **(B)** G and C make 3 hydrogen bonds, while A and T make 2 hydrogen bonds. The result is that the higher the GC content of a strand is, the higher is the dsDNA's "melting" point. The dsDNA with more GC is held together with more hydrogen bonds in the same amount of space. Note that DNA doesn't melt in the same sense that ice melts into water. This question is discussing strand separation of double-stranded DNA. However, most researchers refer to it as "melting." Choice (A) is incorrect because a higher number of AT pairs would lead to a *lower* melting point. Since the composition of DNA varies, so does the melting point. Therefore, choice (C) is incorrect. DNA only melts because it is double-stranded, so choice (D) is incorrect. (LO 3.5)

305. **(B)** Saline solution was used as a control so that Griffith could be sure that any effect was due to the bacteria. Since mice are fairly easy to grow and because previously injected mice have already had something done to them, reusing control mice is not a good strategy. Therefore, choice (A) is incorrect. Griffith likely did some preliminary experiments to learn how to inject mice effectively, but practicing the injection procedure was not the purpose of the injections described in the question. Therefore, choice (C) is incorrect. Choice (D) is incorrect because no information indicates that osmotic conditions were varied or measured. (LO 3.2)

306. **(D)** Immune cells recognize nonself cells and clear them from the body. As a result, blood has very few bacteria in it. The idea presented in choice (A), that competition excluded *S. pneumoniae*, is incorrect. Since Griffith injected bacteria into the blood, bacteria are expected to be found in the blood (not on the teeth) if they could survive. Therefore, choice (B) is incorrect. Blood pH might impact bacterial survival. However, if bacteria are present, it should be possible to isolate them. Therefore, choice (C) is incorrect. (LO 3.2)

307. **(D)** The live rough bacteria were transformed with DNA from the environment that contained genes for a smooth coating and that made the bacteria pathogenic. This DNA originally came from the smooth bacteria that were heat-killed. The idea of a life force has been experimentally disproven by previous experiments, such as those of Louis Pasteur, so choice (A) is incorrect. If some smooth bacteria were still alive after heat treatment, the heat-killed smooth bacteria alone should have killed the mice. Therefore, choice (B) is incorrect. If choice (C) was correct, rough bacteria alone should have been able to express the virulence genes they already had and cause diseases. (LO 3.2)

308. **(B)** Avery, MacLeod, and McCarty destroyed protein with proteinases and still got transformation. Therefore, something else in the mixture must cause transformation and be heritable. Choice (A) is a close second possible answer. A person might use the following logic: if DNA is the hereditary material, then protein cannot be. However, the fact that proteinase did *not* affect transformation is direct evidence that protein is not the genetic material. RNase degrades RNA. Therefore, it doesn't give us any information about protein, which makes choice (C) incorrect. The treatment with no enzyme was the control and was included to ensure that transformation worked, making choice (D) incorrect. (LO 3.2)

309. **(A)** AAGG is the pure-breeding dominant strain, and aagg is the pure-breeding recessive. Choices (B) and (C) do not contain any pure-breeding genotypes. Choice (D) includes only one pure-breeding genotype. (LO 3.14)

310. **(D)** The genotype AaGg is a double heterozygote, which is what is expected for the F1. The genotype aagg is the double recessive. Choices (A) and (B) have the double recessive parent correct but not the F1 parent. Choice (C) has the F1 parent right but not the double recessive parent. (LO 3.14)

311. **(A)** Each gene locus has a $\frac{3}{4}$ chance of showing the dominant phenotype and a $\frac{1}{4}$ chance of showing the recessive phenotype. Therefore, the chance of double dominant (purple–green) is $\frac{3}{4} \times \frac{3}{4} = \frac{9}{16}$, the chance of double recessive (nonpurple–yellow) is $\frac{1}{4} \times \frac{1}{4} = \frac{1}{16}$, and the chances of the other two phenotypes are each $\frac{1}{4} \times \frac{3}{4} = \frac{3}{16}$. Choice (B) has the double dominant and the double recessive reversed. Choice (C) might be obtained if a heterozygote were backcrossed to a homozygous recessive and the two genes were linked. Choice (D) is what would be expected of a heterozygote backcrossed to a homozygous recessive. (LO 3.14)

312. **(D)** Each gene locus has a $\frac{1}{2}$ chance of showing the dominant phenotype and a $\frac{1}{2}$ chance of showing the recessive phenotype. Therefore, the chance of double dominant (purple–green) is $\frac{1}{2} \times \frac{1}{2} = \frac{1}{4}$, the chance of the double recessive (nonpurple–yellow) is $\frac{1}{2} \times \frac{1}{2} = \frac{1}{4}$, and the chances of the other two phenotypes are each $\frac{1}{2} \times \frac{1}{2} = \frac{1}{4}$. Choice (A) shows the correct ratio for self-crossing a double heterozygote and is therefore incorrect. Choice (B) might be obtained if one thought the cross was a double heterozygous self-cross and then got the double dominant and the double recessive switched. Choice (C) might be obtained if the genes were linked. (LO 3.14)

313. **(D)** In this case, one parent had purple stems with green leaves while the other parent had nonpurple stems with yellow leaves. These are the parental phenotypes. The recombinant phenotypes are plants that have purple stems with yellow leaves and those plants that have nonpurple stems with green leaves. There are fewer recombinant phenotypes than would be expected if the genes assorted independently. Therefore, the genes are linked. Random chance would be unlikely to produce these results, so choice (A) is incorrect. Performing a chi-square test would confirm this. Choice (B) would be true if there was a 1:1:1:1 ratio, but this is not what was observed in these results. Choice (C) doesn't explain why there are nonpurple, green leaf plants. (LO 3.15)

314. **(A)** If D = solid and d = polka dot, the diploid plant had a genotype of Dd. When it made haploid spores, it contributed a D allele to half of the spores and a d allele to the other half. When the spores grew into haploid plants, they retained this ratio. Choices (B) and (C) are incorrect because diploid

plants did not mate in this system. Only the haploids produced sperm and eggs. Choice (D) incorrectly defines "haploid." "Haploid" means that an organism contains only one set of chromosomes. (LO 3.15, LO 3.16)

315. **(B)** Most spores sown on day 1 grew into hermaphrodites, while spores sown on day 5 produced predominantly males. This suggests an environmental cause, not a genetic one. Since the plants are haploid, any allele they have will be expressed, so whether it is dominant or recessive is irrelevant. This is unlike the situation in a diploid, where there are usually two copies of each gene, and therefore a dominant allele can mask the presence of a recessive allele. Therefore, choice (A) is incorrect. The ratio of hermaphrodites to males is not consistent, so choice (C) is not correct. There is no evidence of epigenetic control, so choice (D) is incorrect. (LO 3.19)

316. **(C)** Early colonizers may be the only plant in an area. If there are other plants nearby, there may not be very many of them. In the case of a completely isolated plant, producing both eggs and sperm is the only way to ensure that fertilization is possible. Early colonizers experience a lower population density than late colonizers. As a result, the distance to neighboring plants is very high. Therefore, the chance of a sperm reaching an egg to fertilize it is very low as is the chance that a sperm from another plant will reach the plant's eggs. Therefore, making both eggs and sperm increases the likelihood that the plant will be fertilized. Choices (A) and (B) do not answer the question, which was about the early colonizer strategy. Eggs are not abundant early in colonization, so choice (D) is incorrect. (LO 3.15)

317. **(A)** Late colonizers have an abundance of eggs produced by early colonizers that developed into hermaphrodites. In this case, there is a selective advantage to producing lots of sperm rapidly. Since males are much smaller than hermaphrodites and only produce sperm, they are ideally adapted to produce lots of sperm rapidly. Choice (B) is incorrect because egg abundance is not low when late colonizers arrive. Choices (C) and (D) do not answer the question, which was about the late colonizer strategy. (LO 3.15)

318. **(B)** The diploid plants occurred in a 3:1 ratio of solid to polka dotted. In most life cycles studied in biology, the haploids are ignored because they are transient and do not produce multicellular stages. Choice (A) is incorrect because the haploids are not in a 3:1 ratio. Diversifying selection produces trait distributions that are bimodal. Therefore, choice (C) is incorrect. The data do not support the idea that polka dots are present less frequently in males. Additionally, sex-linked traits require there to be sex chromosomes. There is no evidence of chromosomal determination of sex in the data presented, so choice (D) is incorrect. (LO 3.14)

319. **(D)** Chemical *W* has a response of 1, which was the same as that of water. Therefore, chemical *W* has no attractant activity. Choices (A) and (B) incorrectly indicate a strong attractant response, which isn't present. Choice (C)

is incorrect because no evidence is presented that an attractant response is being inhibited. (LO 3.40)

320. **(B)** Attraction is the result of the chemical being tested binding to a receptor. Since both chemicals cause similar reactions, the simplest explanation is that both chemicals *X* and *Y* bind to the same receptor and therefore share charge and shape features. Choice (A) requires two receptors to lead to attraction. This is a more complex mechanism than that described in choice (B). In science, when two explanations fit the observed data, the simpler explanation is more likely to be correct. Since the question asks which choice is most likely to be true, choice (A) is incorrect. Choices (C) and (D) are incorrect because chemicals *X* and *Y* produce strong responses, not weak ones like that for water and chemical *W*. (LO 3.37)

321. **(C)** The concentration of chemical *Y* is higher closer to where chemical *Y* is produced. The mechanism described in choice (C) turns on a flagellum when the flagellum is pointed away from the high concentration of chemical *Y* and turns a flagellum off when the flagellum is pointed toward the source of chemical *Y*. Choice (A) is unlikely to lead to sperm that swim toward the source of the chemical. The sperm would probably rotate in space by the time transcription, translation, and protein transport to the membrane were complete. Additionally, the mechanism described in choice (A) places the flagellum in an orientation that would make the sperm swim away from the source of chemical *Y*. Choice (B) would cause the sperm to swim faster the closer they were to the source of the chemical. As a result, the sperm would swim away from the source of chemical *Y*, not toward it. Choice (D) would cause the sperm to swim away from the source of chemical *Y*. (LO 3.38)

322. **(B)** Since neither of the copies of the *TAS* gene are functional, the signal will not be transduced and no taste will be perceived. Choices (A) and (C) are not correct because a single defective gene is not sufficient to prevent signal transduction. Choice (D) describes someone who has two good copies of the tasting gene. Therefore, that person will be able to taste DA. (LO 3.37)

323. **(C)** Since the only difference in the new protein is that it will bind to a different compound, everything in the signal transduction pathway downstream of binding will be the same. Therefore, this situation should trigger the same response as in a mutation-free protein, which is the perception of butterscotch. Choices (A), (B), and (D) require different downstream effects that will not happen simply by changing the shape of the ligand-binding site. (LO 3.39)

324. **(C)** In order for the protein to span the hydrophobic portion of the membrane, transmembrane domains of the protein need to be hydrophobic. Changing these amino acids to hydrophilic ones would make it difficult for the protein to associate with the membrane. Choice (A) might be true if the change was not large. However, the mutation described affects more than

one amino acid, so choice (A) is incorrect. Since the mutation was not in the binding site, ligand binding should not be affected. Therefore, choice (B) is incorrect. The mutation was not in the kinase domain. Therefore, there is no reason to believe that the protein will be stuck in the active form, making choice (D) incorrect. (LO 3.37)

325. **(D)** If a 700 bp fragment of DNA had an *Eco*R1 site, it could be cut into a 400 bp fragment and a 300 bp fragment. Adding ligase to the PCR mixture would be unlikely to do anything more than denature the ligase on the first cycle of PCR because ligase is not heat stable. Therefore, choice (A) is incorrect. There is no evidence that more than one sample was loaded, so choice (B) is not correct. The question does not mention ligase, so choice (C) is not a reasonable explanation. (LO 3.5)

326. **(C)** If person 2's restriction enzyme worked properly, he or she must not have the mutation since *Eco*R1 did not cut. Therefore, that person should be able to taste butterscotch unless his or her restriction enzyme did not work correctly. Since the person has DNA in both lanes, PCR must have amplified the DNA properly, so choices (A) and (B) are incorrect. Choice (D) is incorrect since an uncut strand indicates no mutation. (LO 3.37)

327. **(B)** Entire deletions of large stretches of DNA are extremely rare. Everyone should have a copy of the *TAS* gene. Choices (A), (C), and (D) would all lead to the results seen. (LO 3.6)

328. **(C)** Since person 4 has a 700 bp band, that person must have a wild-type copy of the *TAS* gene. Since person 4 also has a 300 bp band and a 400 bp band, that person must also have a mutant copy of the *TAS* gene. Therefore, person 4 is heterozygous. A wild-type homozygote would have only a 700 bp piece, so choice (A) is incorrect. A homozygous mutant would have only the 300 bp and 400 bp bands, so choice (B) is incorrect. Since it can be concluded that the person is heterozygous, choice (D) is also incorrect. (LO 3.6)

329. **(A)** Person 4 is heterozygous (Aa). Person 1 is homozygous mutant (aa), and person 2 is homozygous wild-type (AA). Since AA × aa must produce Aa, choice (A) is possible. Crossing Aa × aa cannot not yield AA, so choice (B) is incorrect. Crossing AA × Aa cannot yield aa, so choice (C) is incorrect. Since only one of the choices is possible, choice (D) is incorrect. (LO 3.15)

330. **(D)** Genetic disorders are passed from parents to offspring. They are not contagious. Recessive disorders simply require two alleles rather than one. They are not less potent, so choice (A) is incorrect. Choice (B) is incorrect because there is no risk of spreading a genetic disorder. Since the *Huntingtin* gene is the same in humans as in fish, choice (C) is incorrect. (LO 3.13)

331. **(D)** The ends of chromosomes contain telomeres. When two chromosomes fuse, there should be telomere sequences in the site of fusion. Additionally, having a second region that resembles a centromere suggests two centromeres

as the ancestral condition. Choice (A) would only provide evidence that fusion could have happened. The fact that closely related species have 24 pairs of chromosomes, as stated in choice (B), supports the hypothesis that the ancestral state is 24 separate chromosomes. However, this fact isn't as convincing as the molecular evidence stated in choice (D). Both fusion and cleavage hypotheses explain the presence of homologous genes, so choice (C) is incorrect. (LO 3.31)

332. **(C)** Tetraploids ($4n$) produce diploid ($2n$) gametes. Diploids ($2n$) produce haploid (n) gametes. When a diploid gamete and a haploid gamete combine, a triploid is produced ($2n + n = 3n$). Therefore, choices (A), (B), and (D) are all incorrect. (LO 3.9)

333. **(D)** In meiosis, homologous chromosomes form tetrads and attach to spindle fibers. Motor proteins on the spindle fibers then line up the chromosomes on the metaphase plate. Since there are 3 copies of each chromosome, it is impossible to attach the chromosomes evenly to the spindle fibers. The most likely outcome in this case is apoptosis, resulting in seeds not developing. Choices (A) and (B) are incorrect because triploids do not produce triploid eggs or pollen. Choice (C) is incorrect because homologous chromosomes do not pair in mitosis. In mitosis, each chromosome is composed of 2 sister chromatids, each of which attaches to a spindle fiber. Mitosis could proceed under these conditions. (LO 3.9)

334. **(C)** Transcription happens in the absence of tryptophan. This is the only choice that accomplishes that. Choice (A) is wrong because when tryptophan binds the repressor, the tryptophan activates the repressor, allowing the repressor to bind DNA. Choice (B) correctly states that tryptophan binding to the repressor makes it able to bind DNA. However, this binding prevents transcription. It does not turn on transcription. Choice (D) is incorrect because the repressor does not bind to the operator when tryptophan is absent. (LO 3.21, LO 3.22, LO 3.23)

335. **(A)** This description matches Figures III and IV. Binding of allolactose to the repressor inactivates the repressor so that the repressor cannot bind to the operator sequence of the DNA, so choice (B) is incorrect. When lactose is absent, there is no allolactose to bind to the repressor so the repressor can bind to the operator sequence of the DNA, so choice (C) is incorrect. Choice (D) is incorrect because binding of the repressor to the operator sequence of the DNA prevents transcription; binding of the repressor to the operator sequence of the DNA doesn't activate it. (LO 3.21, LO 3.22, LO 3.23)

336. **(B)** Using one switch to control a suite of genes needed under a single condition is an efficient use of control elements. Eukaryotes don't typically use operons, but this doesn't make operons a bad choice. Bacteria have smaller genomes and therefore need to be a bit more efficient in using the DNA they have. Therefore, choice (A) is incorrect. The downside of operons is that

bacteria can't adjust the transcription of individual mRNA molecules easily, which is why choice (C) is incorrect. Operons don't allow the use of the same DNA sequence to make more than one protein, so choice (D) is incorrect. (LO 3.21, LO 3.22, LO 3.23)

337. **(A)** The *trp* repressor binds to tryptophan, not lactose. The repressor would not bind to the DNA, and the *lac* genes would be transcribed. Any lactose present would be broken down. Choice (B) is incorrect because the repressor would be inactive unless bound to tryptophan. Choice (C) requires the *trp* genes to be present, which they aren't. Choice (D) is incorrect because it relies on the misunderstanding that tryptophan can be made from lactose. (LO 3.21, LO 3.22, LO 3.23)

338. **(C)** Adding tryptophan to the culture would cause the repressor to bind to the DNA, turning off transcription. This would lead to no transcription of the *lac* genes, and the tryptophan concentration would stay the same. Choice (A) is incorrect because the presence of tryptophan would turn off transcription. Choices (B) and (D) are incorrect because no tryptophan would be made. Choice (D) is also incorrect because the presence of tryptophan would turn off transcription. (LO 3.21, LO 3.22, LO 3.23)

339. **(C)** The presence of tryptophan would turn off transcription of the *lac* genes. There would be no change in the concentration of lactose because there would be no transcription of the *lac* genes. Since the concentration of lactose would not decrease, choice (A) is incorrect. Choice (B) is incorrect because no tryptophan would be produced. Choice (D) is incorrect because tryptophan wouldn't increase and lactose wouldn't decrease. (LO 3.21, LO 3.22, LO 3.23)

340. **(B)** The presence of lactose would turn on the *trp* genes that would produce tryptophan, and the concentration of lactose would stay the same. Choice (A) is incorrect because the concentration of lactose would stay the same, not decrease, and the concentration of tryptophan would increase. Since tryptophan would be produced, choice (C) is incorrect. Choice (D) is incorrect because the concentration of lactose would stay the same, not decrease. (LO 3.21, LO 3.22, LO 3.23)

341. **(C)** The presence of tryptophan would have no effect on the control elements, so there would be no change in the concentration of tryptophan. Since there are no *lac* genes present, they cannot be transcribed, making choice (A) incorrect. No tryptophan would be produced, so choices (B) and (D) are both incorrect. (LO 3.21, LO 3.22, LO 3.23)

342. **(B)** Lactose would bind to the *lac* repressor, inactivating it. The *trp* genes would be transcribed, and tryptophan would be produced. Since no lactase would be produced, the concentration of lactose would remain the same. Since the remaining choices do not state these exact results, choices (A), (C), and (D) are all incorrect. (LO 3.21, LO 3.22, LO 3.23)

343. **(D)** The enzyme lactase would be produced when no lactose was present. Therefore, there should already be a high concentration of the enzyme. Adding lactose would turn off transcription. The protein already present would degrade lactose. Enzymes are recycled naturally in cells. How much lactose was broken down would depend on how rapidly the cell broke down the enzyme. Choice (A) might be true if there were enough lactase, but choice (D) is a better explanation of the likely result. Choice (B) is incorrect because enzymes are not used up in the reactions. Choice (C) is incorrect because there would already be high concentrations of enzymes before lactose was introduced. Adding lactose would turn off the production of new proteins, but proteins made prior to the addition would still be present. (LO 3.21, LO 3.22, LO 3.23)

344. **(C)** If the presence of tryptophan turned on tryptophan production, this would be a self-stimulating process resulting in ever more tryptophan, which would stimulate the production of proteins to make even more tryptophan. Choice (A) might be true if it were possible for there to be absolutely no tryptophan present. However, the breakdown of existing proteins would release tryptophan, leading to more tryptophan production. Choice (B) is incorrect because all 20 amino acids are necessary for the bacteria to live, including tryptophan. Choice (D) is incorrect because tryptophan and the genes for lactose production are not present. (LO 3.21, LO 3.22, LO 3.23)

345. **(A)** The mother in the first generation gives the disorder to all of her children. In the second generation, only the females pass on the disorder; they pass it on to 100% of their children. If choice (B) was correct, the females in the first generation would have gotten a dominant allele from their father and not shown the disorder. The reasoning in choice (C) is not sound. Dominant alleles don't always show up in every generation, and the pedigree doesn't match this even if it was true. Since mitochondria are not inherited through the father, choice (D) is incorrect. (LO 3.15)

346. **(A)** The trait must be recessive because neither of the parents shows the trait but one of the sons does. This means that choice (B) cannot be correct. The answer in choice (C) is possible, but the question asked which *must* be true. The gene could be autosomal, so choice (C) is not correct. Since there is enough information to determine that choice (A) is correct, choice (D) is incorrect. (LO 3.14, LO 3.15)

347. **(B)** The trait cannot be dominant because it is impossible for two parents who do not show a dominant trait to have offspring that do. Choice (A) is incorrect because it is possible for the trait to be recessive. Choices (C) and (D) are also incorrect because it is possible that the boy who has the trait could have gotten it from his mother, who would have been heterozygous. This could only be the case if the allele was recessive. (LO 3.14, LO 3.15)

348. **(D)** This trait could be autosomal or sex-linked, dominant or recessive. If the trait was recessive, we could designate the recessive allele as r and the dominant allele as R. The female in the first generation would be Rr, and the male in the first generation would be rr. Their first child would be an rr male. Their second child would be an Rr male. Their third child would be an rr female. Their fourth child would be an Rr female. Therefore, the trait could be recessive. If the trait was dominant, we could designate the dominant allele as D and the recessive allele as d. The female in the first generation would be dd, and her mate would be Dd. Their first child and third child would be Dd while their second child and fourth child would be dd. Therefore, the trait could be dominant. If the trait was on the X chromosome and was recessive, the female in the first generation would be X^RX^r, and her mate would be X^rY. The offspring from left to right would have the genotypes: X^rY, X^RY, X^rX^r, and X^RX^r. If the trait were on the X chromosome and dominant, the female in the first generation would be X^dX^d, and her mate would be X^DY. Since all of these are possibilities but nothing is certain, choice (D), which says that not enough information is provided to make a determination, is correct. Choice (A) is incorrect because the trait is not necessarily recessive. Choice (B) is incorrect because the trait is not necessarily dominant. Choice (C) is incorrect because the trait is not necessarily on the X chromosome. (LO 3.14, LO 3.15)

349. **(B)** This trait cannot be dominant because one of the parents would have to show the trait if it was dominant. The trait is definitely recessive, so choice (A) is incorrect. The trait could be on the X chromosome since the boys could have inherited the trait from their mother, who would have to be a carrier. Therefore, choice (C) is incorrect. Since it is possible to determine that the trait is recessive, there is enough information to determine the mechanism of inheritance, making choice (D) incorrect. (LO 3.14, LO 3.15)

350. **(A)** This trait cannot be recessive, or it would be impossible for the parents to have children without the trait. The trait could be dominant, so choice (B) is incorrect. If the trait were on the X chromosome, all of the offspring could have the phenotypes shown in the diagram, provided that the female in the first generation was heterozygous. Therefore, choice (C) is incorrect. Choice (D) is incorrect because it is possible to determine that the trait must be dominant. (LO 3.14, LO 3.15)

351. **(A)** Ligands bind to receptors because their shape and charge match that of the receptor. Serotonin binds the same way. Choice (B) is incorrect because serotonin is not fired in the sense that a projectile is fired. In order to react with the cytoplasmic domain, serotonin would have to get into the cell somehow, so choice (C) is incorrect. The fact that serotonin is transported out of the cell in a vesicle should indicate that it is not hydrophobic. If serotonin was hydrophobic, it wouldn't need a channel protein. Therefore, choice (D) is incorrect. (LO 3.45)

352. **(A)** In the diagram, vesicles that contain serotonin fuse with the plasma membrane. This is exocytosis. Endocytosis brings vesicles into the cell, which is the opposite direction for secretion. Therefore, choice (B) is incorrect. Direct translation requires ribosomes on the plasma membrane. No ribosomes are shown in the diagram, so choice (C) is not correct. Choice (D) is incorrect because the diagram does not show serotonin moving through a channel protein. (LO 3.44)

353. **(A)** Sertraline blocks the serotonin reuptake channel. As a result, the rate at which serotonin is removed from the synapse is slower and synaptic serotonin levels are higher. Since the serotonin levels are higher, not lower, choice (B) is incorrect. Choice (C) describes the effects of a negative feedback loop and is therefore incorrect. Serotonin is not absorbed by the postsynaptic neuron. There are no channel proteins for this. If there were, sertraline would be expected to block those, too. Therefore, choice (D) is incorrect. (LO 3.39)

354. **(A)** The signal transduction pathway amplifies the signal at every step. Therefore, the earlier in the pathway the doubling is made, the more of an effect it will have. Choices (B), (C), and (D) are all later in the pathway and therefore would have less of an effect. (LO 3.3)

355. **(A)** Constantly phosphorylating Raf would permanently turn on the production of mitosis-promoting genes. This would lead to constant mitosis and cell division and therefore to cancer. If Raf phosphorylation were turned off, mitosis would happen less frequently, so choice (B) is incorrect. Directly phosphorylating MAPK would decrease the frequency of mitosis because there would be no amplification effect of the cascade, so choice (C) is incorrect. If Ras could no longer interact with Sos, there would be no activation of mitosis by this pathway. Therefore, choice (D) is incorrect. (LO 3.3)

356. **(B)** If the antibody was specific to an antigen specifically on cancer cells but not on healthy cells, the drug would be delivered only to the cells that needed to be killed. Piercing each individual cancer cell with a syringe would be incredibly time consuming and difficult, and it is very likely that some cancer cells would be missed. Therefore, choice (A) is incorrect. Choice (C) is incorrect because the antibodies should be highly specific to cancer cells and not be able to bind to lots of different antigens. Choice (D) is incorrect because the antibody must enter the cells to deliver the toxin. (LO 3.38)

357. **(C)** Since the antibody and the toxin are moved into the cell by endocytosis, the toxin is not able to bind to a receptor on the outside of the cell. Preventing transcription would lead to cell death since transcription is absolutely necessary to maintain a living cell. Therefore, choice (A) is a valid choice and is therefore incorrect. Preventing DNA strands from separating would make the DNA impossible to copy. This would prevent the cell cycle from proceeding. Therefore, choice (B) is a good choice and is therefore incorrect. Preventing attachment to centromeres would prevent mitosis from

occurring. This too is a good choice, and choice (D) is therefore incorrect. (LO 3.38)

358. **(B)** The smaller the dose used, the less side effects that will be produced. In comparison to the body as a whole, the doses used are miniscule. Since almost all of the toxin goes into the targeted cancer cells, a high concentration of toxin is delivered to those cancer cells. Choice (A) is incorrect because highly toxic chemicals typically produce many side effects. Choice (C) is incorrect because the chemicals are highly targeted and are not widely distributed. It would be ideal if noncancerous cells could avoid absorbing toxins, but they cannot. Therefore, choice (D) is incorrect. (LO 3.39)

359. **(A)** If this is a case of incomplete dominance, heterozygotes resulting from self-crossing the F1 flowers should produce a 1:2:1 ratio of red:pink:white. Any ratio that significantly differed from this would cause the flower grower to reject the hypothesis of incomplete dominance. Backcrossing the F1 flowers to the white parent would give a ratio of 1:1 pink:white. Therefore, choice (B) is incorrect. Since the expected ratio for self-crossing the F1 flowers is 1:2:1 of red:pink:white, choices (C) and (D) are incorrect. (LO 3.15)

360. **(A)** In order for a dog to be black, it has to have a copy of the dominant black allele (B) and a copy of the dominant coat deposition allele (E). Any genotype with both B and E gives a black phenotype. All dogs with the genotype ee are yellow dogs. Therefore, choices (B), (C), and (D) are incorrect. (LO 3.15)

361. **(A)** Both parent dogs are black, so any combinations that are not black dogs can be ruled out. All dogs with the genotype ee are yellow dogs. Therefore, choices (B), (C), and (D) are incorrect. (LO 3.15)

362. **(D)** The genotype Bbee is a yellow dog, and the genotype bbEe is a brown dog. When these two genotypes mate, they will produce black dogs with the genoytpe BbEe, brown dogs with the genotype bbEe, and yellow dogs with genotypes BbEe and bbEe. Choice (A) is incorrect because both dogs have the genotype BbEe and are thus black. Choice (B) is incorrect because, even though the genotype BBee is a yellow dog and the genotype bbEE is a brown dog, this cross would not be capable of producing brown dogs. All of the dogs would be black with a genotype of BbEe. Choice (C) is incorrect because bbee is a yellow dog, but BBEE is a black dog (not a brown dog). (LO 3.15)

363. **(D)** A single nucleotide deletion in the first exon would cause a frameshift that would change all of the amino acids after it. All of the other mutations described in the answer choices are likely to be less serious. Introns are noncoding. Therefore, a mutation in an intron is likely to have little effect, making choice (A) incorrect. Choice (B) describes a small mutation in a region that is noncoding. Therefore, it is unlikely to have an effect, making

choice (B) incorrect. Choice (C) is incorrect because changes upstream of the promoter region are unlikely to have an effect. (LO 3.6, LO 3.19)

364. **(B)** One functional copy of the receptor is sufficient since a single allele that makes a functional protein would make the cell able to detect and transmit the information about the presence of MSH and trigger eumelanin production. Choice (A) is incorrect because simply stating that an allele is dominant does not explain why a single copy is sufficient. Choice (C) is incorrect because reverse transcriptase is commonly found in retroviruses, not in dogs. Receptor genes encode receptor proteins, not pigment molecules such as eumelanin. Therefore, choice (D) is incorrect. (LO 3.38)

365. **(D)** Reduced surface area, reduced photosynthesis, and reduced growth rate are all physiological consequences. The other choices all provide consequences. However, these are consequences for the humans who grow and eat the bananas, not for the plant itself. Therefore, choices (A), (B), and (C) are incorrect. (LO 3.26)

366. **(B)** Widespread, regular applications of fungicides almost guarantee that mutations that confer fungicide resistance will be selected for in the fungus. Fungicide application should prevent discoloration of the fruit, so choice (A) is incorrect. Eradication is unlikely as long as there are fungi that escape the fungicide, such as those inhabiting wild varieties of bananas. Therefore, choice (C) is incorrect. Fungicide use prevents any selection pressure on bananas to evolve resistance, so choice (D) is not correct. (LO 3.24)

367. **(D)** Vegetative propagation produces clones. Therefore, it reduces genetic diversity. The introductory information states that seeds are not produced by Cavendish bananas. Therefore, seeds cannot be vitally important, making choice (A) incorrect. No information is given concerning the cost of vegetative propagation versus seed propagation, so choice (B) is incorrect. Since vegetative propagation produces clones and reduces genetic diversity, choice (C) is incorrect. Natural selection also decreases diversity, so choice (C) is incorrect for that reason as well. (LO 3.10)

368. **(A)** Wild bananas retain genetic diversity and are likely still exposed to the fungus. As a result, wild bananas would be a naturally selected source for resistance. Cavendish bananas grown in a greenhouse would have the same genetic problems as those grown outdoors. These bananas grown in a greenhouse would not provide resistance genes, making choice (B) incorrect. The fungal genome is unlikely to contain genes for resistance to itself, so choice (C) is incorrect. Choice (D) is incorrect because there is no reason to believe that a species that infects plants would be able to infect animals. Therefore, the animals would not have any selective pressure for genes. (LO 3.28)

369. **(C)** Moving a single gene into the Cavendish banana variety should have little to no effect on taste. In contrast, combining nuclei, as described in

choice (A), would introduce a large number of genes, many of which may have an effect on taste. Choice (B) would involve a wide range of differently tasting bananas, which should have a diversity of different flavors. Choice (D) is physically impossible since Cavendish bananas do not produce fertile eggs. (LO 3.26)

370. **(B)** The information presented mentions the hydrolysis of ATP as the driving force for the sodium-potassium pump, which performs active transport. Choice (A) is incorrect because the sodium-potassium pump is not an example of passive transport. Choice (C) is incorrect because energy is not created. Remember that energy can change form, but it can never be created. The diagram shows the two ions binding to different active sites. Therefore, even though sodium and potassium both have a positive charge, choice (D) is not correct because the sodium-potassium pump does not use the same binding site. (LO 3.45)

371. **(C)** Every cycle of the pump moves a net of 1 positive charge out of the cell, making the inside of the cell negatively charged. The pump does not move single electrons, and there is no buildup due to friction. Therefore, choice (A) is incorrect. Hydrolysis of ATP does not release electrons, and the electron transport chain is not involved in the process shown. Therefore, choice (B) is incorrect. Even though electrolysis of water makes oxygen and hydrogen ions, hydrolysis of ATP does not. Therefore, choice (D) is incorrect. (LO 3.45)

372. **(C)** Since the inside of the membrane becomes negative, moving positively charged potassium into the cell is down the ion's electrical gradient. Similarly, moving sodium out of the cell moves that ion against its electrical gradient. Choices (A) and (B) are incorrect because potassium is not being moved against its gradient. Choices (A) and (D) are incorrect because sodium is moving against its gradient, not down. (LO 3.45)

373. **(B)** When sodium channels open, sodium flows into the cell, making the inside of the cell positively charged. Choices (A) and (C) are incorrect because opening potassium channels would allow potassium to flow out, making the inside of the cell more negative. Choice (D) is incorrect because not all of the sodium ions would move. (LO 3.45)

374. **(C)** When potassium channels open, some of the potassium moves out of the cell, making the inside of the cell less positively charged. Choice (A) is incorrect because potassium would move out of the cell, not into it. Choices (B) and (D) are not correct because moving sodium would make the cell less negatively charged inside, not more. (LO 3.45)

375. **(D)** Action potentials fire when sodium channels open. If the sodium channels are in a conformation in which they cannot open, no action potential can occur. Choice (A) is incorrect because an ion charge separation is created. If the sodium and potassium channels were stuck in the open position, there

would be no charge differential across the membrane. Therefore, choice (B) is incorrect. Potassium channels open after sodium channels. Therefore, inactivating the potassium channels would not prevent action potential, making choice (C) incorrect. (LO 3.45)

376. **(A)** Stimulatory and inhibitory signals are integrated in the neuron. Stimulatory signals cause sodium channels to open, and inhibitory signals cause them to close. Therefore, choice (B) is incorrect. Presynaptic neurons release neurotransmitters. They do not cause a buildup of electrons. Therefore, choice (C) is incorrect. Jumping between nodes of Ranvier speeds up signals, and this is not occurring during time period II. Therefore, choice (D) is incorrect. (LO 3.43)

377. **(C)** The diagram shows the resting potential being reestablished. Choice (A) is incorrect because the rapid opening and closing of ion channels would not have an effect on water. Choice (B) is incorrect because gene expression is not involved in generating action potentials. Translation is too slow to have the effect shown in the diagram, so choice (D) is incorrect. (LO 3.45)

378. **(A)** Antibodies that bind to the receptors block the receptor's active site so that acetylcholine cannot access it. Since acetylcholine cannot access the receptor's active site, it cannot bind to trigger muscle contraction. For choice (B) to be correct, the antibody would have to fit in the ligand-binding site like acetylcholine itself. This is highly unlikely. Since the receptors do not have anything to do with acetylcholine reuptake, choice (C) is incorrect. Choice (D) includes a typical function of antibodies. However, there is no mention in the question of bacteria being present. (LO 3.38)

379. **(D)** Since the receptors that the antibodies bind to are on the muscle cell, the complement system attacks and kills muscle cells. Choice (A) is incorrect because the antibody-producing cell would have to produce antibodies to bind itself. This does not happen. An additional release of acetylcholine might happen if the complement system attacked the motor neuron. However, the antibody does not bind to the motor neuron. Therefore, both choices (B) and (C) are incorrect. (LO 3.38)

380. **(D)** The fact that T_4 needs to bind to helper molecules in both aqueous environments mentioned suggests that it is not soluble in water. Choice (A) is incorrect because carrier proteins float in blood that moves because the heart pumps it. Choice (B) is incorrect because charged molecules are soluble in water and do not interact stably with nonpolar fatty acid tails. Choice (C) is incorrect because there is no evidence that T_4 travels up its concentration gradient. (LO 3.35)

381. **(A)** The diagram shows transcription (as an arrow connecting the DNA to the mRNA in the nucleus) and protein synthesis as the result of thyroid signaling. Choice (B) is incorrect because DNA is not replicated. T_4 molecules

are chemically different from mRNA. While the arrows show one coming from the other, interpret that to mean that one influences the other. One does not turn into the other, so choice (C) is incorrect. The diagram does not mention metabolic rate, so even though T_4 signaling is involved in regulating metabolism, choice (D) is incorrect. (LO 3.36)

382. **(D)** Light reduced the depression rating from the baseline. However, the error bars for morning light and evening light overlap, so it is not accurate to say that they are significantly different. Since morning light and evening light are not significantly different, both choices (A) and (B) are incorrect. Choice (C) is incorrect because both morning and evening light had significantly lower depression ratings than the baseline. (LO 3.40)

383. **(B)** Reversing the treatment groups generated data under all three conditions for all patients and reduced the effects of person-to-person variability. People who volunteer for experiments are often exposed to different treatments. This is not unethical in most cases, so choice (A) is incorrect. Choice (C) is incorrect because a baseline was established for each patient before treatment, by asking participants to fill out a self-diagnostic questionnaire to measure depressive symptoms, not by reversing the treatment groups. Choice (D) is incorrect because the experimenters' goal was to test a hypothesis, not to find the best treatment for a particular patient. (LO 3.40)

384. **(B)** Using a larger sample size would reduce the standard error and allow more resolution of differences. Repeating the experiment with mice would give no information about the effectiveness of light therapy on humans, so choice (A) is incorrect. Removing the baseline measurements would make it impossible to know if any treatment had an effect. Therefore, choice (C) is incorrect. Increasing the number of light treatments would complicate the experiment unnecessarily. Therefore, choice (D) is incorrect. (LO 3.40)

385. **(B)** Cortisol levels rose to their highest level of approximately 0.65 at 20 minutes. Choice (A) is incorrect because it took more than 50 minutes for cortisol levels to return to baseline. Choice (C) is incorrect because the cortisol levels peaked at approximately 20 minutes. The graph gives no information about chronic exposure to stress, so choice (D) is incorrect. (LO 3.40)

386. **(C)** Highly conserved amino acid sequences are usually conserved because changing them causes serious damage to the organism. As a result, the mutant organisms do not survive to pass on their mutated genes. Choice (A) is incorrect because researchers generally do not do mutation analysis on variable regions because these regions are generally unimportant to the protein's function. Conserved regions are not likely to have different functions in different organisms, so choice (B) is incorrect. The choice of mutations was driven by the interest in the mutations, not by what procedure would be easier. Therefore, choice (D) is incorrect. (LO 3.6)

387. **(C)** The graph shows that the wild-type protein has an elongation rate that is intermediate between that of both mutant 1 and mutant 2. Therefore, both choices (A) and (B) are incorrect. The graph does not show a variable rate of expression for the wild-type protein, so choice (D) is incorrect. (LO 3.26)

388. **(C)** The data on the graph for the untrained flies were always nearly 100%. Choice (A) is incorrect because the untrained flies almost always extended their proboscis fully. Choice (B) is incorrect because the trained flies avoided extending their proboscis. Choice (D) is incorrect because the untrained flies extended their proboscis more frequently than the trained flies. (LO 3.43)

389. **(B)** Since the trained flies took at least 25 minutes for their rate of proboscis extension to approach that of untrained flies, the researcher can conclude that the trained flies retained memory of the averse-tasting chemical for at least 25 minutes. Choice (A) is incorrect because the extension rate returned to baseline before 100 minutes. Choice (C) is incorrect because the trained flies retained memory of the averse-tasting chemical for longer than 30 seconds. Since the trained flies did differ from the untrained flies in the first 15 minutes, choice (D) is incorrect. (LO 3.43)

390. **(A)** Fruit flies in the longer training session took a longer time for the rate of proboscis extension to return to baseline levels than did those flies that were only exposed once. Choice (B) is incorrect because there was a difference between the two groups. Choice (C) is the opposite of what happened. Choice (D) is incorrect because the fruit flies were not given the opportunity to seek out quinine. (LO 3.43)

391. **(C)** Turning the taste lobe off led to no learning (100% proboscis extension). Choice (A) is incorrect because associative learning did happen when the taste lobe was on. Choice (B) is incorrect because associative learning happened when the taste lobe was on, but not when it was off. Choice (D) is incorrect because proboscis extension was not 100% when the taste lobe was on. (LO 3.43)

392. **(B)** This region is not important in associating the averse taste with sucrose because turning it on and off gave approximately the same response. Proboscis extension is approximately 30% when this region is off, so choices (A) and (D) are incorrect. Choice (C) is incorrect because if a region that controls learning is turned off, no learning should occur. (LO 3.43)

393. **(D)** When this region is off, almost no learning takes place, as shown by nearly 100% proboscis extension. Therefore, this region is essential for learning. Choices (A) and (C) are incorrect because proboscis extension was nearly 100% when this region was off but was nearly 30% when this region was on. The fact that the data for this region did not differ significantly from that of the whole lobe indicates that this region was important for learning. Therefore, choice (B) is incorrect. (LO 3.43)

394. **(B)** This mechanism would allow the plant to sense blue light as distinct from other colors and also produce a change in gene expression. Choice (A) is incorrect because it does not explain how light changes gene expression. Choice (C) is incorrect because it does not mention how the plant would sense changes in kinetic energy. In addition, blue light has more energy than other colors of visible light and is therefore not cooler. Light is not absorbed through the stomata, so choice (D) is incorrect. (LO 3.36)

395. **(B)** This mechanism would bend the plant toward the blue light. Choices (A), (C), and (D) would all cause the plant to bend away from the blue light. (LO 3.37)

396. **(A)** The ratio of interphase to mitosis for the distilled water control should provide the expected values without the effects of lectin or caffeine. Choice (B) is incorrect because the expected numbers should be equal only if the null hypothesis is a 1:1 ratio. Choice (C) is incorrect because the average of the treatments would change if different treatments were used, as is the case in this scenario. The degrees of freedom are calculated by taking the number of different categories and subtracting 1, so choice (D) is incorrect. (LO 3.6)

397. **(D)** There are 2 categories, so there is 1 degree of freedom. A chi-square value larger than the cutoff means that the observed values are more extreme than expected from random chance, so reject the null hypothesis. Since there is 1 degree of freedom, choices (A) and (B) are incorrect. Since the chi-square value is larger than the cutoff, the null hypothesis cannot be accepted. Therefore, choice (C) is incorrect. (LO 3.7)

398. **(C)** The lectin data are more extreme than that of the 1 mM caffeine. If the 1 mM caffeine is statistically significant from the distilled water, then the lectin must be too. Choice (A) is incorrect because more cells exposed to lectin were counted, not less, and because there was a higher, not lower, % mitotic for cells exposed to lectin. Choice (B) is incorrect because the 5 mM data were further away from that of the distilled water data than the 1 mM data was and therefore should be statistically significant. The distilled water had 86% mitotic cells while 1 mM caffeine had 4.5% mitotic cells and the 5 mM had 4.3% mitotic cells. If a difference of 4.1% (8.6% – 4.5% = 4.1%) is statistically significant, we would also expect a difference of 4.3% (8.6% – 4.3% = 4.3%) to also be statistically significant. Since additional conclusions can be drawn, choice (D) is incorrect. (LO 3.7)

399. **(A)** A round of meiosis produces 4 genetically different nuclei. A round of mitosis that follows this round of meiosis would produce identical copies of each of these 4 distinct nuclei. The result would be pairs of identical spores that are genetically different from the other three pairs. Choice (B) is incorrect because meiosis I could not occur twice and wouldn't result in diploid spores. When crossing-over, two homologous chromosomes switch homologous pieces. Crossing-over does not fragment chromosomes into

smaller pieces, so choice (C) is incorrect. Nondisjunction would cause some nuclei to gain more chromosomes and some to receive less. It would not lead to pairs of identical spores, so choice (D) is incorrect. (LO 3.12)

400. **(C)** Since only half of the spores in an ascus that shows crossing-over are the result of crossing-over, the first step in solving this problem is to multiply the number of map units by 2. This will give the percentage of asci that should show crossing-over. In this case, $26 \times 2 = 52$. Since the students are scoring 200 asci, find 52% of 200, which is 104 ($200 \times 0.52 = 104$). Choices (A), (B), and (D) are incorrect because they don't give the correct number of asci that should show crossing-over. (LO 3.11)

401. **(B)** To find map units, first find the percentage of asci showing crossing-over. The percentage of asci showing crossing-over is $\frac{95}{340} \times 100$, which equals 28%. Since each ascus includes 4 spores produced by crossing-over and 4 produced without crossing-over, divide 28 by 2, which equals 14 map units. Choices (A), (C), and (D) are incorrect because they don't give the correct number of map units. (LO 3.11)

402. **(A)** There is nothing in the LB agar to kill any of the bacteria, so all should grow. Choice (B) is incorrect because there is nothing in the LB agar for the plasmid to provide resistance to. Choice (C) is incorrect because plate 1 has no plasmid and plate 4 has the plasmid, the opposite of what is stated in this answer choice. Choice (D) is incorrect because there is nothing to prevent growth on plate 4. Nothing in the description suggests that bacteria were omitted. (LO 3.24)

403. **(A)** Without the plasmid, the ampicillin should kill bacteria. Only bacteria with the plasmid should grow. Since there is no arabinose to turn on GFP production, the bacteria on plate 5 should not glow. Since the bacteria on plate 5 should not glow, choice (B) is incorrect. Choice (C) is incorrect because the question asks only about plates 2 and 5, not all 6 plates. It is also incorrect because plate 6 should contain glowing colonies. Choice (D) is incorrect because plate 2 would have no growth, and plate 5, which will have growth, will not glow. (LO 3.24)

404. **(B)** Since there is ampicillin in the media, only bacteria with the plasmid (plate 6) will grow. Therefore, there will be no growth on the no pGlo plate (plate 3). Since there is arabinose, GFP will be expressed, and the colonies on plate 6 will glow. Choice (A) is incorrect because the colonies on plate 6 will glow. Choice (C) is incorrect because there will be no growth on plate 3. Choice (C) is also incorrect because plate 6 will have glowing bacteria. Choice (D) is incorrect because there will be no growth or glowing on plate 3. (LO 3.24)

405. **(C)** Without ampicillin, all of the bacteria plated will grow. Since there would be no selection for bacteria that took up the plasmid, none of the bac-

teria would glow. Transformation (taking up plasmid) is a very rare event and will only be seen if the millions of bacteria that aren't transformed are killed by an antibiotic. Choices (A) and (B) describe plates with no growth and are both therefore incorrect. Choice (D) is incorrect because almost none of the bacteria will take up the plasmid, so there should be no glowing. (LO 3.24)

406. **(C)** These bacteria have the plasmid because they did not die on agar with ampicillin. Therefore, they also have the GFP gene. When they are transferred to a new plate with arabinose, the GFP will be expressed and the bacteria will grow and glow. Choice (A) is incorrect because there will be growth. Choice (B) is incorrect because the bacteria will glow. Choice (D) is incorrect because it is possible to predict that the bacteria will grow and glow from the information given. (LO 3.24)

407. **(C)** If the plasmid is 5,371 base pairs long, then there are twice that many nucleotides. To find the approximate molar mass of the plasmid, multiply the total number of nucleotides by the average molar mass of a nucleotide:

$$5{,}371 \times 2 \times 327 = 3.51 \times 10^6 \text{ g/mol}$$

Choice (A) is off by a factor of 10. Choice (B) could result from forgetting that there are 2 bases for every base pair. Choice (D) is high by a factor of 5: $\frac{1.76 \times 10^7 \text{ g}}{5} = 3.52 \times 10^6$. (LO 3.5)

408. **(B)** Transformation efficiency is calculated by first counting the number of transformants and then dividing that number by the amount of DNA used. In this case, we are given the number of transformants and the efficiency, and we are asked to indicate how much DNA was used. Only bacteria that were transformed (took up a plasmid) would have the gene for ampicillin resistance and would thus be able to grow on agar containing ampicillin. Bacteria that were not transformed died. Each colony is the result of a single bacterium multiplying. Therefore, the number of transformants is the same as the number of colonies that formed on the arabinose plate (90 colonies in this case). If there were 90 transformants and the transformation efficiency was 1,800 transformants/μg plasmid DNA, dividing the number of transformants by the transformation efficiency will give the micrograms of plasmid DNA:

$$\frac{90}{1{,}800} = 0.05 \text{ μg plasmid DNA}$$

Choice (A) is incorrect because it is too low. Choices (C) and (D) are incorrect because they are both too high (LO 3.28)

409. **(C)** Inserting DNA into the middle of a gene typically disables the gene. Since the *lacZ* gene spans the CS, inserting DNA in the CS makes the *lacZ* gene nonfunctional. Choice (A) is incorrect because the information does not state what the cDNA inserted in the CS does aside from disrupting the *lacZ* gene. Therefore, we cannot know if it produces anything. Even if it did produce a transcription factor, turning on the *lacZ* gene would lead to the production of a blue color, not the lack of production of blue. Since *lacZ*

causes X-gal to produce a blue color, these colonies would not appear blue. Since choice (B) mistakenly refers to a white color rather than a blue color, choice (B) is incorrect. Choice (D) is incorrect because inserting DNA in the CS does not repair the *lacZ* gene. (LO 3.19)

410. **(D)** The only bacteria that might grow would be ones that took up plasmids that were not digested by the restriction enzyme. None of these bacteria would contain a piece of DNA that disrupts the *lacZ* gene, so they would all be blue. The information provided says that ligase seals the nicks in DNA strands, so it is not reasonable to think that ligase degrades the antibiotic. Therefore, choice (A) is incorrect. White colonies would appear only if the *lacZ* gene was disrupted, so choices (B) and (C) are incorrect. (LO 3.24)

411. **(B)** The diagram of the MCS in Figure I shows that there are two 100 bp pieces on either side of the CS. If a 900 bp piece were inserted into the CS, PCR would amplify one 1,100 bp fragment (900 + 100 + 100). Choice (A) is incorrect because there should only be a single piece of DNA inserted that is 1,100 bp. Choice (C) is incorrect because it ignores the two 100 bp pieces on either side of the CS. Since there should only be a single piece of DNA, three pieces, choice (D), is incorrect. (LO 3.5)

412. **(D)** According to Figure I, the detail of the MCS shows that *Eco*R1 cuts the plasmid in two places. If the circular plasmid is cut with a restriction enzyme at two places, there will be two pieces. In Figure I, the number at the center of the plasmid indicates its total length. For pTrx, this is 3.6 kb. Therefore, the total length of DNA of the plasmid is 3,600 bp, and the insert described in the question is 900 bp long. The diagram of the MCS in Figure I shows that there is a 300 bp piece and a 400 bp piece of plasmid on either side of the CS that will be connected to the 900 bp piece of DNA inserted at the CS. This will make the piece with the insert total 1,600 bp long (900 + 300 + 400). The second piece will be the 3,600 bp plasmid without the 300 bp and 400 bp pieces attached to the insert. Therefore, subtract these pieces from the length of the plasmid to determine that the second piece will be 2,900 bp long (3,600 – 300 – 400). Choice (A) is incorrect because it ignores the second piece that would be created by the rest of the plasmid. Choice (B) is incorrect because it ignores the 300 bp and 400 bp pieces on either side of the insert. Choice (C) is incorrect because it ignores the 300 and 400 bp pieces on either side of the CS as well as the rest of the plasmid. (LO 3.5)

413. **(D)** The more DNA there is, the more dye will bind. More fragments of DNA of the same size bind to more dye. Therefore, the band made up of more DNA shines brighter and looks thicker. Choice (A) is incorrect because it suggests that a brighter band means less DNA, which is the opposite of what is true. Choice (B) is incorrect because the band in lane 1 is brighter than the band in lane M. Choice (C) is incorrect because the two bands are not the same brightness. (LO 3.5)

414. **(A)** The DNA in lane 2 migrated farther on the gel and is therefore shorter than the DNA in lane 1. Recall that longer pieces migrate more slowly through agarose gels. Since the DNA in lane 1 is longer than the DNA in lane 2, choices (B) and (C) are incorrect. Choice (D) is incorrect because electrophoresis is used to separate molecules based on size and charge. Therefore, it is possible to determine the size of the DNA based on how far the piece migrates in the gel. (LO 3.5)

415. **(C)** This is the only choice that gives the correct plasmid size based on the diagram and an explanation that is consistent with the results from the gel. The size of the plasmid is given in Figure I. We would expect a piece that is 3,600 bp long to migrate more slowly than the 3,000 bp standard. The gel in Figure III shows that it did not. The explanation that circular DNA is more compact and runs faster is consistent with the results on the gel. Choice (A) is incorrect because if circular DNA moved more slowly than linear DNA, it would be above the expected site, not below it. Choice (B) is incorrect because it relies on the idea that longer DNA will migrate more quickly on the gel. Since the plasmid is longer than 3 kb, the plasmid should migrate more slowly than the shorter 3 kb standard DNA. This would put it higher than the 3 kb standard, not below it as choice (B) states. Choice (D) is incorrect because the plasmid pTrx is 3.6 kb long. (LO 3.5)

416. **(C)** From the diagram of the MCS in Figure I, we see that the PCR results are each 100 bp from the MCS. As a result, a plasmid with no insert in the MCS would be 200 bp long. Any piece of DNA that is inserted at the MCS will make this amplified fragment longer. To find the size of the inserted fragment using PCR, first determine how long the fragment on the gel is, and then subtract 200 bp to account for the DNA that would be amplified if there were no inserted DNA. The *Eco*R1 sites flank the MCS by 400 bp on one side and 300 bp on the other side. As a result, cutting a plasmid with no insert with *Eco*R1 would produce a 700 bp long piece as well as a second piece that is 2,900 bp long. This should total the 3,600 bp length of the plasmid. To find the size of the inserted fragment using an *Eco*R1 digest, first determine how long the fragment is and then subtract 700 to account for the DNA that would appear even if there were no inserted DNA in the MCS. To summarize, the PCR results should be 200 bp longer than the inserted fragment and the restriction digest results should be 700 bp longer than the inserted fragment. Choice (A) is incorrect because choice (C) is a plausible explanation for the discrepancy. Choice (B) is incorrect because cDNA is made from mRNA, which doesn't have introns. Also, bacteria cannot process introns. Choice (D) is incorrect because the lengths that need to be subtracted are backward. (LO 3.5)

417. **(B)** If the student's 1,500 bp long expected piece had an *Eco*R1 cut site in it, it would be digested into two pieces. If the pieces were approximately the same size, they might appear as an especially broad band. Choice (A) is

incorrect because an extra PCR primer site would lead to two fragments on the gel for the DNA results (Figure II). One of the fragments would be the expected length, and the other would not. This would not explain why the restriction digest fragment was smaller than expected from the PCR results. Choice (C) is incorrect because no RNases were mentioned and because the fragment is made of DNA. Choice (D) is incorrect because no RNA was used. (LO 3.5)

418. **(B)** Damaged DNA shows up in the tail. Since this choice has only a head, there is no damaged DNA. Choices (A) and (C) have tails, so they are incorrect. Choice (D) is incorrect because the assay shows any DNA that is present. (LO 3.7)

419. **(A)** The longer the tail is, the more DNA damage there is. Choice (B) has no DNA damage. Choice (C) has very little DNA damage. Therefore, both choice (B) and choice (C) are incorrect. Choice (D) is incorrect because it shows no head and therefore has no intact DNA. (LO 3.7)

420. **(B)** Extra chromosomes would not have any double-stranded DNA breaks. Choice (A), (C), and (D) would all have damaged DNA. Thus, they would all be useful applications of the comet assay and are therefore incorrect. (LO 3.6)

421. **(B)** A hydrophobic domain would allow the antibody to be anchored in the plasma membrane. Changing that domain to hydrophilic would allow the antibody to dissolve in blood plasma, which is mostly composed of water. Choice (A) is the opposite of what is correct. Antibodies are made of protein, not DNA or RNA, so choice (C) is incorrect. Antibodies bind to antigens. Antigens do not make up antibodies. In addition, T-cells are much larger than antibodies and are not part of them. Therefore, choice (D) is incorrect. (LO 3.6)

422. **(A)** Step I involves deleting segments of DNA. This permanently alters the genome since that DNA is lost. Neither transcription, choice (B), nor RNA splicing, choice (C), alters the DNA. Therefore, the genome is not altered in either of these steps. Choice (D) is incorrect because the genome is edited in Step I. (LO 3.6)

423. **(B)** Step II is transcription, which makes RNA from a DNA template using RNA polymerase. Step I alters the DNA, and Step III is RNA splicing. Therefore, choices (A) and (C) are incorrect. Since Step III does not use RNA polymerase, choice (D) is incorrect. (LO 3.6)

424. **(C)** Step III is RNA splicing. Choice (A) is incorrect because Step I involves deleting DNA from the cell's genome. Choice (B) is incorrect because Step II is transcription. Neither Step I nor Step II involves RNA splicing, so choice (D) is incorrect. (LO 3.6)

425. **(A)** Since the genetic code relies on triplets to specify amino acids, inserting 0, 3, or 6 nucleotides will keep the mRNA in the same frame. Inserting 1, 2,

4, 5, 7, or 8 nucleotides will shift the reading frame and will cause incorrect amino acids to be linked together in the protein. Since three of the nine possibilities do not change the frame and the other six out of the nine do cause a frame shift, approximately two-thirds will contain frameshift mutations. Choice (B) is incorrect because nonsense mutations are mutations to a stop codon. The odds of mutation to a stop codon are approximately $\frac{3}{64}$ because there are three stop codons out of 64 possible triplets. Choice (C) is incorrect because inserting bases does not lead to inversions in which DNA strands reverse their orientation. Choice (D) is incorrect because the odds of an early stop codon are $\frac{3}{64}$. (LO 3.6)

426. **(D)** Since the rejoining occurred in nonimmune cells expressing *rag-1* and *rag-2* genes, immune cells must use the same rejoining process that nonimmune cells use. Choice (A) is incorrect because no rejoining would have occurred if the process was immune-cell specific. Choice (B) is incorrect because the proteins rag-1 and rag-2 make breaks in double-stranded DNA, not in single-stranded DNA. Choice (C) is incorrect because the proteins rag-1 and rag-2 only cut DNA; they do not rejoin it. (LO 3.6)

427. **(A)** Without Step I, the RNA produced would be too long and would result in nonfunctional antibodies. Choice (B) is incorrect because mutations do not occur in order to replace missing functions. Choice (C) is incorrect because normal antibodies would not be produced. Choice (D) is incorrect because two cuts are needed to remove a section of DNA: rag-1 cuts in the V-region and rag-2 cuts in the J-region. Both proteins are necessary, making choice (D) incorrect. (LO 3.6)

428. **(D)** If the binding led to B-cell proliferation, the antibody would attack a self-antigen. However, removing the antigen would be impossible since it is a part of the body. Choice (A) is incorrect because it pairs a hypothetical scenario with its correct outcome. If a B-cell that made antibodies to self-antigens was released from the bone marrow, the B-cell would encounter an antigen that was a normal part of the body and would trigger an inappropriate immune response. Choice (B) is incorrect because it also pairs a hypothetical scenario with its correct outcome. Reediting the DNA would lead to different segments being joined. Therefore, the cell would produce a different antibody. Another common fate of B-cells that bind to self-antigens in the bone marrow is apoptosis (programmed cell death). Since these B-cells undergo apoptosis before entering general circulation, they do not lead to an autoimmune response. Therefore, choice (C) also pairs a hypothetical scenario with its correct outcome and is thus incorrect. (LO 3.21)

429. **(C)** By generating variability around an antibody that binds, some of the antibodies will bind better. These will be selected for. The increased mutation

rate is not driven by the pathogen, and it leads to less disease. Therefore, choice (A) is incorrect. Although some of the B-cells will not be selected to proliferate, sufficient B-cells would already be available to mount an effective immune response. Therefore, choice (B) is incorrect. The variability happens before increased binding affinity, and friction does not increase the mutation rate. Therefore, choice (D) is incorrect. (LO 3.24)

430. **(B)** When they are inhibited, nerves that are normally involved in creating alertness would lead to lack of alertness (drowsiness). Choice (A) is incorrect because antihistamines should relax smooth muscles, not cause them to contract. There is no information provided that suggests that melatonin would be produced, so choice (C) is incorrect. Choice (D) is incorrect because antihistamines decrease, not increase, the activity of alertness-related neurons. (LO 3.39)

431. **(D)** All three of these hypotheses, outlined in choices (A), (B), and (C), would lead to less of a response without any effect on the alertness neurons. If the antihistamine was not able to pass through the blood-brain barrier, it would not be able to reach the neurons involved with alertness and therefore would not cause drowsiness. Therefore, choice (A) is a reasonable hypothesis, but it is not the correct answer because the hypotheses in choices (B) and (C) are reasonable as well. Choice (B) is also reasonable because binding to histamine itself would lower histamine's ability to induce an allergic response, but it also would not affect the H1 receptor. Again, choice (B) is a reasonable hypothesis, but it is not the correct answer because the hypotheses in choices (A) and (C) are reasonable as well. Choice (C) is a third reasonable hypothesis because preventing histamine release would also prevent an allergic response, but it would not affect the arousal center in the brain. Once again, choice (C) is a reasonable hypothesis, but it is not the correct answer because the hypotheses in choices (A) and (B) are reasonable as well. (LO 3.39)

432. **(B)** When dsRNA was injected into the parent, the offspring twitched. Therefore, the offspring had their muscle function protein turned off. Choice (A) is incorrect because turning on the protein would have led to a normal phenotype. Choice (C) is incorrect because the injection was into the parent, not into the offspring. Choice (D) is incorrect because injecting single-stranded RNA did not have an effect, regardless of whether it was message sense or antisense RNA. (LO 3.1, LO 3.21)

433. **(D)** Prior to the invention of RNA interference (RNAi), which is demonstrated in this experiment, it was not possible to turn off specific genes with this level of precision and a lack of impact on the genome. Choice (A) is incorrect because this technique does not allow for direct visualization of DNA. Choice (B) is incorrect because this technique does not remove and replace specific segments of DNA. Choice (C) is incorrect because this experiment did not make permanent changes to the genome. (LO 3.21)

434. **(D)** The strands shown in choices (B) and (C) are both complementary to the sequence shown in the question. Reading from 5′ to 3′ in both choices, you should be able to see that these are the same sequence. Choice (A) looks complementary to the sequence in the question, but it is in the wrong orientation and would not pair correctly. (LO 3.35)

435. **(C)** Since nonsense mutations are the result of mutations that produce stop codons, this type of drug would be effective against those types of mutations. Silent and missense mutations would not include incorrect stop codons and would not be affected by a drug that causes readthroughs. Therefore, choices (A) and (B) are incorrect. Since the mutations described in choices (A) and (B) are incorrect, choice (D) is incorrect as well. (LO 3.36)

436. **(B)** Since the readthrough could cause the ribosome to slip 1, 2, or 3 bases, only one-third of the readthroughs would be in the correct frame and would produce the correct protein. Since the answer is 33%, choices (A), (C), and (D) are incorrect. (LO 3.36)

437. **(D)** *CCA1* gene expression is indicated by the open circles. In the 24–48 hour period, the light comes on at 36 hours, which is the beginning of the shaded region. *CCA1* gene expression begins to rise at about 34 hours. Choices (A) and (C) are incorrect because CCA1 protein levels are rising at these time points. CCA1 protein levels are falling 6 hours before the light is turned on, so choice (B) is incorrect. (LO 3.36)

438. **(A)** The minimum TOC1 protein level is always within 6 hours before the minimum CCA1 protein level. Each of the two proteins rise after they reach a minimum level. Within 6 hours after CCA1 protein levels rise, TOC1 protein levels peak and begin to fall, so choice (B) is not correct. TOC1 protein levels sometimes rise and fall during the periods in which CCA1 protein rises and in which CCA1 protein falls. Therefore, choices (C) and (D) are also incorrect. (LO 3.36)

439. **(C)** TOC1 protein turns on the expression of the *CCA1* gene, which produces CCA1 protein. CCA1 protein inhibits the expression of the *TOC1* gene which produces TOC1 protein. This creates an oscillating pattern which would be maintained for some period of time. However, without light to strengthen *CCA1* gene expression, the timing of the cycles would become disregulated. Since the CCA1 protein inhibits the expression of the *TOC1* gene, we would not expect both to rise at the same time as choice (A) suggests. Although *CCA1* gene expression inhibits the *TOC1* gene, as TOC1 protein levels fall, less CCA1 protein would be produced. This would lead to less TOC1 protein production. As a result, there wouldn't be an indefinite rise of TOC1 protein or an indefinite fall of CCA1 protein. Therefore, choice (B) is incorrect. Although CCA1 and TOC1 protein levels would continue to oscillate for some time, the cycle would not be perpetuated without stimulation by light. Therefore, choice (D) is incorrect. (LO 3.37)

440. **(D)** If the CCA1 protein didn't inhibit the *TOC1* gene, the TOC1 protein levels would rise. Since TOC1 protein stimulates the *CCA1* gene, the CCA1 protein levels would rise. TOC1 protein levels would not fall if the CCA1 protein didn't inhibit the *TOC1* gene, so choice (A) is incorrect. Choice (B) is incorrect because the cycling is largely due to the ability of the CCA1 protein to cause a decrease in the TOC1 protein when levels of CCA1 are high. Since there is no inhibition of either protein, the levels would not fall to zero as suggested by choice (C). (LO 3.37)

441. **(C)** The early plant gets the light. By being prepared to perform photosynthesis before the light comes on, the plant avoids wasting daylight. This also means that choice (A) is incorrect since there is a benefit to expressing these genes before dawn. Moonlight is strongest once per month during the full moon, not once per day. Therefore, choice (B) is incorrect. Nothing in the question suggests that these plants live in an urban environment, so choice (D) is incorrect. (LO 3.32)

442. **(B)** Since CCA1 protein production would be constantly stimulated by the light, the *TOC1* gene would be constantly inhibited by the CCA1 protein. This constant stimulation by light would make CCA1 protein levels rise, not fall as both choices (A) and (C) suggest. If CCA1 protein levels rose, TOC1 levels would fall, which is the opposite of what choice (D) says. (LO 3.37)

443. **(C)** Continuous darkness would prevent the cycles from being synchronized with the 24-hour cycle. However, the inhibition of the *TOC1* gene by the CCA1 protein would allow the periodic rise and fall of each to continue. Choice (A) is incorrect because the CCA1 protein inhibits the *TOC1* gene. If CCA1 protein levels rise, TOC1 protein levels fall. Choice (B) is incorrect because falling levels of CCA1 protein would cause TOC1 protein levels to rise. However, the TOC1 protein stimulates CCA1 protein levels. Higher CCA1 protein levels lead to less TOC1 expression. Therefore, the cycle would continue for some time. Without light, the cycles would not continue indefinitely. Without the stimulation from light, the cycles would most likely lose amplitude without the regular stimulation from sunlight. Therefore, choice (D) is incorrect. (LO 3.37)

444. **(A)** Early photosynthetic organisms would have had to adapt to daily cycles of light and dark. As a result, the genes for regulating the daily cycles of metabolism are probably very ancient. Since modern plants evolved from these ancestors, the genes that control daily cycles evolved from those ancestral genes contained in those ancestral plants. Since the modern DNA sequences derive from a common ancestral DNA sequence, they probably have a high degree of sequence similarity. Choices (B) and (D) are incorrect because evolving two separate sets of genes (either from different times or from different ancestors) to control the same responses is a much more complex explanation that is less likely to be true. Choice (C) is incorrect because the genes probably evolved from common ancestral genes and should show a high degree of similarity. (LO 3.31)

445. **(C)** Sepals form first, and carpels form last. Therefore, gene C would be turned on last, making choices (A), (B), and (D) incorrect. (LO 3.38)

446. **(C)** Without transcription factor B, the plant would not be able to make petals or stamens but could make sepals and carpels. Choice (A) is incorrect because petals would not be made. Choice (B) is incorrect because stamens would not be made. Since the plant would not be able to make petals or stamens, choice (D) is incorrect. (LO 3.37)

447. **(B)** Without transcription factor A, stamens and carpels could be made but sepals and petals could not be made. Choices (A) and (C) are incorrect because there would be no sepals or petals made. Since the plant would not be able to make sepals or petals, choice (D) is incorrect. (LO 3.37)

448. **(A)** Without transcription factor C, stamens and carpels could not be made but sepals and petals could be made. Choices (B) and (C) are incorrect because stamens and carpels could not be made. Since the plant would not be able to make stamens or carpels, choice (D) is incorrect. (LO 3.37)

449. **(C)** Strategies I and II are the only ones that would result in showier flowers with a large number of colorful sepals and petals. Strategy I would result in flowers with only sepals and petals. Petals are the showy parts of the plant. Strategy II would result in colorful sepals. Together, these strategies would produce many colorful organs. Turning on factor E in all plant tissues (strategy III) would cause the plant to produce flowers in inappropriate locations, such where leaves or even roots should form. Choice (A) is incorrect because strategy I is not the only way to produce showier flowers. Choice (B) is incorrect because strategy III would not cause showier flowers. Choice (D) is incorrect because, while strategies I and II would cause showier flowers, strategy III would not. (LO 3.5)

450. **(C)** mRNA peaks between hours 12 and 14, and protein peaks at around hour 20, which is a difference of between 6–8 hours. Choices (A) and (B) are incorrect because their ranges are too short. Choice (D) is incorrect because its range is too long. (LO 3.21)

451. **(C)** Translation of mRNA produces PER and TIM protein and degradation destroys PER and TIM protein. Since the question asks about PER and TIM protein and not RNA levels, only look at the line for PER/TIM protein. When the rates of these two processes are equal, the concentration of protein should not change. The slope of the line for protein concentration should be equal to zero, and the line should look flat. There are two places on the graph where this happens. PER and TIM protein levels flatten out at the minimum at hour 8 and then again at the maximum at hour 20. Choice (A) is incorrect because protein levels are falling at hour 1, so the rate of protein degradation is higher than the rate of translation. Protein levels are rising at hour 13, so the rate of translation is higher than the rate of protein degradation. Hour 1 and hour 13 also correspond to the points when mRNA levels are not

changing, but the mRNA line should be ignored when considering protein concentrations. Choice (B) is incorrect because protein levels are falling at hour 4 and are rising at hour 17. Since the line for protein concentration flattens out twice in the 24-hour period shown, there are two times when the rate of protein degradation equals the rate of protein production by translation. Therefore, choice (D) is incorrect. (LO 3.21)

452. **(A)** Figure II shows that the mRNA concentration is increasing at hour 8. If mRNA levels are increasing, transcription must be occurring. For transcription to occur, both CLK and CYC must be bound to the promoter. Choice (B) states the opposite of the correct explanation and is therefore incorrect. CLK and CYC are both necessary for transcription. There should be no mRNA production without both of them, so choice (C) is incorrect. Choice (D) is incorrect because Figure II shows that CLK and CYC are at their minimum concentration at hour 8. As a result, they would not be bound to PER/TIM. (LO 3.21)

453. **(A)** Since mRNA concentrations are rising, transcription must have been turned on. Therefore, CLK and CYC must be interacting with the promoter and must not be bound to the PER/TIM protein complex. Degradation of the PER/TIM protein complex in the nucleus would result in this effect. Choice (B) is incorrect because it states that mRNA would be produced if CLK and CYC were still bound to the PER/TIM protein complex. Figure I shows that there would be no transcription of mRNA for PER and TIM if the PER/TIM protein complex was bound to CLK and CYC. Degradation of the PER/TIM protein complex in the cytoplasm would not affect transcription in the nucleus, so choice (C) is incorrect. Choice (D) is incorrect because the model shown in Figure I indicates that a buildup of PER and TIM would lead to increased binding of CLK and CYC and therefore a decrease in *per* and *tim* mRNA, not an increase in CLK and CYC or an increase in *per* and *tim* mRNA. (LO 3.21)

454. **(A)** TIM is degraded in the light. If there were no light, there would be no degradation of TIM. This would lead to an increase in the levels of the PER/TIM protein complex earlier. This would lead to more PER/TIM protein being transported into the nucleus at an earlier time, shortening the length of the oscillations. Choice (B) is incorrect because TIM would not be degraded since there would be no light. Choice (C) is incorrect because the rate of PER degradation would not speed up in complete darkness. Sleep/wake cycles are controlled by the oscillating system described in the passage and are entrained by light. Simply turning the light off will not cause the flies to sleep, so choice (D) is incorrect. (LO 3.38)

455. **(B)** TIM would be degraded faster with the light constantly on. This would lead to longer oscillations. Choice (A) is incorrect because TIM is degraded in the light. Choice (C) is incorrect because PER degradation is not affected by light according to Figure I. Choice (D) is incorrect because the sleep/wake

cycle is controlled by the system described in the passage, not simply by the presence or absence of light. (LO 3.38)

456. **(D)** A mutation to a stop codon would lead to a truncated TIM protein that would be nonfunctional. These flies would not have a functional internal clock. Choice (A) is incorrect because an early stop codon would not produce a functional TIM protein. Choice (B) is incorrect because PER cannot act without TIM. Choice (C) is incorrect because the TIM protein would have no functionality, not decreased functionality. (LO 3.37)

457. **(B)** The fruit flies would transcribe the gene for GFP when they would normally transcribe the genes *tim* and *per*. If GFP was degraded like TIM and PER, the fruit flies would glow and then stop glowing each day. If GFP was not degraded, they would simply glow all the time. Choice (A) is incorrect because the gene that codes for GFP is not replacing the genes that code either TIM or PER. If the flies blinked, it would be a daily blink, not a rapid blink, because the cycle occurs on a daily basis. Therefore, choice (C) is incorrect. If GFP had any effect on TIM, it would be to degrade TIM faster because GFP emits light. It is unlikely any such effect would occur. In either case, choice (D) is incorrect. (LO 3.37)

458. **(C)** Signals in the cytoplasm reset the nucleus to a pluripotent state. Choice (A) is incorrect because the genetic material was removed. Choice (B) is incorrect because mRNAs are unlikely to be completely removed. RNA is transcribed and processed in the nucleus. When transplanted, the RNAs that the somatic cell was producing before the transplant should be introduced into the germ cell and prevent reversion to pluripotency. Choice (D) is incorrect because gene expression is situational. In no case will all of the genes that an organism has be expressed. (LO 3.21)

459. **(A)** The frog will be a clone of the donor of the nucleus, so its eyes should be red. The germ cell donor's nucleus was destroyed, so the white eye allele was destroyed with it and would not be present. Therefore, choices (B) and (D) are incorrect. There is a single frog, so it is not possible to have a 3:1 ratio, making choice (C) incorrect. (LO 3.11)

460. **(C)** A good control treatment should be the same as the experimental treatment with the exception of a single factor. Treating somatic cells with UV light would destroy the donor nucleus and would provide no useful information about this experiment. Therefore, choice (A) is incorrect. Transferring germ cell nuclei into somatic cells would not lead to dedifferentiation and wouldn't be helpful. Therefore, choice (B) is incorrect. Choice (D) is incorrect because control treatments are always necessary in order to have something to compare a result to. Without a control, the experiment is invalid. (LO 3.11)

461. **(A)** Mock treatment ensures that the experimental procedure itself had no effect. A simple example of a mock treatment is the use of a placebo

pill in drug studies. The placebo pill contains no active ingredient, but it looks identical to the pill with the active ingredient. Whatever effect the mock treatment (the placebo pill) has can be accounted for. This allows the researcher to separate the effect of the active ingredient from the effect of taking a pill alone. In this experiment, the mock treatment included every component of the mixture used for the other treatments, administered in exactly the same way as in the other treatments. The only difference in the other treatments was the presence of the factors being studied. Choices (B) and (D) are incorrect because the positive control is the treatment that contains all 8 factors. Choice (C) is incorrect because no cells were expected from the mock treatment. (LO 3.11)

462. **(B)** The positive control contained all 8 factors and was known to produce viable cells. Choices (A) and (C) are incorrect because the negative control was the mock treatment. Choice (D) is incorrect because cell development was expected in the positive control. (LO 3.21)

463. **(C)** Very few to no cells developed in mixtures lacking factor 3, 4, 7, or 8. Therefore, all four of these factors are necessary to cause growth. A reasonable hypothesis to draw from this experiment is that factors 3, 4, 7, and 8 are necessary and sufficient to cause growth. An experiment to test this would be to use factors 3, 4, 7, and 8 together to see if they were able to cause growth and then to remove one of the factors at a time to see if any were not needed. Factors 1, 2, 5, and 6 did not have an effect when they were removed from the mix and are therefore not necessary. Therefore, choices (A) and (B) are incorrect. Choice (D) is incorrect because inhibition is not expected. Instead, the production of pluripotent cells capable of developing into a frog is expected. (LO 3.21)

464. **(B)** Since these cells were all produced by mitosis, they all have the same genetic makeup. The differences depend on which genes are expressed and which are not. Gene expression is controlled primarily by transcription factors. The difference between nerve cells and smooth muscle cells, for example, lies in the fact that there are different transcription factors present in nerve cells than what is present in smooth muscle cells. As a result, different genes are expressed in different cells. Since mitosis produces cells with identical nuclei, choice (A) is incorrect. Cell differentiation is not a random process, so choice (C) is incorrect. Choice (D) is incorrect because cells develop as organisms grow; cells do not evolve. (LO 3.21)

465. **(C)** Molecule I is a polypeptide, which is a polymer made of amino acids that are assembled by the ribosome (molecule G) based on the sequence in the mRNA (molecule H). Choice (A) is incorrect because amino acids, not nucleotide triphosphates, are being polymerized. Choice (B) is incorrect because amino acid hydrolysis does not create RNA. Choice (D) is incorrect because the process shown does not elongate actin strands. (LO 3.4)

466. **(A)** Changes in the DNA (molecule D) will affect every mRNA made from it and therefore every protein. Changes in the primary RNA transcript (molecule E) and in the messenger RNA (molecule F) will disappear when the RNA is degraded, so choices (B) and (C) are incorrect. Molecule I is a polypeptide. Changes in molecule I will disappear when the protein is recycled. Therefore, choice (D) is incorrect. (LO 3.4)

467. **(D)** This process is RNA processing. It involves splicing introns out, adding a poly-A tail to the 3′ end, and adding a cap to the 5′ end. Since choices (A), (B), and (C) all occur during this process, choice (D) is the correct answer. (LO 3.4)

468. **(C)** This process is transcription. Choice (A) describes reverse transcription, which is the opposite of what is occurring. Choice (B) describes protein synthesis. Choice (D) describes replication of DNA. (LO 3.4)

469. **(B)** The process labeled C is the transport of the mRNA from the nucleus into the cytoplasm. This does not happen in prokaryotes, like bacteria, because they do not have nuclei. Choice (A) is incorrect because no DNA synthesis is occurring during the process labeled C. Choice (C) is incorrect because adding the poly-A tail happens in an earlier stage. Choice (D) is incorrect because this process takes place in eukaryotes. (LO 3.27)

470. **(A)** Process A is transcription. It is not affected by start codons because they are used during translation, a process that occurs much later. The rate of transcription is affected by transcription factors at the promoter region (choice (B)), how tightly DNA is bound to histones (choice (C)), and the degree of DNA methylation (choice (D)). (LO 3.22)

471. **(B)** Exon selection occurs during RNA processing, which is process B. Process A is transcription. The entire gene, including introns and all exons, is transcribed. Therefore, choice (A) is incorrect. Process C is the export of mRNA from the nucleus. Exons are chosen before process C, so choice (C) is incorrect. If exons are chosen before process C, then choice (D), which contains a process that occurs after process C, is incorrect. (LO 3.4)

472. **(A)** Crossing-over shuffles alleles and produces diversity. It does not create random point mutations, so choice (B) is incorrect. It does not prevent heterozygosity, which is a measure of allele diversity in a population. Therefore, choice (C) is wrong. Crossing-over ensures that nuclei are variable, not that they have the same DNA. Therefore, choice (D) is incorrect. (LO 3.10)

473. **(C)** Most alleles are the result of single nucleotide changes. Choice (A) is incorrect because only small changes are necessary to change the function of a protein. Choice (B) is incorrect because information is stored on chromosomes as DNA sequences. Choice (D) is incorrect because the exact same DNA sequence could not result in different alleles. (LO 3.27)

Grid-In Questions (pages 217–219)

> When working through the grid-in, long free-response, and short free-response questions that involve math, remember that your answer may vary slightly from the answer provided due to rounding or due to reading a graph differently. See page ix of the Introduction of this book for more information regarding how to avoid rounding errors. As a general rule, if you are within 5% of the actual answer, then your answer will be marked as correct.

474. **34** There are 2 A, 1 T, 3 G, and 4 C.

$$2 \times (2 + 1) + 4 \times (3 + 4) = 6 + 28 = 34$$

(LO 3.5)

475. **26** There are 4 A, 3 T, 2 C, and 1 G.

$$2 \times (4 + 3) + 4 \times (1 + 2) = 14 + 12 = 26$$

(LO 3.5)

476. $\mathbf{\frac{1}{4}}$ or **0.25** The AB– man has a $\frac{1}{2}$ chance of providing an A allele. The O+ woman provides only an O allele. Therefore, the child has a $\frac{1}{2}$ chance of having blood type A. The Rh+ woman is heterozygous since her mother was Rh–. Therefore, the woman has a $\frac{1}{2}$ chance of passing on an Rh+ allele. Multiplying $\frac{1}{2}$ by $\frac{1}{2}$ yields $\frac{1}{4}$. (LO 3.14, LO 3.17)

477. $\mathbf{\frac{1}{8}}$ or **0.125** Since there is a blood type O child, the parents must be AO and BO. The first child is also Rh–, so the Rh+ parent must be heterozygous. For the second child to be A–, there is a $\frac{1}{4}$ chance of being AO and a $\frac{1}{2}$ chance of being Rh–. Multiplying $\frac{1}{4}$ by $\frac{1}{2}$ yields $\frac{1}{8}$. (LO 3.14, LO 3.17)

478. $\mathbf{\frac{1}{8}}$ or **0.125** Since they already had a child that is Rh–, the parents must both be heterozygotes at the Rh gene locus. There is a $\frac{1}{2}$ chance that they will produce an AB child, and a $\frac{1}{4}$ chance that the child will be Rh–. Multiplying $\frac{1}{2}$ by $\frac{1}{4}$ yields $\frac{1}{8}$. (LO 3.14, LO 3.17)

479. $\mathbf{\frac{1}{64}}$ or **0.0156** There is a $\frac{1}{4}$ of a chance for each heterozygous locus to produce a homozygous recessive, and there are 3 loci. Multiplying by $\frac{1}{4}$ by $\frac{1}{4}$ by $\frac{1}{4}$ yields $\frac{1}{64}$. (LO 3.14, LO 3.17)

480. $\mathbf{\frac{1}{8}}$ or **0.125** For this cross, there is a $\frac{1}{4}$ chance that the offspring will have the genotype bb and a $\frac{1}{2}$ chance of the genotype Ee. Multiplying $\frac{1}{4}$ by $\frac{1}{2}$ yields $\frac{1}{8}$. (LO 3.14, LO 3.17)

481. $\mathbf{\frac{1}{2}}$ or **0.5** For this cross, only the E locus matters. If the puppy has the genotype ee, it will be yellow. (LO 3.14, LO 3.17)

482. **0** Neither dog has the B allele, so all of the puppies will be either brown or yellow. (LO 3.14, LO 3.17)

483. **16** Based on the model in the information provided, there are $2^4 = 16$ combinations. (LO 3.9)

484. **128** Based on the model in the information provided, there $2^7 = 128$ combinations. (LO 3.9)

485. **216** The current code has $4^3 = 64$ combinations. Adding a pair of nucleotides would give $6^3 = 216$ combinations. (LO 3.9)

Long Free-Response Questions (pages 219–220)

When working through the grid-in, long free-response, and short free-response questions that involve math, remember that your answer may vary slightly from the answer provided due to rounding or due to reading a graph differently. See page ix of the Introduction of this book for more information regarding how to avoid rounding errors. As a general rule, if you are within 5% of the actual answer, then your answer will be marked as correct.

486. This question asks you to create visual representations of processes in the nervous system. Note that this question is worth 10 points total. (LO 3.48, LO 3.49, LO 3.50)

(a) Specialized nerve cells called receptor cells detect and transmit signals from the environment. Identify an external signal that humans can detect. Draw a diagram to illustrate how a receptor cell might detect and transmit that signal. Explain your diagram. (4 points maximum)

You may choose any type of receptor cell. Examples include a rod cell in the eye, cilia in the ear, or a taste receptor on the tongue. (1 point for identifying an appropriate external signal)

Below is a sample diagram that would have received full credit.

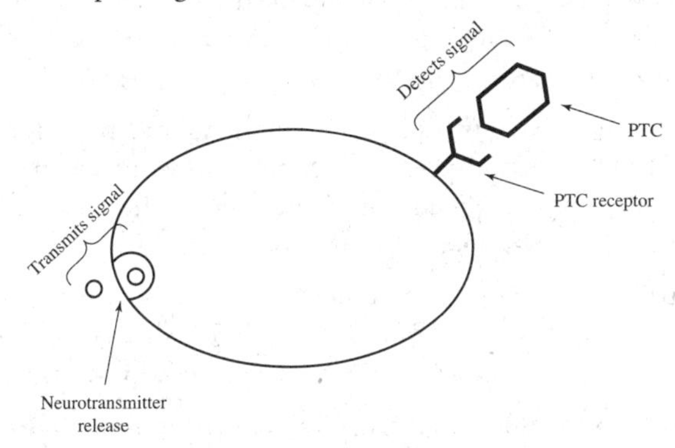

(1 point for creating a drawing that depicts a mechanism for identifying that external signal and 1 point for depicting a mechanism for signal transmission in the drawing)

Below is a sample explanation of the diagram that would have received full credit.

A taste receptor cell (as shown in the diagram) detects the chemical PTC. It can do this because it has a receptor protein that matches the shape of the chemical PTC. Binding of PTC to the receptor causes the receptor to depolarize the membrane and release a neurotransmitter into the synapse of a neuron connected to the brain.

(1 point for clearly explaining the detection step and for clearly explaining the transmission step shown in your drawing)

(b) Identify an example in which the brain must integrate two sources of information to produce a response. Draw an illustration that shows the two sources of information being integrated in the brain in order to produce a response. Describe how that process occurs. (3 points maximum)

The following is a sample response with a diagram that would have received full credit.

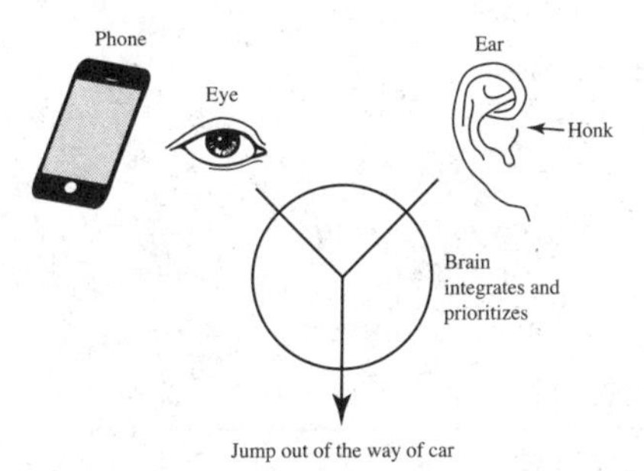

The figure shows a person looking at his phone. If he is walking and texting, his eye is looking at his phone. He then hears the honk of a car's horn. His two sources of information are his ear and his eye. His brain prioritizes the honk as the more important signal and stimulates motor neurons to cause him to move out of the way of the car. The diagram shows sight and hearing as the two sources of information. The sources interact in a circle representing the brain, and there is an output arrow showing the response of the brain.

(1 point for identifying two sources of information; 1 point for drawing a diagram that shows the two sources of information interacting and that leads to an output response; 1 point for writing a description that clearly connects the input, the integration, and the output)

(c) Animals often respond to stimuli with muscle movement. Draw a diagram that shows at least THREE of the steps by which motor neurons stimulate a muscle cell to contract. Describe the process. (3 points maximum)

This part of the question requires you to draw a neuromuscular junction. Your diagram should show the following:

- An action potential arriving at a synapse, stimulating the release of a neurotransmitter
- A receptor on the muscle cell that detects the neurotransmitter
- A release of calcium ions into the intercellular environment, thereby causing muscle contraction

You would receive 1 point each, 3 points total, for including any of the following in your diagram and in your description of the process:

- Action potential transmission along the neuron
- Neurotransmitter release
- Neurotransmitter reception
- Signal transduction from the receptor on the muscle cell
- Calcium release from the T-tubules
- Calcium ions interacting with the troponin/tropomyosin complex
- Movement of actin and myosin relative to one another

The following is a sample response with a diagram that would have received full credit.

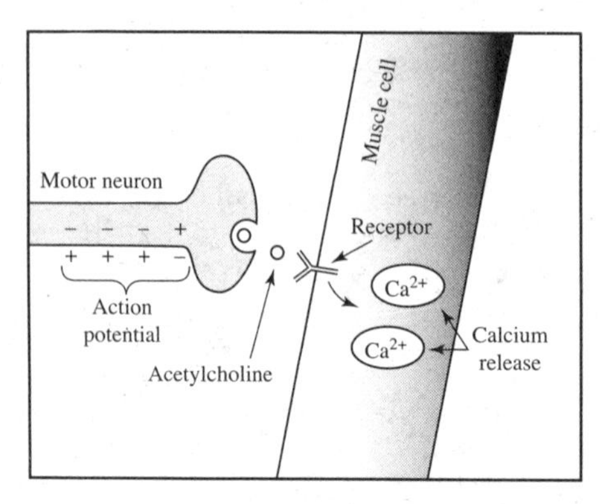

This diagram shows a motor neuron causing a muscle cell to contract. An action potential is propagated along a neuron by voltage gated ion channels opening and closing. The diagram shows this with + and – symbols. When an action potential reaches the end of a motor neuron, the neuron releases the neurotransmitter acetylcholine into the synapse. Acetylcholine receptors on the muscle cell membrane bind to the acetylcholine, causing the receptor to change shape. This shape change causes the release of calcium ions, which cause muscle contraction.

487. This question asks you to analyze data from a series of genetic experiments. Note that this question is worth 10 points total. (LO 3.14, LO 3.40)

(a) Explain how resistance to downy mildew is inherited. Justify your explanation with data from the backcrosses. (3 points maximum)

Resistance is conferred by a dominant allele. (1 point)

The backcross to the cultivated variety results in 1:1 ratio of susceptible:resistant offspring. (1 point)

The backcross to the wild lettuce results in all offspring being resistant. (1 point)

(b) State a null hypothesis for the self-cross data. Calculate a chi-square value, and interpret your result. (4 points maximum)

You would be awarded 1 point for each of the following, up to 4 points maximum:

- Correct null hypothesis (that there is no difference from a 3:1 ratio)

Sample Response: If resistance is conferred by a dominant allele, then R = dominant and r = recessive. The hybrids would be heterozygous (Rr). Crossing Rr × Rr gives a 3:1 ratio of resistant to susceptible plants.

- Correct chi-square value (1.21)

Sample Response: There are 388 + 144 = 532 total offspring. To find the expected number of resistant plants, multiply:

$$532 \times 0.75 = 399$$

To find the expect number of susceptible plants, multiply:

$$532 \times 0.25 = 133$$

Next, subtract each observed value from the expected value, square the result, and divide by the expected value. Then, add the results together.

$$\left(\frac{(388-399)^2}{399}\right)+\left(\frac{(144-133)^2}{133}\right)=\left(\frac{(-11)^2}{399}\right)+\left(\frac{11^2}{133}\right)$$

$$=\left(\frac{121}{399}\right)+\left(\frac{121}{133}\right)=0.30+0.9097=1.2$$

- Correct degrees of freedom (1)

Sample Response: The degrees of freedom is the number of categories minus 1. There are two categories (resistant and susceptible), so there is 1 degree of freedom.

- Correct cutoff value (3.84 at 0.05 p-value or 6.64 at a 0.01 p-value)

Sample Response. According to the Reference Tables, the correct cutoff value is 3.84 at 0.05 p-value or 6.64 at a 0.01 p-value.

- Correct interpretation (that you should accept the null hypothesis)

Sample Response: Since $1.2 < 3.84$, the null hypothesis should be accepted.

Note that an answer that states an incorrect hypothesis can still earn up to 3 points if all of the other parts of the answer are consistent with the stated (incorrect) hypothesis.

(c) Resistance to downy mildew has been shown to be the result of a gene-for-gene system. In this system, resistance occurs when a strain of a pathogen has an avirulence gene that matches a corresponding gene in the host plant. Recognition results in the death of plant cells that the fungus needs in order to live. Describe a molecular mechanism by which this gene-for-gene system might work. (3 points maximum)

The resistance gene might be detecting a molecule on the surface of the fungal cell or a molecule that is secreted by the fungus. (1 point)

The plant would need a receptor protein, which would have a binding site for the signal produced by the fungus. (1 point)

If the signal from the fungus was detected by the receptor, the receptor would change shape and turn on a signal transduction pathway that stimulated apoptosis (cell suicide). (1 point)

Short Free-Response Questions (pages 221–222)

> When working through the grid-in, long free-response, and short free-response questions that involve math, remember that your answer may vary slightly from the answer provided due to rounding or due to reading a graph differently. See page ix of the Introduction of this book for more information regarding how to avoid rounding errors. As a general rule, if you are within 5% of the actual answer, then your answer will be marked as correct.

488. Briefly explain how processes that occur in each of the following contribute to controlling gene expression: the plasma membrane, the cytoplasm, and the nucleus. (LO 3.22)

A maximum of 3 points can be awarded for correctly answering this question.

(a) Receptors on the plasma membrane detect ligands and trigger a signal transduction cascade. (1 point)

(b) The portion of the signal transduction pathway in the cytoplasm that is composed of Raf, MEK, and ERK amplifies the signal and propagates the signal across the cytoplasm to the nucleus. (1 point)

(c) The signal transduction pathway stimulates transcription of genes in the nucleus. (1 point)

489. Label the ladder fragments to show the pieces being 10, 20, 30, 40, 50, 60, 70, 80, 90, and 100 kb long. Draw bands on the gel to indicate the sizes of the DNA pieces for each of the digests listed. (LO 3.5)

A maximum of 4 points can be awarded for correctly answering this question.

Your gel should look like the one below:

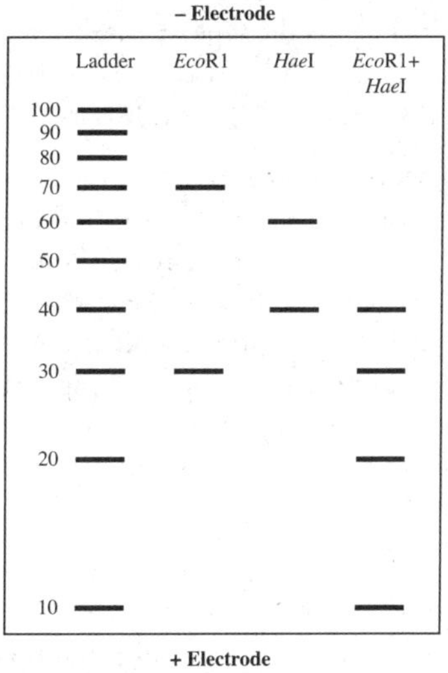

– Electrode

Ladder *Eco*R1 *Hae*I *Eco*R1+ *Hae*I

100
90
80
70
60
50
40
30
20
10

+ Electrode

Positive and negative electrodes should be at the correct sides of the gel. (1 point)

The ladder with smallest pieces should be toward the + electrode. If the electrodes are on the wrong sides, the ladder should be backward. (1 point)

All 4 single-digest pieces should be in their correct locations. (1 point)

All 4 double-digest pieces should be in their correct locations. (1 point)

BIG IDEA 4—INTERACTIONS

Multiple-Choice Questions (pages 223–264)

490. **(A)** Since the chicks were newly hatched, they never had any time to learn feeding behavior. Therefore, the researcher was testing whether pecking at red spots was an innate instinct in these birds, even though they hadn't been exposed to other birds. Choice (B) was not being tested because food was not offered after pecking in the first experiment. In this experiment, there was no way to know if the chicks distinguished among colors, so choice (C) is incorrect. Since they have just emerged from the egg, the herring gull chicks have not yet had any experiences outside of the shell to learn from. Therefore, any behavior the herring gull chicks show must be innate rather than learned, which makes choice (D) incorrect. (LO 4.8, LO 4.18)

491. **(C)** Knowing the effect of the dot on the stick is impossible unless the effect of the stick itself is known, making the unpainted stick the control in this experiment. The dependent variable was the number of pecks per minute, and the independent variable was the type of dot. Therefore, choices (A) and (B) are incorrect. Replication was provided by using 10 chicks per treatment, not by using the unpainted stick, so choice (D) is incorrect. (LO 4.8, LO 4.18)

492. **(C)** The experiment was designed to determine if chicks could learn to peck at a particular colored spot to earn to a food reward. No evolutionary hypothesis was tested in this experiment, so choice (A) is incorrect. To determine the genetics of pecking behavior, the researcher would have to perform crosses or use molecular techniques to look at DNA. This was not done in this experiment, so choice (B) is incorrect. Epigenetics is the modification of gene expression by modifying DNA with acetylation or methylation patterns. This experiment did not address epigenetics in any way, so choice (D) is incorrect. (LO 4.8, LO 4.18)

493. **(A)** All spot colors produced some response above that of the control. The strongest response was toward red spots. Therefore, choice (A) is the best answer. Choice (B) implies a causal relationship that was not established in this experiment. Choice (C) suggests that learning took place. However, the results from Table I show data from newly hatched chicks that couldn't have learned from previous experience, so choice (C) is incorrect. Since red and black spots produced stronger responses than the other colors, choice (D) is incorrect. (LO 4.8, LO 4.18)

494. **(B)** The researchers wanted to study innate behavior, unmodified by learning. Had they been interested in learned responses, older chicks can certainly be made hungry enough to produce a strong enough response to study. Therefore, choice (A) is incorrect. There is no evidence in these experiments that older chicks stop pecking at spots, so choice (C) is incorrect. Neotenous

is a term that means that adults retain traits that are associated with juveniles. This term does not apply in this case, so choice (D) is incorrect. (LO 4.8)

495. **(C)** Pecking behavior is an innate behavior that can be modified by learning, which in this case included teaching the chicks to peck at a particular stimulus in order to be rewarded with food. Since pecking behavior is innate, but can be modified by learning, choices (A) and (B) are incorrect. Choice (D) is incorrect because the unpainted stick seemed to lead to less of a response than any spotted stick. (LO 4.8)

496. **(C)** Without some measure of the variability in the data, it is difficult to determine whether the differences seen in the data are actually meaningful differences, not simply the effect of random chance. The difference between 37 pecks (for those in the blue training group pecking at the blue spot stick) and 38 pecks (for those in the green training group pecking at the green spot stick) is probably due to random chance, but what about the difference between 37 and 42 pecks (for those in the red training group pecking at the red spot stick)? Is it statistically significant? Thus, a measure of variability, like the standard error of the mean, would be useful in this case. Since a measure of variability would be helpful, choice (A) is incorrect. The unpainted stick is a reasonable control, so choice (B) is incorrect. Choice (D) is incorrect because the data presented already use the mean, which is a measure of central tendency. (LO 4.8)

497. **(C)** The forest has 8 fungus species represented, while the nursery has only 4 species represented. Therefore, the forest has more fungus species and higher species richness. Since species richness is the number of species present, it can be clearly measured based on the information in Table I. Therefore, choice (A) is incorrect. Choices (B) and (D) both incorrectly interpret the information from Table I. (LO 4.27)

498. **(C)** In the nursery, species *A* is very common and the other species are a lot less common. In the forest, species *A* is still very common but other species are also fairly common. Choice (A) is incorrect because a high number of an individual species will lead to a lower species evenness. The same number of a particular species does not occur in each environment, so choice (B) is incorrect. Choice (D) is incorrect because species evenness does not refer to even and odd numbers. (LO 4.27)

499. **(A)** According to Figure I, fungicide-sprayed plots had about 900 plants, while unsprayed plots had about 450 plants. The error bars do not overlap, so we can be reasonably sure that these are statistically significant differences. Therefore, choice (B) is incorrect. Choice (C) is the opposite of what the data show. Choice (D) is incorrect because the experiment was clearly designed to answer this question. (LO 4.27)

500. **(A)** Unsprayed plots had more species than the sprayed plots, and the species in the unsprayed plots were more evenly represented. Choice (B) is the opposite of the data displayed in Table II, so it is incorrect. Biodiversity was not about the same in the sprayed and unsprayed plots, so choices (C) and (D) are incorrect. (LO 4.27)

501. **(B)** If species *A* and species *B* were fairly resistant to the pesticide and species *E*, species *F*, species *G*, and species *H* were highly susceptible to the pesticide, this would explain why species *A* and species *B* were present in the sprayed plots yet species *E*, species *F*, species *G*, and species *H* were not. This would provide a causal link. Choice (A) might explain the pattern, but one would need to establish how far the fungicide penetrated into the tissue to argue for a causal link. Therefore, rule out choice (A). Choices (C) and (D) do not provide any information that would support a link between fungicide and decreased diversity. (LO 4.27)

502. **(D)** The data provided in the question demonstrate an inverse correlation (the more pathogens there are, the lower endophytic diversity is, and vice versa). Choices (A) and (B) present competing hypotheses that might explain the correlation, but the information provided does not support either conclusion. We might imagine that a more diverse endophytic community could prevent pathogens from invading the stem. We might also imagine that being infected by a pathogen would reduce endophytic diversity. A good follow-up experiment could be designed to test these alternatives, but choice (D) is still the most reasonable statement to make based on the information provided. Choice (C) is incorrect because the information presented in the passage clearly states that endophytes do not cause disease, like cankers. (LO 4.27)

503. **(C)** Chitinase allows the fungus to dissolve the exoskeleton and invade the ant's body and thus its circulatory system. Fungi do produce chitinases, but chitinases are used to break down cell walls, not to build them. Therefore, choice (A) is incorrect. Additionally, in the context of the question, the important function of chitinases is to dissolve the ant's exoskeleton. Chitinases dissolve chitin; they do not interact with neurons. Therefore, choice (B) is incorrect. Choice (D) is incorrect because the fungus, not the ant, is the organism that produces the chitinases. (LO 4.13)

504. **(D)** If the neurotoxin fits into the active site of the motor neuron, the neurotoxin would cause muscle contraction. Choice (A) is incorrect because if the toxin made holes in the membranes of sensory neurons, sensory processes, such as taste, hearing, or vision, would be affected. Muscle contraction is controlled by motor neurons, not sensory neurons. Blood filtration shouldn't really affect muscle contraction and neither should an increase in blood glucose. Therefore, choices (B) and (C) are both incorrect. (LO 4.9)

505. **(C)** The experiment described in choice (C) directly tests the hypothesis that a vegetation height of 25 cm above the forest floor provides the optimum

conditions for spore production. Choice (A) might be useful in establishing that 25 cm is the standard height, but this design does not measure spore production. The information gathered from choice (B) would be interesting to use when explaining the conditions that lead to the optimum amount of spores. However, this design doesn't directly measure spore production. Sequencing the genomes would give a lot of information about the organisms, but this design also wouldn't directly answer spore production. Therefore, choice (D) is incorrect. (LO 4.11)

506. **(C)** This behavior is controlled by the neurotoxin and increases the fitness of the fungus, which obtains nourishment from the ant in this way. No information presented suggests that the ant absorbs energy or nourishment from the leaf. In fact, the information in the passage states that the ant hangs from the leaf by its mandibles, so no chewing is actually happening. Therefore, choice (A) is incorrect. Similarly, since the ant does not obtain resources from the plant for the fungus to use, choice (B) is incorrect. This behavior has a clear explanation and function, as outlined in choice (C). Therefore, choice (D) is incorrect. (LO 4.8)

507. **(A)** This experiment would establish what, if any, cost was associated with infected ants returning to the hive to die. Choices (B), (C), and (D) would not give any information about what cost there would be to the hive if diseased individuals returned to the hive to die. (LO 4.8)

508. **(D)** A good experiment typically attempts to keep all variables that are not under study constant. In this case, the intensity of the lights needed to be adjusted to be the same since it was the effect of the wavelength of light (depending on the color of the LED) that was being measured. Although plants do grow more slowly under lower light intensity, the reason why the students adjusted the LEDs was to make light intensity a controlled variable in which the LEDs were at the same light intensity. Therefore, choice (A) is not correct. Even if the LEDs varied widely in their light intensity, the purpose for the adjustment was to ensure the same amount of light from the LEDs. Therefore, choice (B) is incorrect. In order to simulate sunlight, as choice (C) suggests, students would have had to use a wide range of different LEDs to get a full spectrum. Since that was not done in this case, choice (C) is not correct. (LO 4.10)

509. **(C)** The increased height in the plants grown under the green light is a typical plant response to poor light conditions. It is called etiolation. The plant uses the resources it has in the seed to seek light. Plants reflect most green light, so choice (A) is incorrect. Choice (B) is not reasonable because the plants grown under the green light were very different from the plants grown under the blue light in the other variables in the table in addition to being nearly 3 times the height of the plants grown under the blue light. Choice (D) is incorrect because visible light has little to no ability to cause muta-

tions. In addition, there is no reason to think that growth differences would be due to mutations. (LO 4.8)

510. **(A)** A positive control is one in which the plants would grow normally. Full-spectrum light is the logical choice. Choice (B) would be a good control to use, but it is a negative control, not a positive control as the question asks for. Choices (C) and (D) would be interesting to include, but they are both additional treatments, not controls. Therefore, both of these answer choices are incorrect. (LO 4.10)

511. **(B)** A negative control is one in which the plants would not be exposed to light. Adding full-spectrum light at the same intensity as the LEDs would be a good positive control because this represents the normal conditions under which the plants grow. Therefore, choice (A) is incorrect. Choices (C) and (D) are both incorrect because each of these only adds another treatment to the experiment (another color of LED). Neither is a negative control. (LO 4.8)

512. **(C)** Most of the dry mass of a plant comes from carbon, and all of the carbon in a plant comes from carbon dioxide in the air. As a result, dry mass is a good measure of carbon fixed. Plants clearly grow laterally in diameter in addition to taller. Carbon that is fixed is used for plant growth in all directions, so choice (A) is incorrect. Leaf color is indicative of the color of light reflected, not absorbed, so choice (B) is incorrect. Choice (D) is a poor choice because carbon dioxide levels will fluctuate significantly with the presence of the students in the room. (LO 4.8)

513. **(C)** Lots of green sea turtles die young, but the adults live a long time. Therefore, the curve should plummet initially and then level off with increasing age. Thus, green sea turtle survivorship is a Type 3 curve, according to this graph. In a Type 1 survivorship curve, young individuals have a relatively low death rate compared to old individuals. The curve stays relatively level and then plummets with old age. This is the opposite of what happens with green sea turtles, so choice (A) is incorrect. In a Type 2 survivorship curve, individuals are equally likely to die at any age. However, green sea turtles have a much greater mortality rate when they are young than when they are old. Therefore, choice (B) is incorrect. Choice (D) is incorrect because a Type 3 survivorship curve accurately describes the situation for green sea turtles. (LO 4.8)

514. **(C)** Although giant sequoias are majestic trees, the amount of seeds and seedling mortality clearly makes this a species with a Type 3 survivorship curve. Choice (A) is incorrect because a Type 1 curve is the opposite of what happens with giant sequoias. For Type 1 to be correct, the seeds would have to have a large chance of survival and the older trees would have to have a much greater chance of dying. This is not the case for giant sequoia trees. Choice (B) is incorrect because a Type 2 curve applies to organisms that have a constant death rate, regardless of the age of the organism. Choice (D) is incorrect because giant sequoia trees do show a Type 3 survivorship curve. (LO 4.8)

515. **(A)** Elephants bear few young, most of which survive. Therefore, their curve begins high and stays high, only dropping off with old age. Choice (B) is incorrect because elephants do not have a constant death rate regardless of age. Choice (C) is the opposite of what is true for elephants. Young animals die infrequently. The death rate increases as elephants get older. Choice (D) is incorrect because elephants do follow a Type 1 survivorship curve. (LO 4.8)

516. **(B)** The population growth rate in the exponential growth equation depends on only the intrinsic growth rate (r_{max}) and the number of individuals in the population (N). The larger N gets, the faster the population grows. Note that the term $\frac{dN}{dt}$ is an expression from calculus. If you have not yet taken calculus, do not be intimidated. This term simply means the change in number (N) divided by the change in time (t). It is the rate of population growth at any point in time. The carrying capacity is not included in the exponential growth equation, so choice (A) is incorrect. The intrinsic growth rate is a constant for any given organism. It is the fastest that an organism is capable of growing under ideal conditions. Therefore, choice (C) is incorrect. The carrying capacity is not included in the exponential growth equation, so choice (D) is incorrect. In addition, the carrying capacity is controlled by the amount of resources present; it does not vary with the population size. (LO 4.12)

517. **(A)** The logistic growth equation does include carrying capacity. When $K = N$, the term $K - N$ in the equation becomes zero, which makes the growth rate also zero. Choice (B) is incorrect because the population growth rate levels off as the population size reaches the carrying capacity in the logistic growth equation. Choice (C) is incorrect because the intrinsic rate of increase does not change. Choice (D) is incorrect because the carrying capacity is the maximum number of individuals an environment can hold. It does not depend on the population size. (LO 4.12)

518. **(C)** A graph of the exponential model would show population levels increasing with the slope continuing to get steeper indefinitely. The logistic model would look very similar to the exponential model at the start. As the population size reaches the carrying capacity, however, the logistic model would plateau. Therefore, human population growth matches both the exponential model and the early part of the logistic model. Although choices (A) and (B) are not entirely incorrect, choice (C) is the best choice because it includes aspects of both models. Choice (D) is incorrect because the data are consistent with aspects of both models. (LO 4.12)

519. **(A)** In separate cultures, both species of paramecia measured in Figure I grew to a larger population size than when that same species was grown in a mixed culture. Choice (B) is the opposite of what is true and is therefore incorrect. Choice (C) is incorrect because paramecia grew better in separate cultures than in mixed cultures, not equally well in both. Choice (D) is incorrect

because there is enough information to determine that paramecia grow better in separate cultures than in mixed cultures. (LO 4.12)

520. **(B)** The two species used in each of the two experiments were competing for the same food source in the mixed cultures. Therefore, the students were trying to determine what effect competition would have on population growth. Neither species was a parasite or a predator on the other, so choices (A) and (C) are not correct. Evolution on the population level is defined as changes in allele frequencies and genotype frequencies. The students did not measure allele or genotype frequencies, so choice (D) is incorrect. If the statement in choice (D) referred to speciation, it would also be incorrect because there is no evidence of one species forming two species. (LO 4.14)

521. **(B)** Since larger individuals need more energy than smaller ones to reach the same number of individuals, the smaller population contains the larger organism. Choice (A) has this backward. The question asks about separate cultures, not mixed ones. Since the two species were grown in different cultures, competition would be impossible. Therefore, choice (C) is incorrect. There is no scale on which it can be measured which organisms are more or less evolutionarily advanced than other organisms. One organism may be better adapted under some conditions, and another may be better adapted in a different environment. Therefore, choice (D) is incorrect. (LO 4.14)

522. **(C)** Competitive exclusion relies on the idea that in a given environment, one strategy is more adaptive than another. A keystone species *increases* diversity with its presence. Removing the keystone species leads to a dramatic decrease in diversity. However, this was not seen in this test tube environment, so choice (A) is incorrect. Energy is lost with increasing trophic levels. However, this doesn't apply to this test tube environment, so choice (B) is incorrect. Niche partitioning is a concept related to competitive exclusion, but it doesn't apply in this case. Therefore, choice (D) is incorrect. (LO 4.14)

523. **(A)** By examining the graph, it is clear that both species grew larger populations in separate cultures than in mixed cultures. Choices (B), (C), and (D) are incorrect because they do not interpret Figure II correctly. (LO 4.14)

524. **(D)** This is a case of niche partitioning. Since one species was more competitive at the surface of the test tube and the other was more competitive at the bottom of the test tube, the food resource was divided, allowing both species to survive. A keystone species increases diversity with its presence. Removing the keystone species leads to a dramatic decrease in diversity. This was not seen in this test tube environment, however, so choice (A) is incorrect. Energy is lost with increasing trophic levels. However, this doesn't apply to this situation, so choice (B) is incorrect. The experiment shown in Figure I shows competitive exclusion, in which one species survives and the other does not. This question asks about the coexistence demonstrated in Figure II, however, so choice (C) is incorrect. (LO 4.14)

525. **(C)** Respiration causes carbon dioxide to increase, and photosynthesis takes up carbon dioxide, causing the concentration of carbon dioxide to decrease. More photosynthesis occurs in summer than in winter. Choice (A) is incorrect because carbon dioxide levels fall in the summer according to this graph. Choice (B) suggests that more fuel is burned to cool (with air conditioners) than to heat, which is incorrect. Warm water actually dissolves *less* gas than cold water. Therefore, choice (D) is incorrect. (LO 4.15)

526. **(B)** The graph shows an increase from just under 355 ppm to just under 370 ppm. The difference is about 15 ppm. Choices (A) and (C) are incorrect because the graph does not show temperature levels. Choice (D) may look promising because the change on the graph looks dramatic. However, it is incorrect because the *y*-axis starts at 350 ppm. (LO 4.15)

527. **(D)** All three of the ways mentioned in choices (A), (B), and (C) would decrease the rate at which atmospheric carbon dioxide increases. While it is true that photosynthesis removes carbon dioxide from the atmosphere and that an increase in the rate of photosynthesis would decrease the rate at which atmospheric carbon dioxide increases, choice (A) is incorrect because it is not the only correct method among the answer choices. Choice (B) is also a correct method because decreasing the rate of respiration would release less carbon dioxide and would thus slow the rate of atmospheric carbon dioxide increase. Again, choice (B) is incorrect because it is not the only correct method among the answer choices. Choice (C) is also a true statement. Increasing the amount of carbon dioxide dissolved in the ocean would remove it from the atmosphere and thus decrease the rate at which atmospheric carbon dioxide increases. Again, since choice (C) is not the only correct method among the answer choices, it is not the correct answer. (LO 4.16)

528. **(B)** To minimize photorespiration, rubisco in the chloroplasts of cells needs to be exposed to high carbon dioxide levels and low oxygen levels. Rubisco can catalyze a reaction in which it combines RuBP with either carbon dioxide or oxygen. Having a high concentration of carbon dioxide and a low concentration of oxygen makes the reaction with carbon dioxide (carbon fixation) more likely and the reaction with oxygen (photorespiration) less likely. A high carbon dioxide, low oxygen atmosphere is a good start, but the carbon dioxide has to get into the leaves through stomata and the oxygen generated by photosynthesis has to exit from the leaf through the stomata. If there is abundant water in the soil, then the plant can leave the stomata open. This maximizes gas exchange so that the oxygen can exit the leaf and carbon dioxide can enter. Choice (A) is incorrect because all three of these conditions would maximize photorespiration. Starting with a low carbon dioxide, high oxygen atmosphere would cause lots of photorespiration. If there was limited water in the soil, the plant would be forced to close its stomata, decreasing gas exchange. This would lead to even higher concentrations of oxygen and even lower concentrations of carbon dioxide in the leaves because photosynthesis releases oxygen and consumes carbon diox-

ide. Choice (C) is incorrect because the low carbon dioxide, high oxygen atmosphere would increase photorespiration. Choice (D) is incorrect because the limited water in the soil would force the plant to close its stomata. This would lead to an increase of oxygen and a decrease in carbon dioxide in the leaf because photosynthesis releases oxygen and consumes carbon dioxide. Closed stomata would prevent gas exchange and the oxygen produced would not be able to exit the leaf and carbon dioxide would not be able to enter to replenish the carbon dioxide used in photosynthesis. (LO 4.13)

529. **(A)** Since C-4 plants are adapted to avoid photorespiration, they outcompete C-3 plants under conditions where photorespiration would be high. These conditions are low carbon dioxide, high oxygen, and limited water in the soil. Choices (B) and (C) both include abundant water in the soil and are therefore incorrect. Choice (B) is also incorrect because it includes a high carbon dioxide, low oxygen atmosphere. Choice (D) also has a high carbon dioxide, low oxygen atmosphere and is therefore wrong. (LO 4.13)

530. **(A)** C-4 plants pay an energy cost to pump carbon dioxide. Therefore, if carbon dioxide levels rise, this cost is less worth paying. C-4 plants will not have the selective advantage, so choice (B) is incorrect. Carbon dioxide is not toxic to plants, so choice (C) is incorrect. Choice (D) is incorrect because both types of plants should show increased growth under higher carbon dioxide levels. (LO 4.13)

531. **(A)** The letters in the sequences indicate amino acids. Choice (B) cannot be correct because DNA and RNA are composed of only 4 different monomers but there are more than 4 different letters shown. Carbohydrates do not contain heritable material and therefore would not likely be shown in a complex sequence alignment like this one. Therefore, choice (C) is incorrect. Choice (D) is incorrect because lipids are not composed of ribosomes or of monomers of any kind. (LO 4.1)

532. **(B)** Conserved regions are those that do not change much from organism to organism. This means that region B is highly conserved. Highly conserved regions are inferred to have important functions since they have not changed over long periods of time. In the case of rubisco, region B has not changed for about 1.5 billion years. Choice (A) is incorrect because it states that region A is more highly conserved than region B, which it is not. The carbon dioxide binding site is likely to be very important and highly conserved, so choice (C) is incorrect. Transmembrane regions may be highly conserved, but rubisco is not membrane bound. It is found dissolved in the stroma. Therefore, choice (D) is incorrect. (LO 4.2)

533. **(B)** Selection pressure keeps functional regions from changing because changes make those regions not work. Since region A changed a lot, likely due to less selective pressure, and since region B did not change a lot, likely due to more selective pressure, choice (B) contains the most accurate state-

ment. Both of these sequences have been around the same amount of time. Therefore, choices (A) and (D) are incorrect. Horizontal gene flow is very rare and is not needed to explain the differences seen here. Therefore, choice (C) is incorrect. (LO 4.8)

534. **(C)** Sequences that are more similar likely have a more recent common ancestor than those that have fewer similarities. Choice (A) is incorrect because the number of letters simply indicates how many amino acids are present. Choice (B) is incorrect because similarities indicate recent ancestry, not more distant ancestry. Choice (D) is incorrect because there is a way to use the information to infer phylogenetic relatedness. (LO 4.8)

535. **(D)** The most compelling experiment would be to turn sonar on and off and see if it affected the diving behavior of whales that researchers could directly observe. Choice (A) suggests a mechanism for why a change in whale behavior might occur, but it is not an experiment that tests whether or not sonar changes the behavior of whales. Since simply observing military ships does not show if those ships are actively using sonar, choice (B) is incorrect. High blood nitrogen levels have been discovered in beached whales, but this discovery doesn't establish a link between military use of sonar and the behavior of whales. Therefore, choice (C) is incorrect. (LO 4.9)

536. **(C)** Without a baseline to determine what normal whale diving behavior is like, observing the whales' behavior after turning on the sonar would be pointless. Choice (A) is incorrect because there is no evidence that the application of the recording device had any effect on the whales. Since the sonar was not turned on in the first 150 minutes, it would have been difficult to determine whether it was working properly. Therefore, choice (B) is incorrect. The graph shows that the whales surfaced several times during this 150 minute period, so if they were waiting for the whales to surface, they would not have had to wait so long. Additionally, sound propagates very well in water. For both of these reasons, choice (D) is incorrect. (LO 4.9)

537. **(C)** Since whales dive to between 100 and 130 meters deep and stay there, it is safe to conclude that that is where their food is. Choice (A) is incorrect because if the whales' food was at the surface, there would be no need for them to dive. If their food was between 0 and 100 meters deep, there would be no reason for them to stay at a depth of 100 to 130 meters deep. We would not expect them to dive below 100 meters. Therefore, choice (B) is incorrect. Choice (D) is incorrect because the data do not show the whales spending much time, if any, below 130 meters. (LO 4.15)

538. **(C)** The first dive on the graph occurs at about 25 minutes and lasts until the whale surfaces again at about 50 minutes. Therefore, the dive duration is about 25 minutes. Choice (A) is approximately how long the whale stays at the surface between dives, but this isn't what the question asks for, so choice (A) is incorrect. Choice (B) might be a good answer for how long it takes a

whale to ascend from the deep water, fill its lungs, and return to the deep water. However, that isn't what the question asked for, so choice (B) is incorrect. Choice (D) doesn't appear to match anything on the graph. (LO 4.8)

539. **(A)** Swimming at 100 meters depth for 20 minutes requires the whales to resurface to obtain the oxygen that they need to complete cellular respiration. Choice (B) is incorrect because decompression sickness (often called "the bends") is caused by diving and surfacing rapidly from depth. If a human diver periodically surfaced like blue whales do, he or she would get decompression sickness. Periodically surfacing would cause, not prevent, decompression sickness. Note that whales do not get decompression sickness. Krill have gills to absorb oxygen from water. They do not need to surface. Therefore, choice (C) is incorrect. Since whale predators are probably aquatic, and because whales use echolocation to "see" under water, surfacing to look for predators doesn't make much sense. Therefore, choice (D) is incorrect. (LO 4.15)

540. **(A)** The graph clearly shows that the whales surfaced just after the sonar was turned on. They appeared to ascend earlier than they would have without the presence of sonar, and they did not dive back down to 100 meters deep for approximately 60 minutes after the sonar was turned off. Although orcas are dangerous to blue whales and sonar may sound like orcas to whales, what the whales thought cannot be determined from the data presented. Therefore, choice (B) is incorrect. Krill are very small and do not make loud, coordinated sounds like sonar. Therefore, choice (C) is incorrect. The data from the graph show the whales surfacing when the sonar was turned on, not diving deep as choice (D) suggests. (LO 4.21)

541. **(A)** Blue whales are BIG. They need to eat tons of food—literally 4 tons a day. The data presented show that 50 minutes of sonar can disrupt feeding for at least an additional 60 minutes after that. Whales would not regularly dive to a depth of 100 meters if there wasn't something (food) at that depth that they need. Therefore, choice (B) is incorrect. Although choice (C) may be true, this experiment did not establish a causal relationship between sonar and beachings. Choice (D) is wrong because we can draw some conclusions from this experiment, as was done for establishing why choice (A) is correct. (LO 4.21)

542. **(B)** The control treatment must be the same as the experimental treatment except for one variable, which is the fungus in this case. Choice (A) is incorrect because the researchers did not want other fungi to grow. No information in the passage suggests that applying sterile fungal growth media to wounds improves wound healing, so choice (C) is incorrect. Choice (D) is incorrect because *S. musiva* grows in growth media. Applying sterile fungal growth media to wounds does not prevent growth of the fungus. (LO 4.11)

543. **(D)** Figure I indicates that cankers were longer in both drought-stressed clones than in the watered clones when they were inoculated with *S. musiva*. Choice (A) is the opposite of what the figure shows, so it is

incorrect. An effect was observed, as described in choice (D), so choice (B) is incorrect. Choice (C) is incorrect because both clones responded the same way. (LO 4.15)

544. **(B)** Stressed NM6 that was inoculated had a higher percentage of stem girdling than watered NM6 that was inoculated, but NE308 had a similar percentage of stem girdling for both treatments (stressed and water that were both inoculated). Choice (A) is the opposite of what Figure I shows, so it is incorrect. Choice (C) is incorrect because an effect was seen, as evidenced by the description in choice (B). Choice (D) is incorrect because stress led to an increased percentage of stem girdling, not a decreased percentage of stem girdling. (LO 4.15)

545. **(C)** No spores were expected on trees that were not inoculated with the fungus. If spores were present on those trees, it would imply that there was some other source for the fungus other than the method used by the researchers. It is not possible to know if a tree is resistant if it is not challenged with the fungus, so choice (A) is incorrect. Drought stress is not measured by fungal sporulation or lack thereof, so choice (B) is incorrect. Choice (D) is incorrect because the fungus was never introduced to these trees. (LO 4.15)

546. **(A)** Fewer spores were produced on NM6 than on NE308. Choice (B) is the opposite of what the table shows, so it is incorrect. Choice (C) is incorrect because the difference was that fewer spores were produced on NM6 than on NE308. Choice (D) is incorrect because it is possible to accurately interpret the results seen in Table I, as evidenced by the statement in choice (A). (LO 4.15)

547. **(C)** NM6 outperformed NE308 in terms of resistance in all of the categories mentioned in choice (C). Choice (A) incorrectly states that NE308 is more resistant in terms of spore production. NM6 is actually more resistant in terms of spore production. Choice (B) is incorrect because significant differences were seen between the clones, as evidenced in choice (C). Choice (D) has all of the relationships backward. (LO 4.15)

548. **(D)** Figure I shows that drought-stressed NM6 and NE308 had longer cankers than watered NM6 and NE308. For the percentage of stem girdling, watered NM6 had less girdling than stressed NM6, but NE308 had approximately the same percentage of stem girdling for both watered and stressed trees. Table I shows that drought-stressed NM6 had more spores than watered NM6 but that NE308 had more spores when watered than when stressed. Since drought stress did have an effect on fungal resistance in the clones, choice (A) is incorrect. Since drought stress did have some effects on the ability of NE308 to resist the fungus, except for percent stem girdling, choice (B) is incorrect. Choice (C) is incorrect because drought stress was detrimental, not helpful, to the trees. (LO 4.15)

549. **(D)** None of the answer choices provided an explanation that is consistent with the results. If choice (A) was true, the percentage of stem girdling should have been higher for NE308 stressed trees, but that is not the case, so choice (A) is incorrect. No evidence is provided that the trees stopped producing antifungal compounds, so choice (B) is incorrect. Choice (C) is incorrect because the fungus did not colonize the entire circumference of the stem. As seen in Figure I, the stressed and watered NE308 bars for those exposed to inoculated media indicate approximately 80% girdling. (LO 4.15)

550. **(D)** The least damage in terms of canker length, the percentage of stem girdling, and sporulation was for watered NM6. Therefore, the best advice to give to the hybrid poplar growers is to plant NM6 and irrigate to supply water during a drought. Choice (A) is incorrect because NE308 is very susceptible to the disease and withholding water would exacerbate the disease. Choice (B) is incorrect because NE308 is very susceptible to the disease. Choice (C) is incorrect because withholding water would increase susceptibility to the disease. (LO 4.15)

551. **(C)** There was no control to compare the results to, so it is impossible to determine if the effect seen was due to the sunbed, the fish, the water, or some combination of these factors. The effect could simply have been due to time. Choice (A) is incorrect because 65 may have been a large enough sample size if the experiment was designed with a control. Choice (B) is incorrect because the intent was to re-create the effect, as described in the passage, without requiring patients to visit the spa in Turkey. Choice (D) is incorrect because the researchers did measure the percent reduction in symptoms, which was the dependent variable. (LO 4.19)

552. **(C)** If patients leaving the spa developed diseases that they didn't have when they came in but that other patrons had, this would suggest that the pathogens may have been transmitted at the spa. Choice (A) is incorrect because the question asks about disease transmission, not about the effectiveness of the treatment for psoriasis. Choice (B) is incorrect because algae are not typically skin pathogens and may be present in perfectly healthy water. Choice (D) is incorrect because simply using the same fish on multiple patients is not evidence of transmission. Patients at the spa also used the same doors, seats in the waiting room, and water, any of which might have harbored skin pathogens. (LO 4.19)

553. **(A)** The reefs with cleaner fish present had a higher percentage of fish larger than 31 mm than the reefs without cleaner fish. The reefs with cleaner fish absent did not have fish that were larger than 50 mm whereas the reefs with cleaner fish present had some fish that were larger than 60 mm. The reefs with cleaner fish absent had a much higher proportion of fish in the two smallest fish size classes (less than 21 mm and between 21 and 30 mm). The presence of the cleaner fish led to more large fish, perhaps because the

fish were healthier, lived longer, and were able to grow bigger as a result of the presence of the cleaner fish. Choice (B) is incorrect because there were significant differences between the reefs with and without the cleaner fish, as evidenced by the observations mentioned above. Choice (C) is incorrect because the lack of cleaner fish had the opposite effect of what choice (C) states. It led to a smaller proportion of bigger fish. Choice (D) is incorrect because the graph shows relative abundance and does not indicate the total number of fish. (LO 4.18, LO 4.19)

554. **(D)** If the cleaner fish caused other fish to grow longer to greater sizes or faster, a higher proportion of large fish would be seen on reefs with the cleaner fish, as evidenced in Figure I. If the reefs with the cleaner fish absent produced more offspring, there would be a higher proportion of small fish on those reefs, as seen in the figure. Since all of these statements are reasonable hypotheses, choices (A), (B), and (C) individually cannot be correct, meaning that choice (D) must be the correct answer. (LO 4.19)

555. **(D)** Approximately 1,100 fish were supported in the presence of the cleaner fish, while only 400 were supported in their absence. Choice (A) is incorrect because clear differences were present. Fish were still present on reefs with the cleaner fish absent, so choice (B) is incorrect. Choice (C) has the relationship shown in Figure II backward. (LO 4.18)

556. **(B)** The cleaner fish were associated with both more fish and more types of fish. Choice (A) states the opposite of what the graphs show, so it is incorrect. Choice (C) is incorrect because the species richness was *higher* in the presence of cleaner fish. Since differences were seen in the graphs, choice (D) is incorrect. (LO 4.27)

557. **(B)** Mutations that promoted survival in the presence of a predator were selected for. Choice (A) is incorrect because there is no predation pressure in an environment without a predator. Choice (C) is incorrect because predation does not cause mutations to form. Choice (D) is incorrect because a lack of predation selected for the wild-type phenotype. (LO 4.26)

558. **(A)** The presence of predators decreased the number of bacteria. Choice (B) is incorrect because the bacteria were given the same amount of media. Since there were fewer bacteria in the microcosms with predators, even with the same amount of media (resources), there should have been less competition, not more, in the presence of predators. Choice (C) is incorrect because there were fewer, not more, bacteria in the microcosms with predators. Also, the bacteria were not parasitic on the predators. Choice (D) is incorrect because the bacterial density was greater in the microcosms without predators. (LO 4.26)

559. **(C)** Random mutation is the source of new alleles. Choice (A) relies on genes being transferred from the predator to the prey. This is unlikely. Choice (B) relies on genes being transferred from humans to the bacteria, which is also

unlikely. Natural selection is a force that decreases diversity by removing some alleles from the gene pool. It does not create new alleles, so choice (D) is incorrect. (LO 4.26)

560. **(C)** In the initial experiment, the new phenotypes arose as a result of a mutation followed by selection pressure from a predator. Removing the predator removed the selection pressure. Without the selection pressure from the predator, the wild-type exhibited higher fitness, so the population reverted to the wild-type. Choice (A) is incorrect because there would be selection pressure back to the original type, designated as the wild-type in this case. Choice (B) is incorrect because antiherbivory compounds are not mentioned in the question. Choice (D) is incorrect because there is competition in the absence of a predator. This competition favors those bacteria that do not have antipredator alleles. (LO 4.26)

561. **(B)** Since the selection pressure of the predator was not present, the original type was more fit and was selected for. Choice (A) is incorrect because bacteria are unable to turn these genes on and off. Since there were no dead, smooth bacteria present from which to take up genes, choice (C) is incorrect. Choice (D) is incorrect because the statement in choice (B) is a reasonable explanation for this phenomenon. (LO 4.26)

562. **(D)** *S. mutans* switches to lactic acid fermentation when there is no oxygen. The acid dissolves enamel. Choice (A) is incorrect because *S. mutans* does not grow as well in the presence of oxygen as does *S. sanguinus*. Choice (B) is incorrect because *S. mutans* does not produce hydrochloric acid and lactic acid does dissolve enamel. Choice (C) is incorrect because carbon dioxide does not dissolve enamel. If it did, enamel would dissolve as humans exhale. (LO 4.14)

563. **(D)** Decreasing the depth of the biofilm prevents the environment from becoming anaerobic, reducing the amount of lactic acid produced. Choice (A) would not help prevent tooth decay. Choice (B) is incorrect because toothbrushes cannot break DNA strands. Choice (C) is incorrect because a lower pH would lead to more enamel decay, not less. (LO 4.14)

564. **(C)** *S. mutans* prefers a lower pH than does *S. sanguinus*. The pH levels typically found in the human mouth will not kill bacteria or prevent them from growing. However, pH levels can favor one species over another. In this case, a higher pH will favor *S. sanguinus*. Choice (A) is incorrect because it discusses a lower pH whereas the question asks about the effect of a higher pH. Increasing the pH favors bacteria that do not cause tooth decay, like the *S. sanguinus*, so choice (B) is incorrect. The higher pH will not prevent all bacteria from growing, so choice (D) is incorrect. (LO 4.19)

565. **(C)** The assumption that the concentration of algae led to decreased transmittance is likely correct. However, no information suggests that this was

checked and thus it is just an assumption. Choice (A) is incorrect because the students assumed that the change in transmittance was due to algae, not bacteria. Choice (B) is incorrect because the fact that the algae perform photosynthesis was not really in doubt. Choice (D) is incorrect because this is a scientific fact that was not in doubt. (LO 4.11)

566. **(B)** If the students measured the percent transmittance of samples with known concentrations of algae, they could use these values to determine the actual concentration of algae in their samples. Choice (A) is incorrect because the blank would not allow the students to compare their results to actual algae concentrations. Choice (C) is incorrect because absorbance has the same issues as transmittance as long as the students do not know how many algae per mL produce a particular amount of transmittance or absorbance. Choice (D) is incorrect because this method will not connect transmittance to algae concentration. (LO 4.11)

567. **(B)** The school power plant exhaust was enriched in carbon dioxide according to the passage. This would explain why the algae that were fed this air grew faster. Choice (A) is incorrect because the algae fed natural air did not grow faster. Choice (C) is incorrect because there were differences between natural-grown and exhaust-grown algae, as seen on the graph. Choice (D) is incorrect because increased algae growth would lead to increased carbon fixation, not the reverse. (LO 4.15)

568. **(B)** If the effect that the students observed in their first experiment was due to CO_2 and not due to other components of exhaust air, bioreactor II (exhaust air with enough CO_2 removed to match the level of CO_2 in environmental air) should have algae growth similar to that in environmental air. Additionally, bioreactor III (environmental air with additional pure CO_2 to raise the concentration of CO_2 to the same level as that of the exhaust air) should have algae growth similar to that in exhaust air. If the increased growth were due to CO, NO_x, N_xO, CH_4, SO_2, and/or VOC, bioreactor II would have growth similar to that in the exhaust air and bioreactor III would have growth similar to that in environmental air. Choice (A) is incorrect because removing CO_2 from the environmental air would not test the effect of CO, NO_x, N_xO, CH_4, SO_2, and/or VOC. Choice (C) is incorrect because bioreactor IV (pure CO_2 at 1 atmosphere of pressure) would not be directly comparable to any of the bioreactors used in the initial experiment. In addition to excluding CO, NO_x, N_xO, CH_4, SO_2, and VOC from the mixture, using pure CO_2 also excludes other components of environmental air, such as N_2 and O_2. Choice (D) is incorrect because repeating the experiment alone would not add additional information about the effect of CO_2 or the effect of CO, NO_x, N_xO, CH_4, SO_2, and VOC. (LO 4.15)

569. **(C)** The data indicate that lettuce grew most rapidly in the highest concentration of coffee grounds. The growth rates varied for the different treat-

ments, so choice (A) is incorrect. The lettuce didn't grow most rapidly in the lowest concentration of coffee grounds, so choice (B) is incorrect. It is possible to determine an answer from this graph, so choice (D) is incorrect. (LO 4.15)

570. **(A)** Whether or not food is healthier is a difficult thing to measure and therefore would not be a testable hypothesis for a future school experiment. The hypotheses in choices (B), (C), and (D) specify things that are possible to measure like plant's height, mass, harvest date, and growth rate. Since these choices do present testable hypotheses, they are all incorrect. (LO 4.11)

571. **(A)** The slope was the steepest for the treatment without any coffee between day 21 and day 35. Therefore, the rate of growth was fastest in 0 g/100 mL between those days. Therefore, choices (B), (C), and (D) are incorrect. (LO 4.14)

572. **(B)** There was a large difference between brand F and the control. Note that their error bars do not overlap at all. There was apparently no significant difference between brand E and the control. Note that the top of the error bar for brand E nearly overlaps with the average for no mascara, and the bottom of the error bar for no mascara nearly overlaps with the average for brand E. The fact that these error bars overlap so much suggests that no significant difference exists. No brands promoted growth, since no brands caused more growth than the control, so choice (A) is incorrect. Choice (C) is incorrect because some brands, such as brand F, did appear to inhibit growth. Brand F is probably the most toxic brand, so choice (D) is incorrect. (LO 4.15)

573. **(C)** Similar relative rankings of the brands in the paramecium assay and in the standard assay would make the paramecium test a good replacement of the standard assay. Choice (A) is incorrect because the standard assay is not measured in terms of cell growth. Choice (B) is incorrect because comparing more paramecium species would not give any information about the test's validity in determining the effect of mascara on the cornea of the eye. Choice (D) is incorrect because consumer feedback depends on many factors other than the toxicity of the product. It is also a good idea to know if a product is toxic before selling it. (LO 4.8)

574. **(C)** If the same factors that irritate the human eye and the eyes of the rats used in the standard method also inhibit the growth of paramecia, this test might be useful. Choice (A) is incorrect because paramecia are single-celled organisms that grow in water, not in the eyes of humans or rats. Choice (B) is incorrect because this assay only requires the paramecia to be exposed to mascara, not to metabolize it. Choice (D) is incorrect because the assumption is that an increased growth rate of paramecia would indicate *decreased* toxicity. (LO 4.8)

575. **(D)** Autoclaving kills the organisms that inhabit soil. If the presence of these organisms does have an effect, killing them should remove that effect.

Choice (A) is incorrect because no measures were reported on the mineral content of the soil. Choice (B) only asks about pathogenic organisms, which would be detrimental to plants, and is therefore a more specific question than the one posed in choice (D). Remember, the question asked for the broadest question that these students could attempt to answer with this experiment, so choice (B) is incorrect. Choice (C) is incorrect because the students only varied the amount of unsterilized soil they added. If they wanted to know what type of soil provided the optimum growing conditions, they would have needed to vary many other soil factors, such as the amount of nitrogen, the amount of potassium, the amount of phosphorus, and/or particle size. (LO 4.19)

576. **(C)** Plants that were grown in soil with 0% unsterilized soil were only 6 cm, which is much shorter than any of the plants that had at least 1% unsterilized soil. Choice (A) is incorrect because pathogens are detrimental to plant growth. If any were present in the unsterilized soil, the plants grown in the sterilized soil should have been taller than those grown in the pathogen-contaminated unsterilized soil. Choice (B) is incorrect because completely sterile soil did not produce taller plants. Choice (D) is incorrect because the presence of nitrogen-fixing bacteria was not tested. In addition, their presence typically leads to increased growth. (LO 4.18)

577. **(A)** Unsterilized soil produced taller plants. Therefore, the organisms that were killed when the soil was sterilized were probably beneficial. Since the organisms in the unsterilized soil were clearly beneficial to the health and growth of the plants based on the graph, choice (B) is incorrect. Some soil-borne pathogens do cause damage to *Brassica rapa*, but this study did not detect or indicate any. Therefore, choice (C) is incorrect. There was no test for the presence of nitrogen-fixing bacteria, so choice (D) is incorrect. (LO 4.18)

578. **(A)** Plating the autoclaved soil on sterile growth media would isolate the soil as the source of any microorganisms that grew. Choice (B) is incorrect because DNA survives autoclaving and can be amplified from dead organisms. If there were organisms present in the soil before autoclaving, their DNA would still be in the soil after autoclaving, even if they were dead. Therefore, confirming the presence of DNA would not confirm the presence of microorganisms. Choice (C) is incorrect because a simple examination under the microscope would be unlikely to reveal any microorganisms present. Choice (D) is incorrect because it introduces many alternative sources of microorganisms, including the paper towels, oatmeal, tap water, and air. (LO 4.11)

579. **(B)** If the drop in the lynx population levels was due to a lack of food, providing the lynx with alternative food sources when the snowshoe hare population falls should remove that effect and prevent the cycling. Choice (A) is incorrect because a lack of cycling might simply be caused by the fact

that the animals are in enclosure. Choice (C) is incorrect because this experiment would not test the hypothesis stated in the question, which is that the lynx population levels fall due to a lack of food. Choice (D) would not test the effect of a lack of food on the lynx population. (LO 4.19)

580. **(B)** If the drop in the snowshoe hare population was due to toxins in the snowshoe hares' food, providing an alternative food source should prevent the crash. Removing the lynx from the environment would not test the effects of these toxins, so choice (A) is incorrect. Sampling the plants for toxins after the snowshoe hare population crashes would give no information about why the population crashed, so choice (C) is incorrect. Choice (D) is incorrect because simply removing the hares from this environment would give no information about the toxicity of the plants or the effect of plant toxicity on the snowshoe hare population levels. (LO 4.16)

581. **(D)** Rabbits are not predatory toward mice, so there would be no reason for the mice to avoid the rabbit urine. Therefore, the rabbit urine was simply the control in this experiment. Choice (A) is incorrect because there is no reason to believe that mice would be attracted to rabbit urine. Choice (B) is incorrect because rabbit urine and bobcat urine were not placed near each other. Choice (C) is incorrect because the mice were not given a choice between different types of urine. The mice were each placed into a box with either one or the other type of urine. (LO 4.19)

582. **(A)** The infected mice did not avoid the bobcat urine as much as the uninfected mice did. Choice (B) is incorrect because the mice did not have an aversion to rabbit urine that could be lost. Choice (C) is incorrect because the mice were never given a choice or a chance to select their preference between the different types of urine. Additionally, the uninfected mice would probably have preferred rabbit urine to bobcat urine, given the results from the graph. Choice (D) is incorrect because there is no evidence that mice are actually attracted to cat urine. (LO 4.23)

583. **(C)** *T. gondii* must infect cats to complete its life cycle. Infecting a mouse and causing it to lose its aversion to cat urine would be advantageous to the pathogen. Choice (A) is incorrect because there is no evidence of mice without an aversion to cat urine being more likely to obtain food. Choice (B) is incorrect because the explanation presented is a physiological explanation, not an evolutionary one. An evolutionary explanation must demonstrate how one behavior increases the fitness of an organism. Choice (D) is incorrect because no evidence is provided that suggests that mice are attracted to cat urine from cats infected with *T. gondii*. (LO 4.24)

584. **(A)** Based on the results shown in Table I, which show that the number of people infected is a lot higher for people with a lot of cats than for those with no cats, it is clear that there is a strong correlation between an infection

with *T. gondii* and the number of cats owned. Any statement stronger than this one is beyond what is supported by the data. Choice (B) is incorrect because it suggests a causal relationship that would need to be tested by other experiments. Choice (C) is also incorrect. It would explain the data, but there is no information about cause and effect presented. Choice (D) is incorrect because choices (B) and (C) suggest causal relationships from data that show only a correlation. (LO 4.25)

Grid-In Questions (pages 264–267)

When working through the grid-in, long free-response, and short free-response questions that involve math, remember that your answer may vary slightly from the answer provided due to rounding or due to reading a graph differently. See page ix of the Introduction of this book for more information regarding how to avoid rounding errors. As a general rule, if you are within 5% of the actual answer, then your answer will be marked as correct.

585. **4.0** To test for the effect of water on the presence of spores in clone NM6, determine the expected value if there was no difference in the presence of the spores between watered and stressed NM6. To do this, add the observed values, and divide by the number of categories:

$$\frac{(4+12)}{2} = \frac{16}{2} = 8$$

Next, subtract the expected value from each observed value, square the result, and divide by the expected value. Then, add the results together:

$$\frac{(12-8)^2}{8} + \frac{(4-8)^2}{8} = \frac{(4)^2}{8} + \frac{(-4)^2}{8} = \frac{16}{8} + \frac{16}{8} = 2 + 2 = 4.0$$

(LO 4.19)

586. **3.2** To test for the effect of a particular clone on the absence of spores in the stressed trees, determine the expected value if there were no difference in the absence of spores between the different clones. To do this, add the observed values, and then divide by the number of categories:

$$\frac{(20+33)}{2} = \frac{53}{2} = 26.5$$

Next, subtract the expected value from each observed value, square the result, and then divide by the expected value. Finally, add the results together:

$$\frac{(20-26.5)^2}{26.5}+\frac{(33-26.5)^2}{26.5}=\frac{(-6.5)^2}{26.5}+\frac{(6.5)^2}{26.5}$$

$$=\frac{42.25}{26.5}+\frac{42.25}{26.5}$$

$$=1.59+1.59$$

$$=3.18$$

Round to the nearest tenth to get 3.2.

(LO 4.19)

587. **4,143** All of the fish caught in the original sample were marked and then returned to the lake surrounded by the forest. Therefore, the total number of marked fish in the lake surrounded by the forest was 254 fish (value M in the equations that follow). If these fish swam around and mixed with all of the unmarked fish in that lake, then when the students resampled the lake, the ratio of marked fish that they recaptured (value $m = 16$ in the equations that follow) to the total number of fish in the second sample (value $n = 261$ in the equations that follow) should equal the ratio of all marked fish in that lake to the total number of fish in that lake (N). Mathematically, let:

N = the population size of the lake

M = the number of fish caught in the first sample that were marked and then released

n = the number of fish caught in the second sample

m = the number of marked fish captured in the second sample

In this case:

$$\frac{M}{N}=\frac{m}{n}$$

Rearrange this equation to give:

$$N=\frac{nM}{m}$$

Substitute the values from the table:

$$\frac{261\times 254}{16}=\frac{66{,}294}{16}=4{,}143$$

(LO 4.19)

588. **1,180** All of the fish caught in the original sample were marked and then returned to the lake surrounded by the athletic fields. Therefore, the total number of marked fish in the lake surrounded by the athletic fields was 163

fish (value M in the equations that follow). If these fish swam around and mixed with all of the unmarked fish in that lake, then when the students resampled the lake, the ratio of marked fish that they recaptured (value $m = 21$ in the equations that follow) to the total number of fish in the second sample (value $n = 152$ in the equations that follow) should equal the ratio of all marked fish in that lake to the total number of fish in that lake (N). Mathematically, let:

N = the population size of the lake

M = the number of fish caught in the first sample that were marked and then released

n = the number of fish caught in the second sample

m = the number of marked fish captured in the second sample

In this case:

$$\frac{M}{N} = \frac{m}{n}$$

Rearrange this equation to give:

$$N = \frac{nM}{m}$$

Substitute the values from the table:

$$\frac{152 \times 163}{21} = \frac{24{,}776}{21} = 1{,}180$$

(LO 4.19)

589. **8** The minimum value in 2002 was 370 ppm. The maximum value in 2003 was approximately 378.2. Subtracting the minimum value in 2002 from the maximum value in 2003 gives the answer 8.2, which rounds off to 8 ppm as the nearest part per million. (LO 4.14)

590. **2** The value in January 2001was approximately 370.1, and the value in January 2003 was approximately 374.8. Subtracting the January 2001 value from the January 2003 value gives 4.7 ppm. Divide 4.7 ppm by 2 years to get the rate of increase per year, which is 2.35. This rounds off to 2 ppm as the nearest part per million per year. (LO 4.14)

591. **50** The value for the sprayed bar appears to be about 900, while the value for the unsprayed bar seems to be about 450. Divide 450 by 900, which is 0.5. Then, multiply 0.5 by 100, which is 50. Depending on how you read the graph, a range of values, from 45% to 50%, would be acceptable answers for this question. (LO 4.14)

592. **120** The line for *P. aurelia* in separate culture rises from about 150 on day 1 to 390 on day 3. Therefore, there is a difference of 240 over 2 days, or about 120 individuals per mL per day. Note that, depending upon the values you noted

for day 1 and day 3 based on your analysis of the graph, a range between 110 and 125 individuals would be acceptable answers for this question. (LO 4.14)

593. **36** The line for *P. burseria* in mixed culture rises from about 25 on day 1 to 240 on day 7. Therefore, there is a difference of 215 over 6 days, or about 36 individuals per mL per day. Note that, depending upon the values you noted for day 1 and day 7 based on your analysis of the graph, a range between 30 and 40 individuals would be acceptable answers for this question. (LO 4.14)

594. **290** The carrying capacity is the maximum number of individuals that a population can sustain. Therefore, this is where the line plateaus, which is at about 290 individuals/mL. Note that any range from 280 to 295 mL would be acceptable answers for this question. (LO 4.14)

595. **575** The carrying capacity is the maximum number of individuals that a population can sustain. Therefore, this is where the line plateaus, which is about 575 individuals/mL. Note that any range from 550 to 590 would be acceptable answers for this question. (LO 4.14)

Long Free-Response Questions (pages 268–270)

When working through the grid-in, long free-response, and short free-response questions that involve math, remember that your answer may vary slightly from the answer provided due to rounding or due to reading a graph differently. See page ix of the Introduction of this book for more information regarding how to avoid rounding errors. As a general rule, if you are within 5% of the actual answer, then your answer will be marked as correct.

596. This question asks you to evaluate the results of an experiment on species diversity in suburban lawns and discuss how humans interact with their environment. Note that this question is worth 10 points total. (LO 4.18, LO 4.27)

    (a) Compare the species richness and the diversity among the three different management techniques. Explain how the choice of management techniques might have led to the differences. (3 points maximum)

    Both the species richness and the diversity are the highest in the low-input DIY treatment, lowest in the professionally managed treatment, and in the middle for the high-input DIY treatment. (1 point)

    Pesticide use is a mechanism for reducing species richness and diversity because it kills certain species of plants. (1 point)

    The fertilizer applications, used in both the professionally managed and high-input DIY treatments, negatively impacted plants that harbor nitrogen-fixing bacteria, such as clover. (1 point)

(b) Identify ONE factor that might have led to the observed differences in the species richness and diversity. Describe a controlled experiment to test that factor. (5 points maximum)

One point is awarded for identifying an appropriate factor to test. Any of the following factors would be acceptable:

- Fertilizer, pesticide, or herbicide application
- Irrigation (watering)
- Human or animal use of the lawn

One point is awarded for including any of the following experimental design components in your description of a controlled experiment, with 4 points maximum possible:

- Stating a hypothesis
- Describing how to manipulate the independent variable
- Specifying at least two variables to keep constant
- Describing how to measure a response variable
- Using multiple replicates
- Using mathematics to analyze the data

Below is a sample response that would have received full credit.

One factor that might have led to the reduced species diversity in the professionally managed and high-input lawns is the use of fertilizer. My hypothesis is that fertilizer use decreases plant diversity. To test this hypothesis, I will need a 4 m by 4 m area of lawn that receives the same amount of rain and sunlight across the whole area. I will divide the plot into 16 square meter plots. I will then count the number of each type of species present in each plot at the beginning of the experiment, and I will calculate species diversity. I will randomly select 8 of the plots to spray with fertilizer once a week for 8 weeks. I will treat the other 8 plots exactly the same way, but I will not mix any fertilizer into the spray (the spray will only contain water). The water-only plots will be my control. After 2 months, I will count the number and type of each species on each plot to compare to my initial data.

(c) Discuss the impact that social views on aesthetics might have on the diversity and stability of an ecosystem. (2 points maximum)

The monoculture of grass using professional management techniques may be attractive to the human eye, as evidenced by the high attractiveness rating for the professionally managed treatment and the low attractiveness rating for the low-input DIY treatment in the experiment. However, using pesticides and other professionally managed techniques leads to decreased biodiversity. (1 point)

Decreased biodiversity will decrease the ecosystem stability. (1 point)

597. This question asks for you to graph and analyze some data on photosynthesis. Note that this question is worth 10 points total. (LO 4.8, LO 4.16)

(a) Graph the data using the grid provided. (3 points maximum)

You would be awarded 1 point for each of the following:

- Properly orienting the axes on the graph (making sure that the carbon dioxide concentration is on the *x*-axis and that photosynthesis is on the *y*-axis)
- Correctly labeling the graph, including the axes, and including both a legend and a title (or a caption)
- Correctly plotting the data

The graph below would earn full credit.

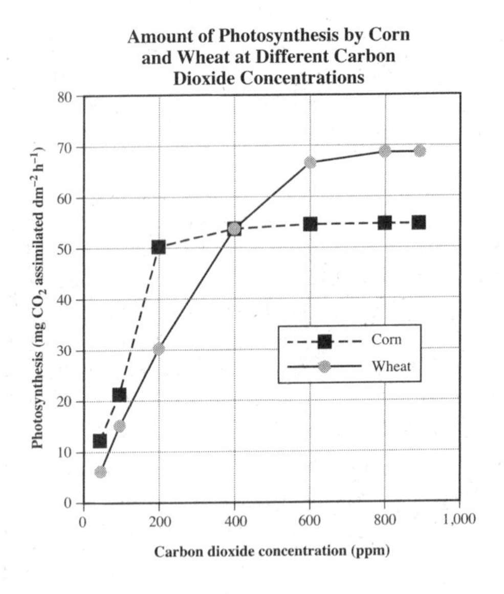

(b) Describe the shape of the curve for corn. Explain the physiological reason for the shape of the curve for corn between 0 ppm and 200 ppm and between 400 ppm and 900 ppm. (3 points maximum)

Photosynthesis increases for corn between 0 and 200 ppm of carbon dioxide. Between 400 ppm and 900 ppm, photosynthesis stays relatively the same for corn. (1 point)

As the carbon dioxide increases, the plant is able to fix more of the gas. (1 point)

The plateau occurs because something other than the carbon dioxide becomes a limiting feature. This could be some other nutrient or light. (1 point)

(c) Compare the responses of the corn and the wheat to increased carbon dioxide. (2 points maximum)

Corn shows a more rapid increase at lower carbon dioxide concentrations than wheat. (1 point)

Corn reaches a plateau before wheat does. (1 point)

(d) Global atmospheric carbon dioxide is currently at approximately 410 ppm. Use the data from the graph to predict the effects on corn growth and wheat growth if the carbon dioxide levels continue to increase. (2 points maximum)

At 410 ppm of carbon dioxide, corn is currently on the plateau. Increasing the carbon dioxide will not increase the photosynthesis for corn. (1 point)

Wheat can still grow from higher carbon dioxide levels until approximately an 800 ppm carbon dioxide concentration when wheat plateaus as well. (1 point)

Short Free-Response Questions (pages 270–271)

When working through the grid-in, long free-response, and short free-response questions that involve math, remember that your answer may vary slightly from the answer provided due to rounding or due to reading a graph differently. See page ix of the Introduction of this book for more information regarding how to avoid rounding errors. As a general rule, if you are within 5% of the actual answer, then your answer will be marked as correct.

598. Perform a chi-square test on the data. Specify the null hypothesis that you are testing, and enter the values from your calculations in the table. Explain whether the null hypothesis is supported by the chi-square test at the 0.05 level. (LO 4.8)

A maximum of 4 points can be awarded for correctly answering this question.

(a) The following must be included in your answer:

- A correct null hypothesis (1 point)
- Correct expected values (1 point)
- Correct chi-square values (1 point)

Your table should read as follows:

Null Hypothesis: Catnip extract has no effect on fruit flies. Therefore, the fruit flies will land on both treated (catnip extract) and control (water) banana slices in a 1:1 ratio.			
	Observed (o)	**Expected (e)**	$\frac{(o-e)^2}{e}$
Catnip Extract	27	35.5	2.035
Water	44	35.5	2.035
Total			4.07

The null hypothesis for this experiment is stated in the table above and should produce a 1:1 ratio of fruit fly landings on banana slices treated with catnip extract to fruit fly landings on banana slices treated with water. To find the expected value, add the observed values, and divide by the number of categories:

$$\frac{(27+44)}{2} = \frac{71}{2} = 35.5$$

Next, subtract the expected value from each observed value, square the result, and divide by the expected value. Then, add the results together:

$$\begin{aligned}\frac{(27-35.5)^2}{35.5} + \frac{(44-35.5)^2}{35.5} &= \frac{(-8.5)^2}{35.5} + \frac{(8.5)^2}{35.5} \\ &= \frac{72.25}{35.5} + \frac{72.25}{35.5} \\ &= 2.035 + 2.035 \\ &= 4.07\end{aligned}$$

(b) 4.07 is larger than the 0.05 cutoff (3.84) on 1 degree of freedom. Therefore, the null hypothesis should be rejected. (1 point)

599. Briefly explain the rationale for the timing of the virus application. Identify ONE potential environmental risk of this practice, and explain what could be done to manage the risk you identified. (LO 4.12)

A maximum of 3 points can be awarded for correctly answering this question.

(a) By timing the virus application to shortly after the petals fall, the grower is able to attack the insect when it is present on the surface of the fruit. Spraying earlier would not be effective because the spray application would be before the codling moth's eggs were laid. Spraying after that time would not be effective because the larvae would have already burrowed into the fruit and the virus spray would not penetrate into the fruit where the insect larvae would be. (1 point)

(b) Two possible risks and their management options are listed below. One point is awarded for identifying one potential environmental risk of this practice, and one point is awarded for explaining the management options for that risk. Other risks and their management options are possible, as long as the risk that you identify is reasonable. Remember, only the first risk that you identify would be scored, so do not identify more than one risk.

One potential environmental risk of this practice is the infection of nontarget organisms. A management option for that risk is to choose a virus, such as papaya ringspot virus, that has high host specificity.

Another potential environmental risk of this practice is that the insect may develop resistance. A management option for that risk is to provide reservoirs of susceptible insects by releasing susceptible males into the environment.

600. State a hypothesis that the students were testing. Interpret the data in terms of the hypothesis. (LO 4.8)

A maximum of 3 points can be awarded for correctly answering this question.

(a) The hypothesis that the students were testing is that exposure to alcohol had an effect on the behavior of the fish. (1 point)

(b) The fish that were exposed to the highest percentage of alcohol spent more time at the top of the tank than the control fish (those that were exposed to 0% alcohol). (1 point)

The question stated that it has been established by previous research that anxious fish avoid the top of the tank. Since the graph shows that those fish that were exposed to the highest percentage of alcohol spent a lot of time at the top of the tank, it can be concluded that higher alcohol exposure led to lower anxiety for the fish. (1 point)

Reference Tables

Equations, Formulae, and Terms

Standard Error of the Mean	Mean
$SE_{\bar{x}} = \frac{s}{\sqrt{n}}$	$\bar{x} = \frac{1}{n}\sum_{i=1}^{n} x_i$
Standard Deviation	**Chi-Square**
$s = \sqrt{\frac{\sum (x_i - \bar{x})^2}{n-1}}$	$\chi^2 = \sum \frac{(o-e)^2}{e}$

Chi-Square Table

	Degrees of Freedom							
p value	1	2	3	4	5	6	7	8
0.05	3.84	5.99	7.82	9.49	11.07	12.59	14.07	15.51
0.01	6.64	9.21	11.34	13.28	15.09	16.81	18.48	20.09

s = sample standard deviation (i.e., the sample-based estimate of the standard deviation of the population)

$\bar{x}$ = mean

n = size of the sample

o = observed results

e = expected results

Degrees of freedom are equal to the number of distinct possible outcomes minus one.

Laws of Probability

If A and B are mutually exclusive, then $P(\text{A or B}) = P(\text{A}) + P(\text{B})$

If A and B are independent, then $P(\text{A and B}) = P(\text{A}) \times P(\text{B})$

Hardy-Weinberg Equations

$p^2 + 2pq + q^2 = 1$	p = frequency of the dominant allele in a population
$p + q = 1$	q = frequency of the recessive allele in a population

Mode = value that occurs most frequently in a data set
Median = middle value that separates the greater and lesser halves of a data set
Mean = sum of all data points divided by number of data points
Range = value obtained by subtracting the smallest observation (sample minimum) from the greatest (sample maximum)

Metric Prefixes					
Factor	Prefix	Symbol	Factor	Prefix	Symbol
10^9	giga	G	10^{-3}	milli	m
10^6	mega	M	10^{-6}	micro	μ
10^3	kilo	k	10^{-9}	nano	n
10^{-2}	centi	c	10^{-12}	pico	p

Water Potential (Ψ)

$\Psi = \Psi_P + \Psi_S$ Ψ_P = pressure potential Ψ_S = solute potential

The water potential will be equal to the solute potential of a solution in an open container because the pressure potential of the solution in an open container is zero.

The Solute Potential of a Solution

$$\Psi_S = -iCRT$$

i = ionization constant (For sucrose this is 1.0 because sucrose does not ionize in water.)
C = molar concentration
R = pressure constant (R = 0.0831 liter bars/mole K)
T = temperature in Kelvin (273 + °C)

Dilution—used to create a dilute solution from a concentrated stock solution

$$C_iV_i = C_fV_f$$

i = initial (starting) C = concentration of solute

f = final (desired) V = volume of solution

Gibbs Free Energy

$$\Delta G = \Delta H - T\Delta S$$

ΔG = change in Gibbs free energy ΔH = change in enthalpy

ΔS = change in entropy T = absolute temperature (in Kelvin)

$$pH = -\log_{10} [H^+]$$

Rate and Growth		
Rate	$\frac{dY}{dt}$	dY = amount of change dt = change in time B = birth rate D = death rate N = population size K = carrying capacity r_{max} = maximum per capita growth rate of population
Population Growth	$\frac{dN}{dt} = B - D$	
Exponential Growth	$\frac{dN}{dt} = r_{max}N$	
Logistic Growth	$\frac{dN}{dt} = r_{max}N\left(\frac{K - N}{K}\right)$	

Temperature Coefficient Q_{10}	$Q_{10} = \left(\frac{k_2}{k_1}\right)^{\frac{10}{T_2 - T_1}}$	T_2 = higher temperature T_1 = lower temperature k_2 = reaction rate at T_2 k_1 = reaction rate at T_1 Q_{10} = the factor by which the reaction rate increases when the temperature is raised by ten degrees
Primary Productivity Calculation	mg O_2/L × 0.698 mL/mg = mL O_2/L mL O_2/L × 0.536 mg C fixed/mL O_2 = mg C fixed/L	

Surface Area and Volume		
Volume of a Sphere	$V = \frac{4}{3}\pi r^3$	r = radius l = length h = height w = width A = surface area V = volume Σ = sum of all s = length of one side of a cube
Volume of a Rectangular Solid	$V = lwh$	
Volume of a Right Cylinder	$V = \pi r^2 h$	
Surface Area of a Sphere	$A = 4\pi r^2$	
Surface Area of a Cube	$A = 6s^2$	
Surface Area of a Rectangular Solid	$A = \Sigma$ (surface area of each side)	